BassBox™ 6 Pro

High-performance speaker box design software for Microsoft® Windows®

User Manual

Fifth Edition (for version 6.0.23 and later)

by D.E.Harris

BassBox 6 Pro User Manual, 5th Ed

© 2013 by D.E.Harris

ISBN-10: 1494773333
ISBN-13: 978-1494773335

All rights reserved.
No part of this book may be reproduced or distributed in any electronic or printed form without permission. Please do not participate in or encourage piracy of copyrighted materials in violation of the author's rights. Purchase only authorized editions.

Trademarks
BassBox is a trademark of Harris Technologies, Inc.
Windows and Microsoft are registered trademarks of Microsoft Corporation.
Other trademarks belong to their respective owners.

Warning and Disclaimer
This book is designed to provide information about the BassBox 6 Pro computer program (version 6.0.23 and later) for speaker enclosure designers. Every effort has been made to make this book as complete and as accurate as possible, but no warranty of fitness is implied.

The information is provided on an "as is" basis. The author and publisher shall have neither the liability nor responsibility to any person or entity with respect to any loss or damages arising from the information contained in this book.

Neither the BassBox 6 Pro software nor a license to use the software are included with this book. The software and license to use the software must be purchased from its manufacturer, Harris Technologies, Inc., or one of the manufacturer's authorized distributors. See the manufacturer's website at www.ht-audio.com for more information.

Technical Support
Technical support for the BassBox 6 Pro software is not available from the author of this book. All technical support requests should be sent to the software manufacturer, Harris Technologies, Inc. (Harris Tech). Harris Tech provides support two ways: General support topics are available on their website at www.ht-audio.com. These topics answer many common questions. Support is also available via email to support@ht-audio.com. Support is available only in the English language.

How to contact Harris Tech:
Email: support@ht-audio.com (technical questions)
         sales@ht-audio.com (sales questions)
Website: www.ht-audio.com

Contents

Getting Started 7
What is BassBox Pro? 7
System Requirements 11
User Requirements 11
Starting BassBox Pro 12
Quitting the Program 14
Overview 15
Beginning a New Design 18
Getting Help 19

Box Designer's Guide

1 What Sound Is 23
Frequency: The Identity of Sound 23
Pressure: The Loudness of Sound 25
How Loud is Too Loud? 27
Sound Pressure Levels versus Acoustic Power Levels 27
Summary 28

2 How a Speaker Works 29
Drivers 29
Crossover Network 33
Box 35
Summary 43

3 How to Choose a Box 45
Closed Boxes 45
Vented Boxes 48
Bandpass Boxes 51
Passive Radiator Boxes 56
Choosing a Box 58
Summary 62

4 Constructing a Box 63
Avoiding Standing Waves 63
Diffraction 64
Materials 64
Construction 65
Vent Placement Tips for Bandpass Boxes 67

Sample Designs

1. **Closed Box Example** **71**
2. **Vented Box Example** **99**
3. **Bandpass Box Example** **117**
4. **Passive Radiator Box Example** **139**

BassBox Pro Reference

1. **Menus** **159**
   - File Menu 159
   - Edit Menu 160
   - Graph Menu 161
   - Test Menu 163
   - Tools Menu 163
   - Help Menu 164
   - Graph Popup Menu 164
2. **Design Wizard** **169**
3. **Driver Properties** **171**
   - Driver Locator 173
   - Description 183
   - Configuration 186
   - Parameters 193
   - Dimensions 201
   - Response 203
   - External 210
4. **Box Properties** **213**
   - Description 214
   - Box Design 215
   - Damping 230
   - Vents 234
   - Passive Radiator 240
   - Internal 243
   - Parts List 250
5. **Room / Car Acoustic Properties** **251**

Contents

BassBox 6 Pro

6 Evaluating Performance **255**
    Graph Modes 256
    Graph Features 257
    Normalized Amplitude Response 269
    Custom Amplitude Response 271
    Maximum Acoustic Power 273
    Maximum Electric Input Power 276
    Cone Displacement 278
    Vent Air Velocity 279
    System Impedance Response 281
    Phase Response 283
    Group Delay 285

7 Saving / Opening a Design **287**

8 Printing a Design **291**
    Options 292
    Title Block 294
    Logo 295
    Graphs 296
    Sample 298
    Exporting & Printing a Graph 304

9 Clearing / Closing a Design **305**

10 Editing the Driver Database **307**
    Adding a New Driver 309
    Changing an Existing Driver 313
    Deleting an Existing Driver 313
    Editing Companies 313
    Compacting the Database 316
    Repairing the Database 316

11 Testing Drivers & Passive Radiators **317**
    Test Equipment 317
    Testing Drivers 318
    Testing Passive Radiators 322

12 Tools **327**
    Design Wizard 327
    Wavelength Calculator 327
    Start X•over Pro 328

13 Preferences **329**
    General 329
    Print 332
    Driver 335
    Box 337
    Graph 340
    Constants 342

Appendix A: Command Shortcuts **343**
Appendix B: Glossary of Terms **345**
Appendix C: The Driver Shapes in BassBox Pro **349**
Appendix D: The Box Shapes in BassBox Pro **353**
Appendix E: Suggested Reading **356**
Appendix F: Driver Parameter Worksheet **358**
Appendix G: Acoustic Response Worksheet **359**

Index **360**

Getting Started

This *User Manual* will show you how to use BassBox Pro to design bass (low-frequency) loudspeaker enclosures. It does not describe how to install the software. **Please refer to the separate *Installation Instructions* included with the software if you have not yet installed the program.** (Note: BassBox is pronounced: bās·bäks.)

What is BassBox Pro?

BassBox Pro is a state-of-the-art speaker enclosure design program. It can help you design a wide variety of speaker boxes for many different applications including home hi-fi, home theater, car, truck, van, pro sound reinforcement, recording studio monitors, stage monitors, PA, musical instruments, etc.

BassBox Pro helps you design a speaker box in three ways: 1) it models how a speaker will sound; 2) it models the maximum loudness and power limits of a speaker; and 3) it helps you calculate the box dimensions.

How a Speaker Sounds

Modeling the sound of a speaker is often called the "small-signal" analysis because it examines the speaker at small or low power levels. BassBox Pro does this with the following performance graphs:

- Normalized Amplitude Response (often referred to as the "frequency response").
- System Impedance Response.
- Phase Response.
- Group Delay.

Maximum Loudness and Power Limits

Modeling the loudness of a speaker is often called the "large-signal" analysis because it examines the speaker at large or high power levels. BassBox Pro does this with the following performance graphs:

- "Custom" Amplitude Response for a desired input power or voltage.
- displacement- and thermal-limited Maximum Acoustic Power.
- displacement- and thermal-limited Maximum Electric Input Power.
- Cone Displacement for a desired input power or voltage.
- Vent Air Velocity for a desired input power or voltage.

The performance graphs of BassBox Pro help you evaluate a speaker design and decide if it fulfills your design goals.

Box Dimensions

BassBox Pro provides eighteen different box shapes for closed, vented and passive radiator boxes and two different shapes for two- and three-chamber bandpass boxes. The overall dimensions of the box can be entered as internal or external dimensions and a three-dimensional wireframe drawing of the box is created. For most box shapes, two wall thicknesses can be specified in case a thicker material is used for the front to accommodate flush driver mounting. After the dimensions have been entered or calculated, a parts list is made available with a two-dimensional drawing of each box part, including cut angles.

Features

User Interface

- BassBox Pro employs a modular user interface that remembers user settings. Extensive use of tab sheets helps to concentrate more information into fewer windows.
- The program has a "real-time" feel. Changes made to a design are immediately reflected in other property windows and displayed in a mini "thumbnail" graph in the main window. Changes made to box dimensions are immediately visible in the box drawing.
- Many default settings can be adjusted by the user. For example, the user can control whether numbers will be rounded or displayed with double precision and whether the context-sensitive "balloon" help is on or off.
- English or metric units can be used (or a combination of both).
- An extensive, illustrated on-screen manual and help system are provided.

Speaker Design

- A "Design Wizard" helps ease speaker design. It can begin with an existing box or driver.
- Perform both small-signal and large-signal analyses.
- Open and compare up to ten different designs at the same time.

- Boxes can be closed, vented, bandpass or use passive radiators. Vented boxes can have a 12 dB/octave active high-pass equalization filter for B_6 designs. Bandpass boxes include double- and triple-chamber designs (triple chamber designs require multiple drivers).
- The program can quickly "suggest" a box for a driver. Suggested closed and vented boxes can be tailored by the design priorities of the user.
- A large variety of box shapes are available. Single-chamber boxes include the following shapes: barrel, cone, truncated cone, cube, cylinder, domed cylinder, truncated cylinder, ellipsoid, square prism, optimum square prism, regular polygon prism, slanted front prism, truncated edge prism, four-sided pyramid, three-sided pyramid, truncated pyramid (trapezoid), sphere and wedge. Multi-chamber boxes include these shapes: two-chamber bandpass cylinder, three-chamber bandpass cylinder, two-chamber bandpass prism and three-chamber bandpass prism.
- Boxes can be dynamically drawn to scale in real-time as the dimensions are entered/edited and the dimensions can be either internal or external. A box parts list is also available with a two-dimensional drawing for each part, complete with cut angles.
- Box leakage losses (QL) and vent losses (QLv) are utilized in all relevant box models and the amount of internal damping or "fill" can also be specified for all box types.
- A variety of vents can be modeled, including vents with flared ends, and the effects of vent "pipe" resonance can be displayed in many graphs.

Graphs

- Examine the performance of a design with nine graphs: 1) normalized amplitude response; 2) custom amplitude response for a specified input power or voltage; 3) maximum displacement-limited and thermal-limited acoustic power; 4) maximum electrical input power; 5) cone excursion for both drivers and passive radiators; 6) vent velocity; 7) system impedance; 8) phase response; and 9) group delay. *Note: Some graphs may not be available depending on the box type and amount of driver parameters that are available. For example, a closed box does not have a vent and so the vent velocity graph cannot plot a closed box design.*
- Two graph modes are available. For VGA resolutions, a single graph window displays graphs one-at-a-time. For SVGA and XGA resolutions, a separate window with two size options is available for each graph so graphs can be viewed simultaneously.
- The graphs can display the estimated piston band on-axis amplitude rise of a driver and the estimated diffraction response shelf (of select box shapes) that can arise from the front-panel circumference of the box.
- Other options include two vertical and two horizontal scales, seven graph memories, a cursor and a Graph Properties window. The width of plot lines and the grid intensity are adjustable. A graph can also be exported via the Windows clipboard for use in other programs such as word processors and page layout programs.

Parameters
- Both Thiele-Small and electromechanical parameters are used and can be entered manually or imported from a CLIO measurement system, a DATS (Dayton Audio Test System) or WT3 (Dayton Audio Woofer Tester 3), a LAUD version 312 (or later) measurement system, or the WT2 (Smith & Larson Audio Woofer Tester 2) measurement system. Parameters may also be imported from other BassBox 6 Pro or Lite "bb6" design files.
- An "Expert Mode" includes a self-analyzing feature for driver parameters. This feature automatically checks the parameters and places a green indicator beside the correct ones, a yellow indicator beside the marginal ones, a red indicator beside the incorrect ones and a grey indicator beside the ones that it is unable to test. The user can adjust the sensitivity of the analysis.
- Dual voice coil drivers are also accommodated with separate parameters for individual, parallel and series voice coil wiring.
- "Net" driver parameter values are displayed for multiple driver designs. These designs can select from three types of mechanical configurations (standard, isobaric and bessel) and four types of electrical configurations (parallel, series, series-parallel and separate). Standard and isobaric configurations can also be push-pull.
- The external series resistance and amplifier source resistance can be entered and their effects included in the graphs.
- The parameters of external passive networks, such as a crossover network or filter, impedance equalization circuit and L-pad can be imported from X•over Pro or entered manually in BassBox Pro and used to display the system response on the graphs.
- Detailed driver dimensions can be entered for a wide variety of driver shapes and types (including coaxial two-way, coincident two-way and three-way drivers).
- An extensive driver database is provided. It includes the parameters of thousands of drivers and can be searched by manufacturer, model name, driver parameters or certain box parameters. The user can add, edit or remove drivers from the database. The driver database is compatible with X•over Pro.

Acoustical
- Acoustic measurements for both drivers and listening environments can be imported from many popular measurement systems (including Brüel & Kjaer, CLIO, IMP, LMS, MLSSA, OmniMic, Sample Champion, Smaart, TEF®-20 and TrueRTA). The listening environment has two modes: architectural and automotive.
- The automotive acoustic mode allows the user to add a 12 dB/octave boost filter to the graphs at a specified start frequency to model the bass rise in many automotive interiors.
- The architectural acoustic mode includes options for a 3, 6 or 12 dB/octave LF boost filter and a 3, 6 or 12 dB/octave LF shelf to model the bass rise in a variety of acoustical environments.

Printouts

- Printouts can include the box and driver parameters, a three-dimensional box drawing with internal and external dimensions, a parts lists with two-dimensional box part drawings, dimensions and cut angles, an English fraction-to-decimal conversion table, a wiring diagram with an external network parts list and nine graphs.
- A custom logo or graphic can be imported and printed in the title block. Many bitmap file types are supported.

Miscellaneous

- A convenient wavelength calculator is provided.
- A built-in test procedure is provided to assist with the measurement and calculation of driver and passive radiator parameters (requires test equipment). The procedure uses the delta volume method.
- BassBox Pro can share design files with BassBox Lite.
- Older BassBox 5.1, 5.0, 4.0 and 3.0 files can be opened.

System Requirements

BassBox 6 Pro requires Microsoft® Windows® and has broad compatibility with 32-bit versions of Windows 8*, 7, Vista, XP, 2000, NT4, Me and 98. It also runs under many 64-bit versions of Windows. The program requires a minimum of 36 MBytes of free hard disk space and a CD-R compatible CD-ROM, DVD or Blu-ray drive. Also, the Arial and Symbol TrueType fonts must be installed (both fonts are a standard feature of Windows). For best results, a single Windows user account with administrator privileges should be used to install and run the program.

*Windows RT, the tablet version of Windows 8, is not supported.

User Requirements

BassBox 6 Pro is a versatile computer program that can serve a wide range of speaker box designers. Yet, it is no substitute for experience and creative insight when it comes to speaker enclosure design. Some basic familiarity with speaker boxes and drivers is advised.

It is this author's hope that this *User Manual* will similarly serve a wide range of BassBox 6 Pro users. It includes general information, sample designs, hints and tips for new users and in-depth information about the operation of BassBox Pro for experienced designers. However, this manual does not pretend to be a tutorial on speaker enclosure design nor a tutorial in the use of Microsoft Windows. For additional information on speaker design, please refer to the Suggested Reading list in Appendix E. For additional information on Windows, please refer to Microsoft and/or your computer vendor.

BassBox 6 Pro

Getting Started

Starting BassBox Pro

BassBox Pro can be started from the Windows Start menu. If the program was installed using the default settings, the menu path is "Start > Programs > HT Audio > BassBox 6 Pro" as shown below:

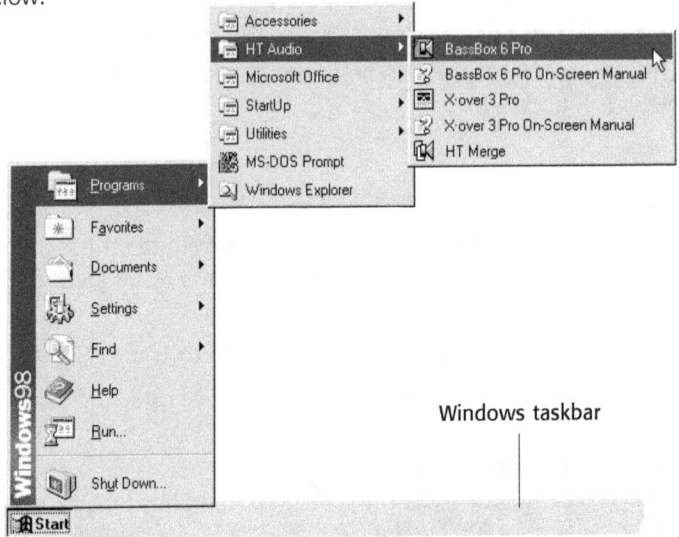

Windows taskbar

The first time that BassBox Pro is run after being installed, a "Welcome" window will appear (shown on the top of the next page). It will walk you through the following steps:

- Select a default acoustical environment.
- Select the way numbers and their units are displayed.
- Select "Normal" or "Expert" mode for the driver parameters.
- Select BassBox or "Classical" box models.
- Read selected portions of the on-screen manual to learn about the program.

Use the "Next" button at the bottom of the window when prompted to advance through each of these steps. A "Finish" button will appear at the end. Clicking (🖱) it will save the configuration and the "Welcome" window will close.

The next time you run BassBox Pro, its standard title window will appear (shown on the bottom of the next page).

Notes: If you previously used the "Cancel" button to close the "Welcome" window without saving an initial configuration, then the "Welcome" window will continue to appear in place of the program's standard title window. On the other hand, if the "Finish" button was clicked but the "Welcome" window continues to appear anyway, it may be necessary to give BassBox Pro administrator privileges every time you launch it. This can be done by right-clicking (🖱) on the program and selecting "Run as administrator".

Getting Started

BassBox 6 Pro

BassBox Pro "Welcome" window

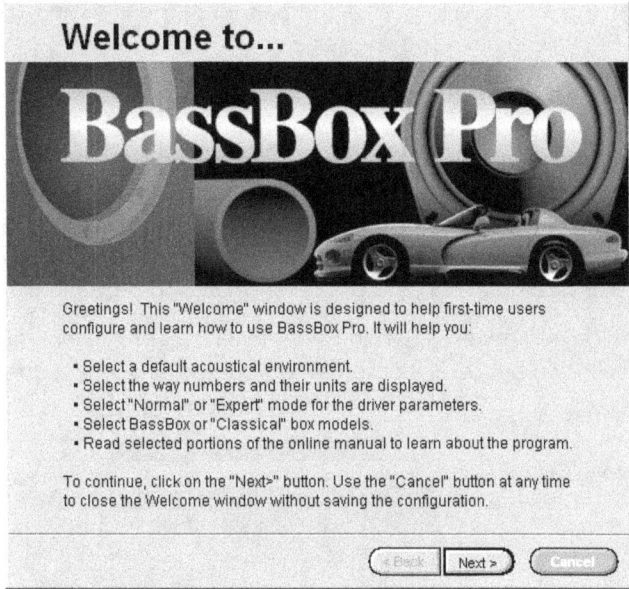

The title window includes the program version number, copyright notice, licensee name, company name and program serial number. To open this window later while the program is running, select "About BassBox Pro..." from the Help menu.

BassBox Pro title window

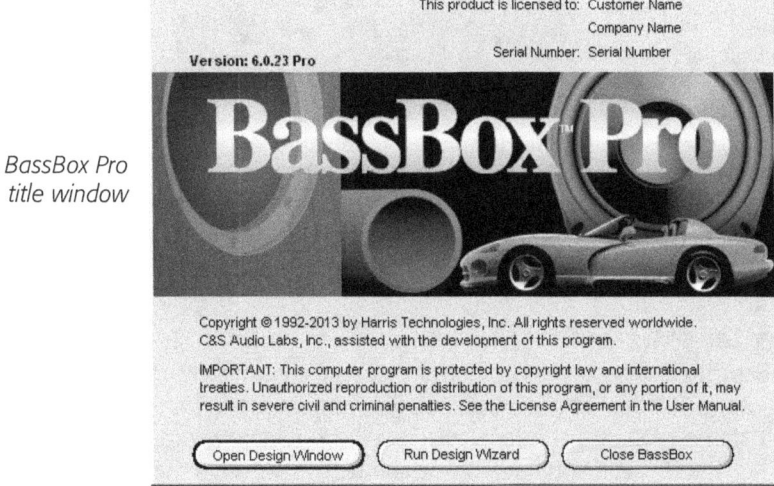

Notice the three buttons located at the bottom of the title window. They determine what BassBox Pro will do next. The choices are:

- **Open Design Window** – opens the BassBox Pro main window shown below. From it you can begin a new design or open an existing design (see "Beginning a New Design" on page 18).
- **Run Design Wizard** – launches the BassBox Pro Design Wizard to help you design a box (see Chapter 2 in the *BassBox Pro Reference* section later in this manual). After the design is finished, it will be loaded into the BassBox Pro main window.
- **Close BassBox** – stops the loading of BassBox Pro and closes the title window.

These choices will be discussed in more detail later in this manual. The main BassBox Pro window is shown below. It is empty when BassBox Pro is first opened:

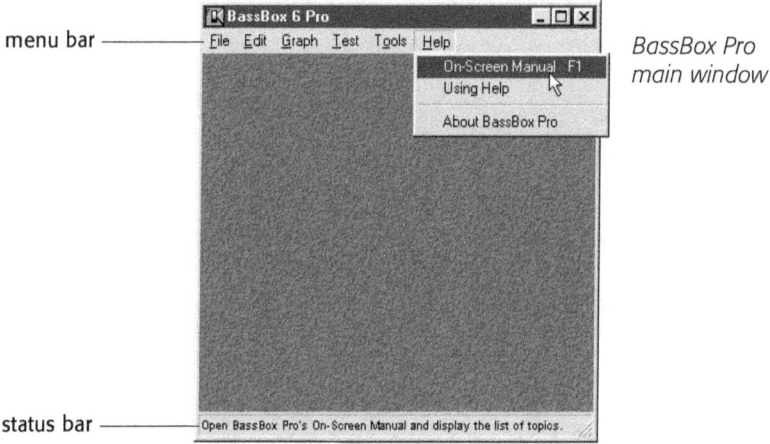

BassBox Pro main window

Like most Windows applications, the main window has a menu bar where many commands can be found. Additional command buttons are displayed in each design panel when a design is open. This is explained in more detail later. For example, use the File menu to begin a new design or open an existing design. Use the Help menu (shown above) to open the on-screen manual.

Quitting the Program

To close BassBox Pro from the title bar of the main window, click the Close button at the far right end of the title bar. To close BassBox Pro from its menu bar, select "Quit" from the File menu or use the keyboard shortcut Ctrl+Q. The keyboard shortcut works only when the program has the focus. If you have made changes to a design that have not been saved, you will be given an opportunity to save them before the program terminates.

Overview

The heart of BassBox Pro is its main window which includes design panels. Each speaker design is given its own design panel and up to ten speaker designs can be open and viewed at the same time as shown below.

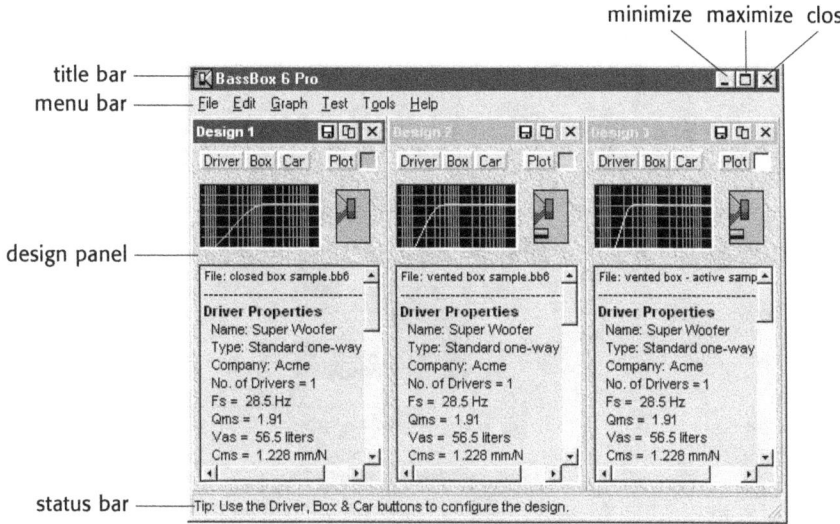

If necessary, a horizontal scroll bar will be added to the bottom of the main window to facilitate left/right scrolling so the design panels of all open speaker designs can be accessed.

The main window can be resized and the height of the design panels will be automatically adjusted to fit. The width of the design panels is not adjustable.

The major components of the main window are:

Title bar Lists the program name and includes the Minimize, Maximize and Close buttons. Like other Windows applications, the color of the title bar also indicates which window has the focus.

Minimize button Removes BassBox Pro from the desktop without closing it. It can be restored by clicking on the BassBox Pro button on the Windows taskbar.

Maximize button Causes the main window to fill the screen. Clicking the same button again will restore the main window to its former size.

Close button Clicking the close button has the same effect as selecting "Quit" from the File menu ([Ctrl]+[Q]). It will close the program. Before the program is closed, you will be given the opportunity to save unsaved changes to all open speaker designs.

Menu bar The menus of the menu bar contain many of the commands for the program (see Chapter 1 of the *BassBox Pro Reference* section of this manual). Many commands also have keyboard shortcuts which are listed beside them in the menus. A list of keyboard shortcuts is also provided in Appendix A.

Design panel Displays information about a single speaker design in the main window and provides buttons which allow design-specific tasks to be performed. The components of a design panel are labeled and explained below:

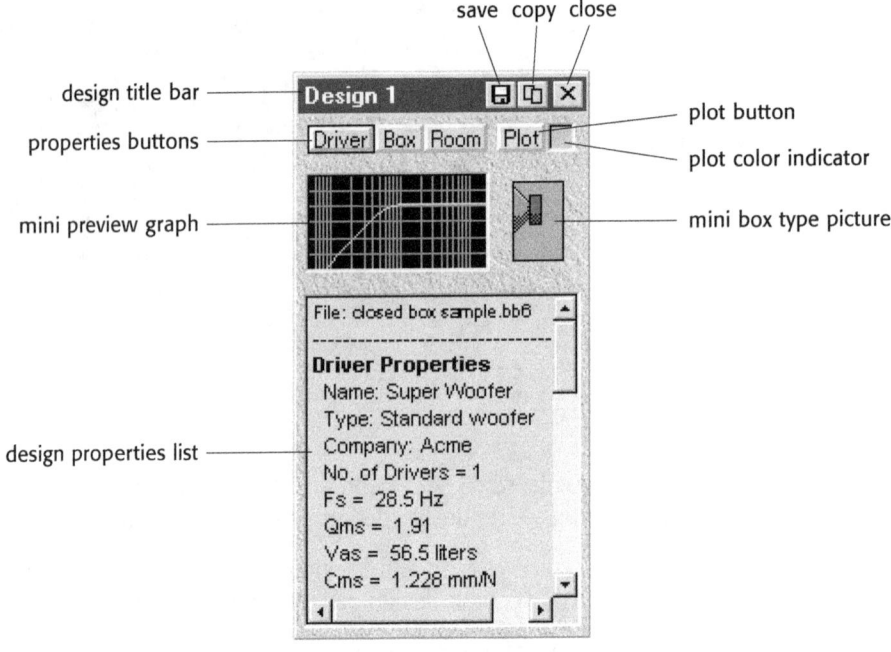

A single design panel.

Design title bar Lists the number of the design and includes the Save, Copy and Close buttons. The design title bar indicates which design has the focus.

Save button Saves the design to its existing BassBox design file. If the design is new, you will be prompted for a file name and a new design file will be created.

Copy button Copies the selected design. It does this by first creating a new design and then copying all the information from the selected design to it.

Close button Closes the selected design after first checking to see if changes need to be saved. *Note: BassBox Pro does not allow a design number to be skipped. When a design is closed, all designs after it will be renumbered.*

Driver properties button Loads the driver information from the selected speaker design into the Driver Properties window. If it is not open, the Driver Properties window will be opened. (Keyboard shortcut: Ctrl+D.)

Box properties button Loads the box information from the selected speaker design into the Box Properties window. If it is not open, the Box Properties window will be opened. (Keyboard shortcut: Ctrl+B.)

Room/Car properties button Loads the architectural/automotive acoustic information from the selected speaker design into the Room/Car Acoustic Properties window. If it is not open, the Room/Car Properties window will be opened. (Keyboard shortcut: Ctrl+A.)

Plot button Plots the response of the design in one or more of the full-size graphs. The way it functions varies slightly depending on the graph mode. The graph mode is selected in the Graph menu of the main window. The default graph mode can be set in the "Graph" tab of the Preferences window. (Keyboard shortcuts: Ctrl+[design number] to plot all open graphs. Ctrl+Alt+[design number] to plot only the selected graph in the single window mode. For example, use Ctrl+1 to plot Design 1 in all open graphs. Use Ctrl+Alt+1 to plot Design 1 in only the selected graph in the single window mode.)

Single Window Mode
This is the graph mode that displays the graphs one-at-a-time in a single window. Clicking the "Plot" button will cause all graphs to be plotted. Holding down the Alt key while the "Plot" button is clicked (🖰) will plot in the selected graph only.

Individual Window Mode
This is the graph mode that displays each graph in its own individual window, making more than one graph visible at the same time. Clicking the "Plot" button will plot in the open graphs only. The Alt key will have no effect in this mode.

Plot color indicator Displays the color used for the plot lines the next time the "Plot" button is clicked. Clicking it will cause the color to advance through the graph color palette (red—orange—yellow—chartreuse—green—cyan—blue—magenta—white—grey—and 2 user custom colors). Changing the plot color will only affect future plotting. It will not change the color of existing plot lines.

Mini preview graph Automatically displays the normalized amplitude response of the speaker design from 5 Hz to 2 kHz. Its vertical scale is 9 dB/division. It replots automatically whenever the design is changed. In this way it provides real-time feedback of the design while you work on it. You can force it to replot by left clicking (🖰) on it. Right clicking (🖰) on it will display a popup menu containing some of the same options of the full-size graphs. Changing any of these options will affect <u>all</u> BassBox Pro graphs.

BassBox 6 Pro

Getting Started

Mini box type picture Displays a small picture of the box type used in the design. It helps you quickly identify the type of speaker design. *Note: The mini box picture is not drawn to scale.*

Design properties list Lists most of the design's properties, including the driver and box properties. This list is provided solely for information—it cannot be edited (use the Driver and Box Properties windows to change the design).

Status bar Located at the bottom of the main window, the status bar displays tips, instructions or the status of a function.

Beginning a New Design

There are two ways to begin a design. The first and easiest way is to use the Design Wizard. It will act as an assistant to step you through the process of designing a speaker. The Design Wizard can be started from the BassBox Pro title screen by clicking (🖰) on the "Run Design Wizard" button (see Chapter 2 of the *BassBox Pro Reference* section of this manual). It can also be run from the main window by selecting it from the Tools menu.

The second way to begin a new design is to select the "New Design" command from the File menu of the main window or use the keyboard shortcut (Ctrl)+(N). An empty design panel will be created as shown below.

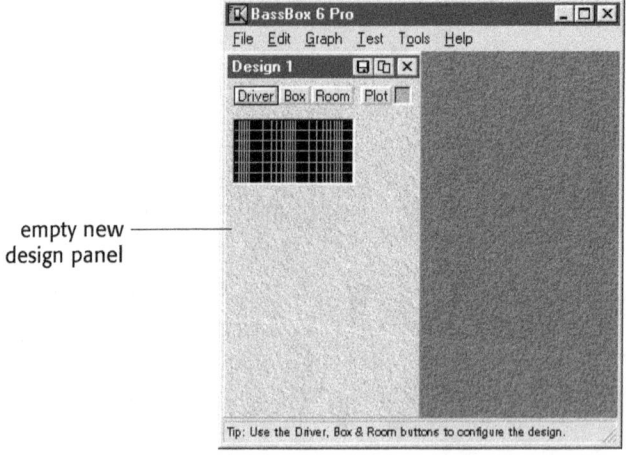

BassBox Pro main window

empty new design panel

Next, enter the driver and box information. This is done with the Driver and Box Properties windows. Simply click on the "Driver" button in the Design Panel (or use (Ctrl)+(D)) and the Driver Properties window will open so you can enter the driver information. Or you can begin with the box and click on the "Box" button (or use (Ctrl)+(B)) and cause the Box Properties window to open. It doesn't matter which you start with. You may want to begin with the driver if you already have one in mind. Begin with the box if it already exists.

After the driver and box information have been entered, it is time to configure the acoustic environment if the speaker will be used inside an automotive vehicle. To configure the acoustic environment select the "Car" button (or use [Ctrl]+[A]). *Note: If BassBox Pro is set for an architectural environment, the button will be labeled "Room". You can change this setting in the "General" tab of the Preferences window.* Because of their relatively small size and resulting "cavity effects", most vehicle interiors have a 12 dB/octave rise in bass response that begins around 50 Hz. Including this +12 dB/octave acoustic response of the vehicle can have a dramatic effect on the overall bass response of the design. Clicking the "Car" (or "Room") button will open the Acoustic Properties window where the acoustic settings are controlled.

To finish a new speaker design, plot its response in the main graphs and evaluate its performance (see Chapter 6 of the *BassBox Pro Reference* section of this manual). Does the speaker have the desired frequency or amplitude response? Will it be loud enough? Will the driver have enough excursion at the maximum power of your amplifier? If the box is vented, is the vent large enough to avoid vent turbulence noise and yet small enough to still fit in the box. If the box is a bandpass, is the vent also small enough to avoid vent resonance peak problems? These are some of the questions that the graphs will help you answer. If necessary make changes to your design and replot the graphs. Once you're finished, you can print the design with the "Print Design…" command in the File menu.

The *Sample Designs* section of this manual includes four sample speaker designs. It is best to begin with Chapter 1 ("Closed Box Example") if you are new to speaker design sbecause it is the most detailed and lays a fouindation for the other design samples..

Getting Help
BassBox Pro provides four different help systems. Each one is described next:

Balloon Help This help system gets its name from the small message boxes which pop up like small text balloons whenever the mouse pointer ( ) is paused over one of the many objects in the various windows of BassBox Pro. A sample is shown below:

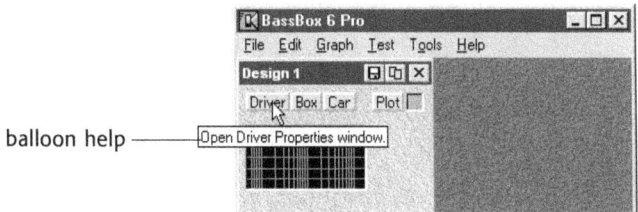

By the way, if you don't need to use balloon help, you can disable it. For details see Chapter 13 of the *BassBox Pro Reference* section of this manual.

BassBox 6 Pro

Getting Started

Status Bar A status bar is provided at the bottom of the main window. A description of each menu command is displayed in the status bar when a menu command is selected. It is also used to display tips, instructions and the status of some functions.

Context Sensitive Help Pressing the F1 or "Help" key will open BassBox Pro's on-screen manual to a relevant topic. For example, pressing F1 while viewing the "Dimensions" tab of the Driver Properties window will open the on-screen manual to the topic on this subject (shown at right). Scroll through the topic to find information about a specific feature.

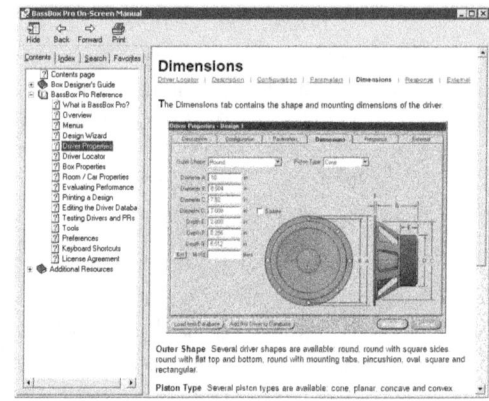

BassBox Pro on-screen manual

On-Screen Manual Select "On-Screen Manual" from the Help menu of the main window to open it to its "Contents" page. From there you can select any of the major topics, including topics in the *Box Designer's Guide* section which are not accessible with context sensitive help.

The on-screen manual can be opened even when BassBox Pro is not running. A direct link to it is included with BassBox Pro in the Start menu as shown on page 12.

Technical Support
When the preceding help systems do not provide the assistance you need, then software support is available in two ways. First, *Technical Notes* are provided at the Harris Tech website and can be accessed 24 hours per day, 7 days per week. The *Technical Notes* answer some common questions. The web address to the BassBox 6 Pro tech note index is:

www.ht–audio.com/pages/support/BBxPSupport.html

Second, support is also available from Harris Tech via email at **support@ht–audio.com**. Support is available only in the English language. Please remember to include your BassBox Pro serial number when you contact them.

Box Designer's Guide

This is the place to start if you are new to speaker box design. This *Box Designer's Guide* begins by explaining what sound is, how a speaker works and how to choose a box type. It answers questions like "Why does a speaker need a box?" and "Why do most speakers have more than one size driver?"

Chapters

1. What Sound Is 23
2. How a Speaker Works 29
3. How to Choose a Box 45
4. Constructing a Box 63

Guide: 1 What Sound Is

BassBox 6 Pro

1 What Sound Is

Like waves in water, sound is waves in air. But unlike water, air is invisible so you cannot see the ripples that are made by sound waves—at least not with the naked eye.

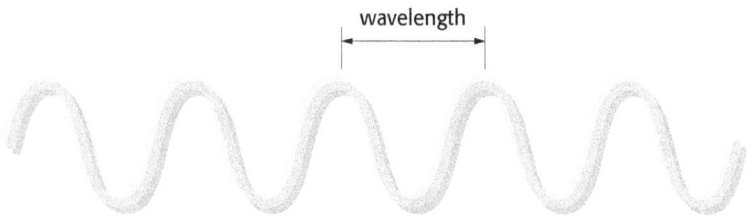

frequency (Hertz) = number of waves / second

Frequency: The Identity of Sound

If you have ever sat on a beach and counted the incoming waves of an ocean or large lake, then you already understand "frequency" and "wavelength". Frequency is simply the number of waves that arrive in a specified amount of time. The distance between the waves is the wavelength. Whether their wavelengths are long or short, sound waves all travel at about the same speed. So the frequency will be lower for long wavelengths and higher for short wavelengths. Incidentally, the speed of sound in air is 1130 feet per second or 344 meters per second.

With sound, frequency is measured in "hertz" (abbreviated "Hz"). *Note: "H" is capitalized in "Hz".* Hertz is the number of waves per second. We identify sound primarily by its frequency. The bass notes from a bass guitar or a big drum produce sound waves that are spaced far apart so they have long wavelengths and low frequencies. The treble notes from a cymbal or a tiny bell produce sound waves that are spaced closely together so they have short wavelengths and high frequencies.

When you look at a "frequency response" graph (they are called "amplitude response" graphs in BassBox Pro) you will usually see a horizontal frequency scale along the bottom as shown in the example below.

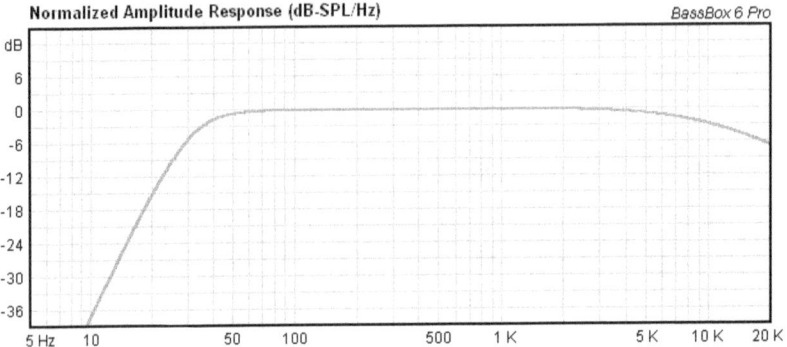

The low frequencies usually begin on the left end of the scale (5 Hz) and progress to the high frequencies on the right end (20 K). Notice that some numbers on the frequency scale are followed by a "K" . "K" means thousands and it can be combined with "Hz" to form "kHz" (pronounced "kilohertz"). So 1000 Hz equals 1 kHz and 20000 Hz equals 20 kHz. *Note: By itself, "K" is capitalized but a small "k" is used when it is added to Hz.*

Human hearing can detect sounds from approximately 20 Hz to 20000 Hz (20 kHz). However, few people can actually hear all the way to 20 kHz. Most high-fidelity speakers attempt to accurately reproduce sound waves in this audible frequency range. Sound waves whose frequency is less than 20 Hz are called <u>sub</u>sonic because they are below our range of hearing. But even though we cannot hear subsonic sounds we can "feel" some of them because they can cause tactile vibrations. Sound waves whose frequency is greater than 20 kHz are called <u>ultra</u>sonic because they are above our range of hearing and we usually cannot detect them with our ears. One possible exception occurs with ultrasonic sounds that are extremely loud—they can sometimes be perceived as a feeling of pressure in the ear— but this isn't very helpful because other phenomena can also create feelings of pressure in the ear.

Is there a benefit for a speaker to reproduce subsonic sound waves (less than 20 Hz)? The answer depends on the purpose of the speaker and the environment where it will be used. If the speaker will be used in an apartment building the answer is an emphatic "no"! unless you want angry neighbors. The answer is also "no" if the speaker will be used for live sound reinforcement because you will want to avoid low-frequency feedback with microphones. However, the answer may be "yes" if the speaker will be a part of an extended-range sound system for a movie theater and you want the audience to "feel" a dinosaur as it thunders across the screen. In this case a powerful subwoofer that can reproduce subsonic sounds may be desired—assuming that the theater has soundtracks that include subsonic sounds.

Is there any benefit for a speaker to reproduce ultrasonic sound waves (greater than 20 kHz)? Not really, unless you want to impress your dog which hears up to 30 kHz.

The table below shows the frequency ranges of some common acoustical music sources:

Piano:	20 Hz — 4096 Hz	Viola:	128 Hz — 1152 Hz
Human voice:	32 Hz — 4500 Hz	Trumpet:	160 Hz — 960 Hz
Bass viol:	40 Hz — 240 Hz	Clarinet:	160 Hz — 1536 Hz
Bass tuba:	43 Hz — 341 Hz	Violin:	192 Hz — 3072 Hz
Cello:	64 Hz — 682 Hz	Flute:	256 Hz — 2394 Hz
Kettle drum:	85 Hz — 170 Hz	Piccolo:	512 Hz — 4606 Hz
French horn:	100 Hz — 683 Hz		

sound level meter

Pressure: The Loudness of Sound

The loudness of sound is determined by the pressure that is created by sound waves when those waves push on a surface—in this case our ear drums. This is why engineers and technicians refer to the loudness level as the "sound pressure level" (SPL) and measure it with a sound pressure level meter. However, our amazing ears respond to sound pressure in a fascinating way—instead of responding to sound pressure in a linear fashion they respond to it in a logarithmic fashion. Let's explore this in more detail...

The unit used by most scientists and engineers to measure pressure is the pascal (Pa) which equals about 69 pounds per square inch (PSI). If our hearing were linear, a one pascal increase in pressure would be just as noticeable whether the change was from 1 to 2 Pa or from 10 to 11 Pa or from 100 to 101 Pa. But this is not the way our hearing works. As the sound pressure rises it takes a larger and larger increase before our ears notice a change in level—the level must be multiplied. This means that a change from 1 to 2 Pa is perceived similarly as a change from 10 to 20 Pa. This allows us to hear a phenomenally wide range of sound levels from a pressure of 0.00002 Pa (the quietest sound level we can hear) to a pressure of 100 Pa (a sound level so loud that it causes most people to feel pain). By the way, 100 Pa is 5000000 (five million) times more pressure than 0.00002 Pa!

Fortunately, sound levels are rarely described in pascals or we would have to be experts in the use of decimals. Instead of pascals, sound levels are measured in a unit called the "decibel" (dB). *Note: The "B" is capitalized in "dB".* As mentioned previously the threshold of our hearing begins with sound levels that have a small pressure of only 0.00002 Pa. This equals 20 dB. And a high pressure of 100 Pa causes most people to feel pain. This equals 134 dB. The table below lists some typical sound levels:

30 dB	Voice: very soft whisper
40 dB	Quiet residential area at night
50 dB	Average residence with television
60 dB	Shopping mall
70 dB	Voice: typical conversation level
70 dB	Freight train (from 100 ft or 31 m)
75 dB	Vacuum cleaner
82 dB	Pneumatic drill
90 dB	Subway train (from 20 ft or 6 m)
95 dB	Automobile on a freeway
110 dB	Very loud radio
115 dB	Circular saw
120 dB	Jet takeoff (from 200 ft or 61 m)
165 dB	Turbojet engine with afterburner
195 dB	Saturn rocket launch

What is a decibel? A decibel is the logarithm of a ratio. It can be any kind of a ratio but in audio it is usually a pressure ratio, a voltage ratio or a power ratio. A 2:1 pressure or voltage ratio equals a 6 dB change in level and a 2:1 power ratio equals a 3 dB change in level.

Let's put the decibel in perspective. For most people, a 10 dB change in sound level will sound twice as loud as the level before the change. A 3 dB change in sound level is very small but most everyone can hear it. A 2 dB change in sound level is very, very small and some people will have trouble hearing it. A 1 dB change in sound level is super small and very few people (unless their hearing is trained or special test signals are used) will be able to hear it.

Another way to view the decibel is with its relationship to amplifier power and driver cone excursion. To increase the sound pressure level just 3 dB, you have to double the amplifier power and increase the cone excursion by 50%. Suppose that you have a speaker with a sensitivity of 90 dB and a cone excursion of 1 mm (0.04 inch) when driven with 1 watt from an amplifier. How loud will the speaker be and what will its cone excursion be if it is driven with 2 watts? Answer: 93 dB and 1.5 mm. The table at the top of the next page further illustrates this progression:

Amplifier Power	Sound Level	Cone Excursion
1 watt	90 dB	1 mm (0.04 in)
2 watts	93 dB	1.5 mm (0.06 in)
4 watts	96 dB	2.25 mm (0.09 in)
8 watts	99 dB	3.38 mm (0.13 in)
16 watts	102 dB	5.06 mm (0.20 in)
32 watts	105 dB	7.59 mm (0.30 in)
64 watts	108 dB	11.39 mm (0.45 in)
128 watts	111 dB	17.09 mm (0.67 in)
256 watts	114 dB	25.63 mm (1.01 in)
512 watts	117 dB	38.44 mm (1.51 in)
1024 watts	120 dB	57.67 mm (2.27 in)

The important point here is that small changes in dB require large changes in power and excursion. This should make the advantages of a high-efficiency speaker obvious—because not many speakers can handle 1000 watts of power and produce 57 mm (2¼ in) of cone excursion.

How loud is too loud?

The answer to this question depends on the type of sound and the duration. Most of us realize that really loud sounds can cause permanent hearing damage. This is especially true of very sudden sounds, like gun fire, because the natural protective mechanisms of our ears don't have time to respond. But the threshold of pain is quite high at 134 dB. Will hearing damage occur at lower levels? Yes. If you listen to an average sound level of only 85 dB for 8 hours per day, you will eventually suffer permanent hearing damage.

Sound Pressure Level   Acoustic Power Level

Sound Pressure Levels versus Acoustic Power Levels

Sound pressure level measurements are always referenced to a distance and direction from the sound source. This is because the sound pressure level decreases as you move farther away from the source. And unless a sound source is perfectly omnidirectional (produces a uniform sound level in all directions), it will be louder or quieter at different directions. Most speakers are louder on axis—that is, they are louder in the direction they are pointed. This is why speaker sensitivity ratings are usually specified at 1 watt and 1 meter. This means that the speaker will produce the specified sound pressure level at a distance of 1 meter when it is driven with 1 watt of power. If no direction is specified, it is assumed that the direction is straight ahead or "on axis".

Unfortunately, this method of specifying loudness has its share of shortcomings. This is because the off-axis sound can have a dramatic effect on the overall quality of the sound in many acoustical environments. For example, sound emanating from the sides, top, bottom and back of the speaker may reflect off side walls, ceiling, floor and the rear wall to alter the character and apparent loudness of the speaker. Depending on the delays between the direct on-axis sound and the various reflected sound waves, the overall sound may seem to stay the same, be improved or made worse. Is there another way to measure loudness? You bet!

Another way to measure loudness is to measure the total energy from a sound source in all directions. This is the "acoustic power level" measurement and its value does not change when the distance or direction from the sound source is changed. Its only reference is the amount of power driving the source, usually 1 watt. Like sound pressure, acoustic power can use the decibel as its unit of measurement. The acoustic power level offers two advantages: 1) It provides a good overall picture of the total energy radiating from a speaker and therefore provides a better idea of its loudness in a reflective or reverberant acoustical environment, and 2) It provides a convenient way to compare one speaker to another. BassBox Pro includes a Maximum Acoustic Power graph to make these comparisons easy.

Summary
Sound is an exciting phenomenon to study and much more can be said about it. But this chapter is designed to be an introduction only. A few highlights are listed next: Sound is a wave and bass notes have long wavelengths and treble notes have short wavelengths. We recognize different sounds primarily by the frequency of their waves. Frequency is measured in hertz (Hz) and one Hz equals one wave per second. Loudness is determined by the pressure level created by sound waves and is measured in units called decibels (dB). Our threshold of hearing is a mere 20 dB and our threshold of pain is an enormous 134 dB. A small 3 dB change in sound level requires that the amplifier power double and that the cone excursion of the driver increase 50%.

Guide: 2 How a Speaker Works   BassBox 6 Pro

2 How a Speaker Works

The word "speaker" is the shortened form of the word "loudspeaker" and it refers to a device that converts electrical signals into sound waves that we can hear. A speaker has several parts:

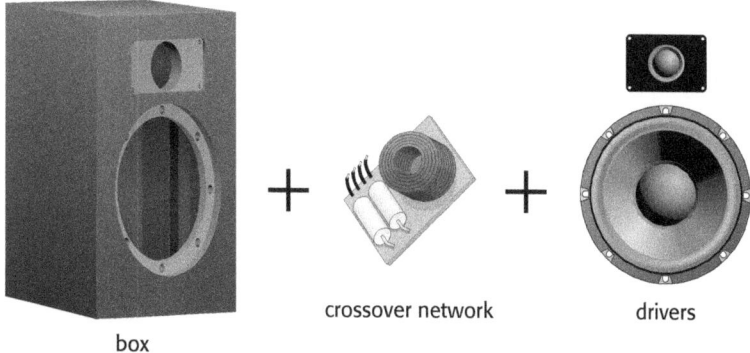

box   crossover network   drivers

- **Box** – houses the drivers and, if present, a passive crossover network.
- **Crossover network** – divides the audio signal between the drivers.
- **Driver** – converts the electrical audio signals into sound waves.

Sometimes people refer to the drivers as speakers. Yet, the speaker is really the entire system, including box, crossover network (if present) and drivers. Let's examine each part beginning with the drivers.

Drivers

There are many different kinds of drivers but they all do basically the same thing: create sound waves. By far, the most common type of driver is the moving coil electrodynamic piston driver. It has a moving part called a diaphragm that acts like a piston to pump air and

 **BassBox 6 Pro**   *Guide: 2 How a Speaker Works*

thereby create sound waves. A common diaphragm for a woofer is a paper cone. A common diaphragm for a tweeter is a fabric dome. The moving coil piston driver is the type that BassBox Pro models so our discussion will focus on them.

Why do drivers come in so many different sizes? Because it is nearly impossible to make one piston driver that can reproduce sound waves over the entire 20 Hz to 20 kHz frequency range of human hearing. To produce low frequencies a driver needs to have a large diaphragm and enough mass to resonate at a low frequency. To produce high frequencies a driver needs to have a small diaphragm with a low mass. Obviously, these requirements are in opposition and so drivers are usually designed to produce only a portion of the sound. This gives rise to multi-way speaker systems like the two-way system shown previously. It uses a tweeter for the high frequencies and a woofer for the low frequencies.

Besides their different sizes, there is another very significant difference between a tweeter and a woofer. The tweeter usually has a sealed back while the woofer usually has an open back. With a sealed back, the tweeter creates sound waves on the front side only. A woofer creates sound waves on both its front and back side. We'll see later that box design is dominated by the woofer because of this difference.

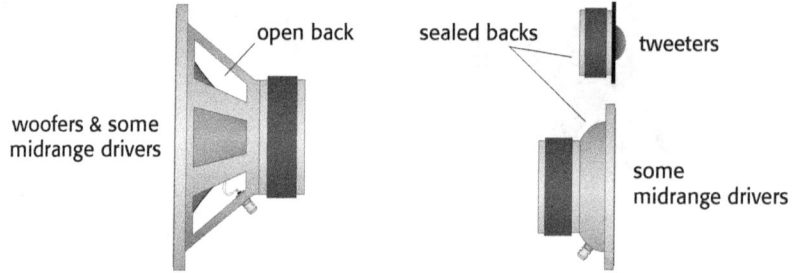

Tweeters are the smaller drivers since they produce the highest frequencies with the shortest wavelengths. Woofers are the largest drivers since they produce the lowest frequencies with the longest wavelengths. Are there other driver sizes? Yes, there are also midrange drivers of various sizes that reproduce middle frequencies between the tweeter and woofer. Midrange drivers are used in multi-way speakers with three or more driver sizes. Some have open backs and some have sealed backs.

A cutaway view of a typical woofer is shown on the next page:

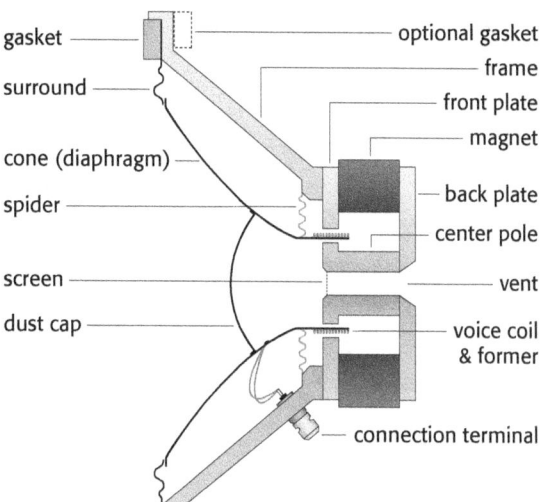

Let's examine each of these parts. They are organized below according to their function.

Motor parts:

- **back plate, center pole, front plate** – these three parts are usually made of iron or a similar permeable material to form the magnet circuit with the magnet. The front plate and center pole piece form the "gap" of the magnet circuit. All three parts, along with the magnet and frame, serve to dissipate heat away from the voice coil. Some drivers (usually small ones such as tweeters and midrange drivers) include ferrofluid in the gap to further cool the voice coil and provide resonance damping.

- **magnet** – provides a stationary magnetic field to oppose the alternating electromagnetic field of the voice coil and thereby cause the attached diaphragm to move inward and outward. Most drivers use a ring shaped magnet that is made of a ferrous ceramic material.

- **screen, vent** – some drivers include a rear vent to prevent pressure from building behind the cone in the magnet assembly and to provide cooling of the voice coil. A screen is usually provided to prevent debris from entering through the vent.

- **voice coil & former** – the voice coil is a coil of wire, usually copper or aluminum, through which the electrical audio signal flows. The flowing current of the audio signal alternates, creating an electromagnetic field which is opposed by the permanent magnetic field of the magnet circuit. This causes the voice coil and diaphragm to move. Some drivers have two (dual) voice coils to provide various wiring options, including the ability of simultaneously connecting two different signals to the driver.

Voice coils can be "overhung" or "underhung" as shown below:

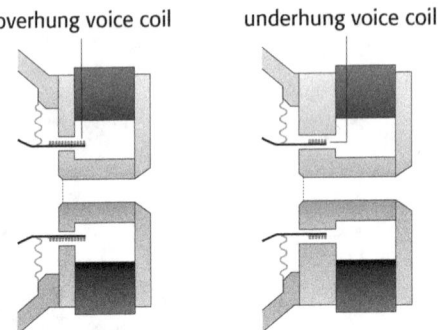

An overhung voice coil is taller than the height of the gap while an underhung voice coil is shorter than the height of the gap. Overhung voice coils are the most common. Underhung voice coils can offer the advantage of a more linear motor strength (BL product) over their excursion range (Xmax) but they are usually more expensive—especially when a large excursion is desired. Finally, the voice coil is wound around the former which serves as a heat-resistant spool for the wire.

- **connection terminal** – provides a way to make an electrical connection to the voice coil. A variety of terminal types are used, including simple push-on terminals or gold-plated 5-way binding posts. The positive terminal should be labeled "+" or with a red dot. When the driver is wired so that a positive signal flows to the positive terminal, the cone should move outward. If it moves backward, the terminal labels are reversed. Drivers with dual voice coils will have two sets of terminals.

Diaphragm parts:
- **cone** – also called the "diaphragm", moves like a piston to pump air and create sound waves. The mass of the moving parts (the cone, dust cap, voice coil and former) and the compliance of the suspension (surround and spider) control the resonance (Fs) of the driver which in turn controls its low-frequency response.
- **dust cap** – covers the hole in the center of the cone. This has several benefits: it reduces the amount of dust and dirt that can get into the gap of the magnet, it reduces the leakage losses (QL) through the driver, it adds strength to the cone while helping to maintain its shape and it can add mass to the cone to help lower the driver's resonance (Fs). Some dust caps include a screen or vent to allow airflow and aid cooling of the voice coil.

Suspension parts:
- **spider, surround** – these two parts form the suspension of the driver. The suspension fulfills several purposes: it centers (both axially and front-to-back) the voice coil

in the gap of the magnet circuit and it exerts a restoring force to keep it there, it limits the maximum mechanical excursion (Xmech) of the diaphragm and voice coil, it determines the compliance (Cms and Vas) of the driver and together with the mass of the moving parts determines the resonance (Fs) of the driver. Ideally, the suspension should provide a linear restoring force on the diaphragm and voice coil over its full range of excursion.

Frame parts:

- **frame** – also called the "basket" or "chassis", provides a rigid structure to which the driver components are mounted. It must be made with a high degree of precision so that all of the driver components will align properly. The frame can also aid the motor parts in dissipating heat away from the voice coil. It is commonly made of stamped steel, cast aluminum or plastic.

- **gasket, optional gasket** – most drivers include a front gasket to provide a smooth and flat mounting surface. However, since most drivers are mounted using the back side of the mounting flange, a rear (optional) gasket is often desired. The driver should have an airtight seal to the box.

Crossover Network

Most speakers must use more than one size driver because it is extremely difficult for one driver to accurately reproduce sound waves over the entire 20 Hz to 20 kHz frequency range of human hearing. The most common multi-way speakers use two drivers, a tweeter and a woofer. This requires the electrical audio signal to be divided into a high-frequency part and a low-frequency part before the signals reach the drivers. This is very important because most tweeters will be damaged if they are driven with a low-frequency signal. The illustration below shows the sound being divided between the tweeter and the woofer:

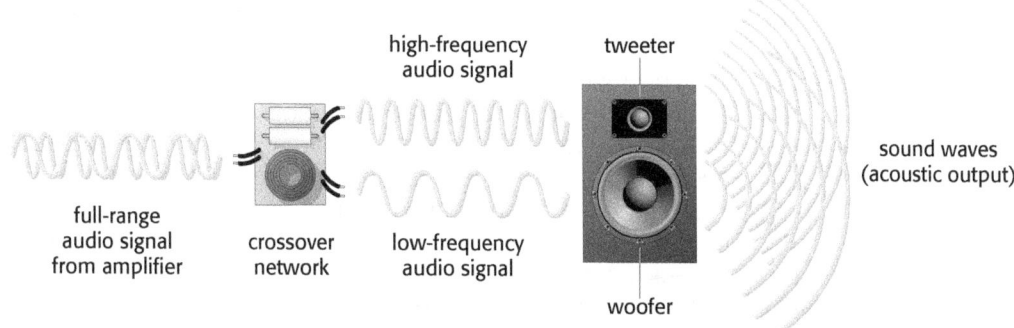

The job of the crossover network is to divide the audio signal. For this reason, crossover networks are sometimes called "dividing networks". The frequency where the sound is divided is called the "crossover frequency". Ideally, a crossover frequency is chosen which protects the tweeter, allowing it to produce only those frequencies that it can reproduce the best, and allows both the response and coverage pattern of the woofer to blend well with the tweeter. *Note: The "coverage pattern" is the shape of the listening area where a driver will provide a relatively uniform direct sound pressure level.*

If the speaker has more than two sizes of drivers the crossover network would also divide the audio signal into one or more additional midrange frequency bands.

There are two places where a crossover network can be placed in the audio system: after the amplifier or before the amplifier. Here are some points to consider for each location:

After the amplifier
- Generally the most common and least expensive location for a crossover network.
- Uses passive components that do not require an external power supply and so they are referred to as "passive crossover networks".
- Uses large components that can handle the full power delivered to the speaker.
- Is very sensitive to the impedance response of the drivers.
- Can be mounted inside or outside of the speaker box. When a crossover network is mounted in or on the speaker box, it is considered a part of the speaker.
- Only one amplifier channel is required per speaker because the audio signal is divided after it has been amplified.
- Are often easier for the hobbyist to construct at home because a printed circuit board and power supply are not required.

Before the amplifier
- Generally a more expensive location for a crossover network but it can produce higher fidelity and offer more adjustability.
- Usually uses active components that require an external power supply and so they are referred to as "active crossover networks". *Note: Passive crossover networks can also be used before the amplifier but this is rarely done because they are more difficult to design.*
- Uses smaller components since they are located "upstream" of the amplifier outputs.
- Is not affected by the impedance response of the drivers.
- Must be located between the preamplifier and power amplifier(s), usually in an equipment rack or cabinet. Because it is not located with the speaker, it is usually considered a separate component and not a part of the speaker.
- Requires a separate amplifier channel for each driver or crossover network filter and thereby raises the overall cost of the audio system.
- Are usually more difficult to construct because a printed circuit board and case (chassis) are often desired and an external power source is required.

The selection and design of a crossover network is a large subject and Harris Tech offers software to assist with the design of passive crossover networks. (Please visit our website at www.ht-audio.com for more information.) In this chapter, the important point to remember about crossover networks is that they divide the audio signal so that each driver in a multi-way speaker will receive only frequencies that they can handle and reproduce well.

Box

At first glance, it is obvious that the speaker box provides a place to mount the drivers. But is this the primary purpose of the box? The answer might be "yes" for tweeters because they have sealed backs and are not affected by the compliance of the air inside the box or, if present, the resonance of the box. *Note: "Compliance" is "springiness" or "mushiness". It describes how easily something can be compressed. A large volume of air is more "mushy" or compliant than a small volume of air. "Resonance" is the ability to prolong or amplify something. A bell resonates because it continues to make a sound long after it has been struck. Normally a closed box should not resonate. However, boxes with vents or passive radiators are designed to resonate.*

The answer to the preceding question is definitely "no" for drivers with open backs like woofers. A box provides two very important functions for an open back driver:

- Make it possible for an open-back driver to work efficiently.
- Shape the low-frequency response of an open-back driver.

Let's examine each of these functions in more detail:

Making a Driver Work Efficiently

In many ways the diaphragm of a speaker driver is very similar to the piston of a car engine. In the car engine, the piston needs a cylinder so it can create pressure. This is illustrated below:

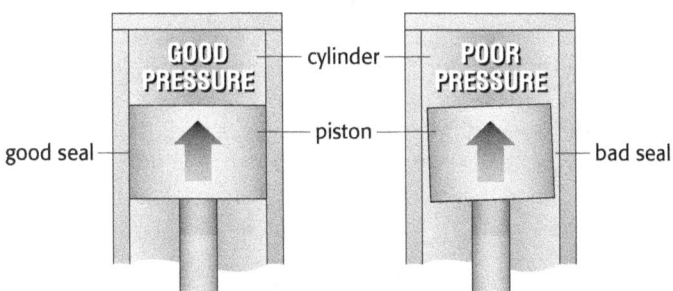

Guide: 2 How a Speaker Works

The piston in the engine moves up and down inside the cylinder. As it moves upward, it compresses a mixture of gasoline and air so that they will create a powerful explosion when the pressurized mixture is ignited by the spark plug. As the engine gets older the piston and cylinder wear and the seal between them begins to leak. When this happens, the cylinder loses pressure because the mixture of gasoline and air can leak past the piston. This causes the engine to lose power. In other words, it loses its ability to work efficiently.

In similar fashion, the diaphragm of a speaker driver needs to compress air so that it can create sound waves. But if the diaphragm of an open-back driver like a woofer is not sealed, the air can slip past it, reducing its ability to compress the air to create sound waves. So it is very important to seal the back side of the diaphragm with a box.

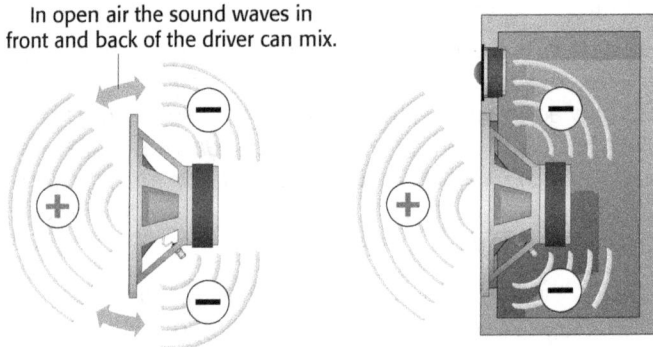

In open air the sound waves in front and back of the driver can mix.

Notice above that the woofer creates sound waves on both sides of the diaphragm. When the sound waves on the front are positive (the cone moves outward), the sound waves on the back will be negative or inverted. If the two should meet, they would cancel each other and you'd be left with no sound or very little sound. With this in mind, here's another way to describe this function of the box: The box prevents the sound waves which emanate with inverted polarity from the back side of the driver's piston from canceling the sound waves which emanate from the front side of the driver's diaphragm.

This need to isolate sound waves between the front and back of an open-back driver does not exist at all frequencies. It occurs only for lower frequencies whose wavelength is relatively large when compared to the piston diameter of the diaphragm. Why? Because the diaphragm becomes directional as the wavelengths become shorter and so the sound will naturally not mix between the front and back even when the driver is in open air. This is why a driver in open air lacks bass or sounds "thin". It produces mid and high frequencies without the rear sound waves canceling the front ones, but the low frequencies are diminished or cancelled.

Since the greatest effect of the box is to control the low-frequency response of open-back drivers, this is the focus of box design. And this leads us directly to the second function of the box:

Shaping the Low-Frequency Response
Enclosing the back side of an open-back driver in a box does more than just prevent the rear low-frequency sound waves from mixing with and canceling the front. The box also has a strong influence on the quality of the sound. Let's look at three examples. The graph below shows the amplitude response of three different boxes, each with the same model driver.

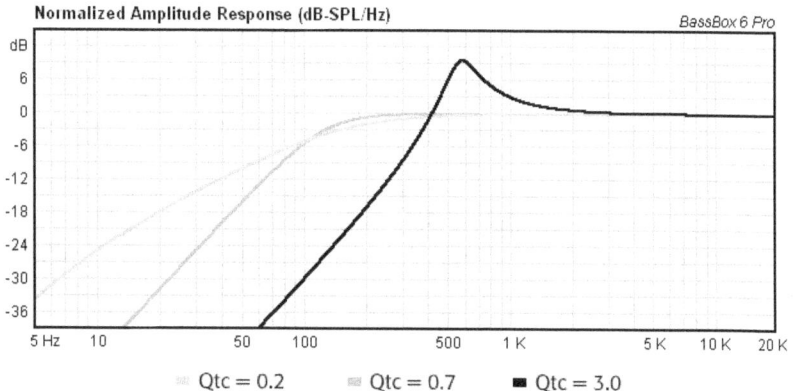

The medium grey curve is considered by many box designers to be ideal. This is because the response is "maximally flat", meaning that the response stays level for as long as it can before dropping off. The reason why a "flat" response is considered the best is because it allows the speaker to produce sound waves that more closely match those of the original audio signal. This is what "high fidelity" is all about. Of course, there are times when you do not want a flat response. For example, an electric guitar speaker is used to create sound—not reproduce sound—and so a non-flat response may be desired in order to give the guitar a desired sound quality.

The light grey curve was created by putting the same driver in a box that was over 100 times bigger than the maximally flat box. In fact, the box was so big that it makes the driver behave as if it were mounted in a really huge wall. Speaker designers call this an "infinite baffle" design. While the use of such a large box extends the low-frequency response, it does so at the expense of loudness because the level begins to decrease much earlier than the maximally flat box.

The black curve was created by putting the same driver in a box that was about 25 times smaller than the maximally flat box. Such a small box has a very dramatic effect. It created a 9 dB response peak and caused the response to drop off very early. The large peak indicates that the speaker will have a large resonance. This means that the box will "ring" like a bell at frequencies within the peak. This will make the speaker sound louder at those frequencies but it will do so at the expense of transient response because any sudden sounds (like the beat of a drum) will no longer sound as distinct as they originally did.

Mass & Compliance

To understand why these three boxes had the effect they did we need to learn a little more about the driver. There is a very good analogy that has been used by many to describe key characteristics of a driver. It is the analogy of the "weight and spring" (illustrated below).

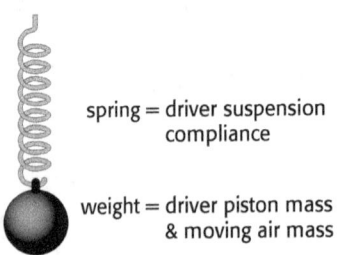

spring = driver suspension compliance

weight = driver piston mass & moving air mass

In this analogy, the spring represents the suspension of the driver (its surround and spider). The driver suspension acts like a spring to pull the piston back to the center whenever it is pushed or pulled by the motor (the voice coil and components of the magnet circuit). The weight on the end of the spring represents the mass of the diaphragm (the cone and dustcap) and the mass of the air that moves with the piston. Once the diaphragm is set in motion, the inertia of the mass exerts a force against the suspension.

Back to our analogy: If the weight is set in motion it will eventually settle into a particular up/down rhythm. This is its "resonance". No matter how fast or slow the weight was originally moving, it will eventually settle into the rhythm of its resonance frequency when there are no more external forces pushing or pulling it. There are two ways to change the frequency of its resonance: The length of the spring can be changed to make it more or less stiff and/or the weight can be changed. Both of the changes shown below will have a similar effect—they will lower the resonance frequency.

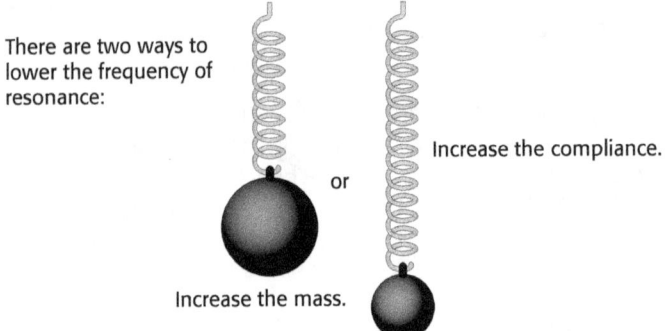

There are two ways to lower the frequency of resonance:

Increase the mass.

or

Increase the compliance.

This is why larger drivers usually have lower resonance frequencies—their larger diaphragms typically have more mass.

How does the box affect these characteristics? It has very little effect on the mass of the diaphragm. But it has a very significant effect on the compliance of the driver. In our preceding analogy, mounting a driver in a box causes the spring to shorten. This is because the volume of air inside the box has its own compliance which serves to "stiffen" or reduce the compliance of the driver. This "stiffening" of the overall compliance causes the driver resonance frequency and "Q" to increase. This introduces another characteristic: Q.

Damping & Q
Before we define Q, let's think about a car again. All automobiles have a suspension which prevents bumps and dips in the road from jarring the passengers. However, a car cannot use a spring suspension by itself because the car would continue to bounce (resonate) after each bump or dip. The bouncing needs to be suppressed and this is what the shock absorbers or dampers do. They suppress the unwanted bouncing after the springs have absorbed the initial impact of the bump or dip. In similar fashion, a speaker driver needs to be damped or its diaphragm will tend to vibrate excessively at its resonance frequency.

Several things serve to damp the driver. These include the suspension (surround and spider), motor (voice coil and magnet circuit) and amplifier output resistance. Achieving an optimal amount of damping can sometimes be a challenge and it almost always requires a consideration of the type of box in which the driver will be used. If the driver is not designed properly, it may be underdamped or overdamped. An underdamped driver will have ripples in its response and it will reproduce transient signals poorly. An overdamped driver will have a reduced low-frequency response.

"Q" is the term used by speaker designers and engineers to describe how well damped a driver is. However, Q is not damping. Instead, it is resonance magnification which is the exact <u>opposite</u> of damping. So a higher amount of damping results in a lower value of Q and visa versa. Let's return to our earlier graph:

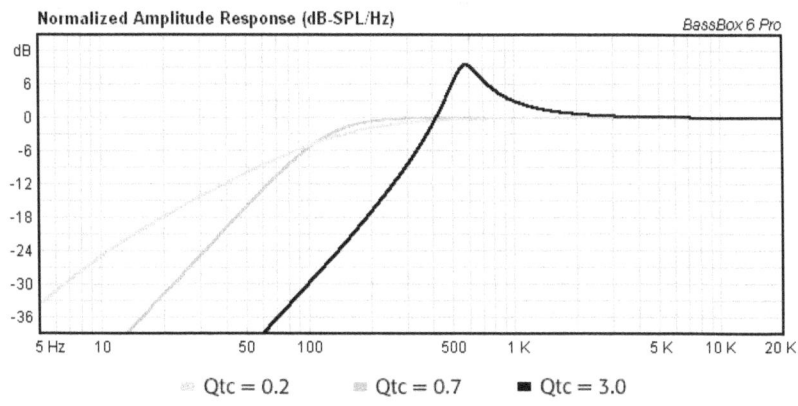

The driver which was used to create this graph has a total Q (Qts) that is super low at only 0.19. This low Q value results primarily from an extremely strong magnet and this driver by itself would be considered severely <u>over</u>damped. To get a maximally flat response from it required the overall compliance to be stiffened in order to counteract the overdamping. This was accomplished by putting the driver in a relatively small box having an air volume with a much smaller compliance than the driver suspension. This caused the overall compliance to decrease considerably and the total Q (Qtc) to increase to 0.7 as shown in the medium grey curve in the graph. A Q of 0.7 is considered by many to be the ideal value because it balances driver damping with a smooth low-frequency response and a reasonably good transient response.

The light grey curve was created with a huge box whose volume was over 100 times larger than the maximally flat box. As a result, the box has a very high compliance and so it could not stiffen the compliance of the driver much at all. In fact, the total Q was 0.2 which is just slightly greater than the driver, itself. Because the total Q is allowed to remain so low, the design is considered overdamped. Notice how it produces less low-frequency amplitude between 95 and 400 Hz. You might consider its greater output below 95 Hz to be an advantage but this is probably not the case because the driver would have to move a tremendous amount of air at these low frequencies and its maximum excursion (Xmax) may not be large enough to handle this. If it is not, it will at best have higher distortion and at worst become damaged.

The black curve was created with an ultra small box whose volume was over 25 times smaller than the maximally flat box. As a result, the box has a very low compliance and so it stiffened the compliance of the driver by a very large amount. In fact, the total Q was 3.0 which is quite high and causes the box to be underdamped. Notice how it produces a 9 dB response peak that is centered at 580 Hz. The peak occurs at this frequency because the low compliance of the box has in effect shortened the spring, raising the resonance to 580 Hz. The peak is 9 dB high because the driver lacks sufficient damping to overcome its resonance. This means that the driver will ring like a bell at 580 Hz. Needless to say, it will also have poor transient response at this frequency.

Note: The compliance of the driver and box are in series with each other but they sum in parallel like capacitors and so the sum is always less than the <u>lowest</u> of the driver or box compliance values. In other words, the lowest compliance dominates the system. With the light grey curve, the box had a compliance that was much higher than the driver and so the driver's compliance was dominant. Therefore the total compliance was not influenced much by the box. However, the box of the black curve had a compliance that was much lower than the driver, and being the lower value it became the dominant one and caused the total compliance to appear to be much lower.

Guide: 2 How a Speaker Works   BassBox 6 Pro

Vents & Passive Radiators

So far, our description of a speaker box has been that of a "closed" box. This is a box that has no openings to the outside and the only devices that radiate sound waves are the drivers. We've learned that the box causes both the resonance frequency and Q of the woofer to shift upward. By itself, the closed box is assumed to have no natural resonance of its own. At least this is the ideal, and whether or not it has resonance problems will depend on the construction details.

Would we ever want the box to intentionally resonate? And if we did, what would be its effect on the sound? The answer to the first question is yes. This is what vents and passive radiators do and their effect on the low-frequency response can be very dramatic.

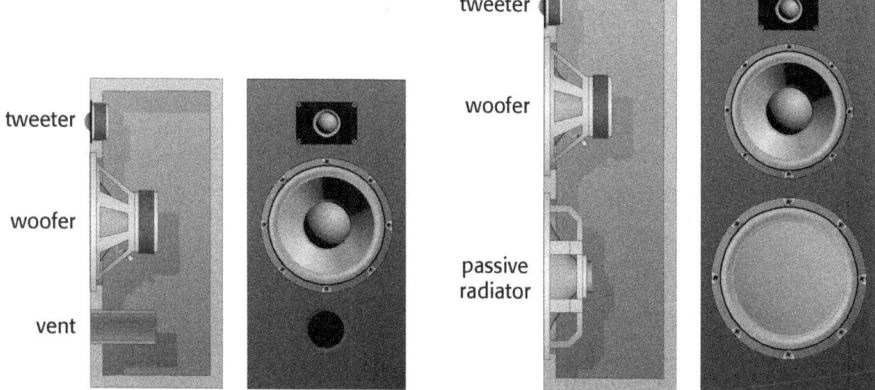

A vent is an opening in the wall of the box that allows air to flow in and out. As long as the opening is modest in size, it will not affect the box's ability to trap the unwanted sound waves that emanate from the rear of the woofer and it will not affect the compliance of the air in the box. The vent is usually constructed with a tube that is mounted in a round hole. The vent tube dimensions are calculated to cause the box to resonate at a desired frequency and the vent creates its own sound waves at this frequency. A sample vented box is shown at left in the illustration above.

A passive radiator is like a woofer without a motor. They used to be called "drone cones" and they can be made by removing the magnet assembly from a woofer. Because it has no magnet circuit the diaphragm is allowed to move freely. The mass of the passive radiator's diaphragm and the compliance of its surround cause it to resonate much like a driver. The net effect is very similar to a vented box. The mass of the diaphragm is adjusted in order to change the resonance frequency of the box. A sample passive radiator box is shown at right in the illustration above.

Using vents or passive radiators it is possible to tune the box resonance to a frequency that will extend the low-frequency response of the speaker. Sample response curves are shown below:

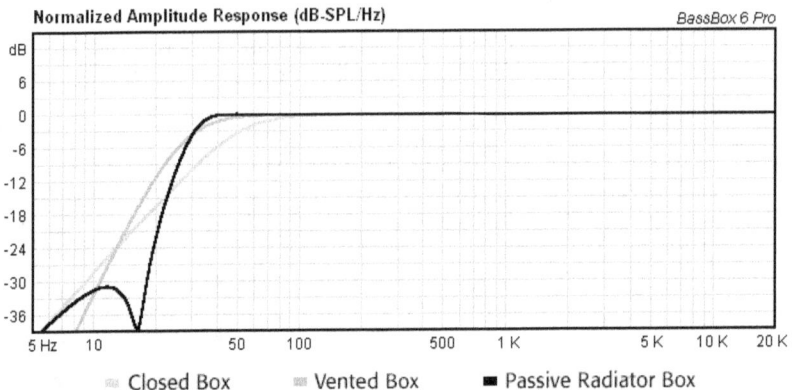

The curves in the above graph were created with a different driver than the previous graphs. The driver used here has a total Q that is equal to 0.38 and is therefore more universally acceptable to various box types. The light grey curve is a maximally flat closed box design. It is included for comparison. Notice that its −3 dB point is approximately 50 Hz. *Note: The low-frequency limit of a speaker's response is usually specified as the point where the response has dropped 3 dB. This is also called the "half-power frequency" because it takes a doubling of power to increase the sound pressure level 3 dB (this was explained in Chapter 1). So a decrease of 3 dB equals a halving of the power.*

The medium grey curve was created by placing the same driver in a maximally flat vented box. Two things are apparent. First, the vented box extends the low-frequency response lower than the closed box. It moved the −3 dB point down to 30 Hz. Second, the low-frequency response rolls off much faster than the closed box (a vented box has a 24 dB per octave cutoff rate which is double the 12 dB per octave cutoff rate of a closed box).

How does a vented box work? The air inside the vent acts like a piston and vibrates in response to the movement of the woofer's diaphragm. However some of the vent's sound waves lag behind those of the woofer, creating a phase shift. At and above the resonance frequency of the woofer, which was shifted upward by the box, the vent's sound waves have the same phase as those of the woofer and so they reinforce each other. At the box resonance frequency, the vent also damps the woofer so that its diaphragm moves very little while the air velocity in the vent reaches a maximum. Below the box resonance frequency, the phase of the vent's sound waves quickly shift 180 degrees so that they are out of phase with the woofer's sound waves. This does two things: First, the sound waves of both the woofer and vent begin to interfere and cancel each other below the box resonance. This produces the rapid low-frequency cutoff rate which can protect the woofer from excessive excursion. Second, it "unloads" the woofer which can make it susceptible to excessive excursion at ultra-low frequencies in spite of the rapid cutoff rate mentioned previously (however, this problem can be easily overcome with a subsonic or low-frequency high-pass filter).

The black curve was created by placing the same driver in a maximally flat passive radiator box. In general, a passive radiator performs very similarly to a vent with the exception that the passive radiator has a suspension and therefore a compliance and excursion limit that the vent does not have. However, if the passive radiator compliance and excursion limit are both very large, it will behave very close to a vent. And like a vent, the passive radiator will begin to move out of phase with the driver below the box resonance frequency, causing a rapid low-frequency cutoff rate similar to a vented box and unloading the woofer at ultra-low frequencies (although its compliance prevents it from unloading the woofer as much as a vent).

If a passive radiator is very similar to a vent, why does the example above have a huge notch in the response at approximately 16-17 Hz? This is a distinguishing feature of many but not all passive radiator boxes. It occurs at the resonance frequency of the passive radiator and results from the out-of-phase behavior of the passive radiator. However, if we plotted the response farther, we would see that it eventually assumes a nearly 24 dB per octave cutoff rate like a vented box. That being said, it is possible to tune a passive radiator box differently than a vented box and use the passive radiator's compliance and resonance to create a more gradual cutoff rate that is more similar to a closed box. However, these latter designs tend to have highly damped passive radiators which are not very efficient.

The next chapter "How to Choose a Box" will discuss the pros and cons of the different box types, including a box type that we haven't discussed yet, the bandpass box.

Summary

A speaker is usually comprised of three parts: a box, an optional passive crossover network and one or more drivers. This chapter has covered many of their key characteristics and explained how they work together to create sound. This summary will highlight some of the more prominent points:

Drivers:
- A single size electrodynamic driver cannot accurately reproduce the entire 20 Hz to 20 kHz frequency range of human hearing and so most high-fidelity speakers use more than one size driver.
- The low-frequency response of a driver is usually limited by its resonance (Fs) and size.
- The high-frequency response of a driver is usually limited by the size and mass of its diaphragm and by the inductive reactance of its voice coil.

 **BassBox 6 Pro**

- A driver behaves like a weight on a spring because it has mass and compliance. The driver's suspension has the compliance and is similar to the spring and the driver's diaphragm has the mass and is similar to the weight.
- A driver's resonance frequency can be lowered by adding mass to the diaphragm and/or making the suspension more compliant.

Crossover networks:
- Divide the audio signal into separate frequency bands so that each driver receives the portion that it can best reproduce.
- Come in two varieties: the more common passive crossover network which is located after the amplifier and the more expensive active crossover network which is located before the amplifier(s).

Boxes:
- A driver needs a box to prevent sound waves emanating from the back of the diaphragm from mixing with sound waves which emanate from the front. Without a box the low-frequency response will diminish because the front and back sound waves tend to cancel each other.
- A box will raise the resonance and total Q of a driver.
- A box can be used to shape the low-frequency response.
- By itself, a well constructed closed box should not resonate.
- By itself, a box with a vent or passive radiator is designed to resonate. This resonance is used to augment and shape the low-frequency response.

3 How to Choose a Box

The previous chapter provided an overview of the parts of a speaker, including the box. It also introduced some of the differences between a closed box, vented box and passive radiator box. This chapter will explore these boxes and the bandpass box in more detail and provide some guidelines for their selection.

Let's begin with a summary of each box type.

Closed Boxes

A "closed box" is a completely sealed box with only one chamber and no vents or passive radiators (see the illustration below). It is the simplest box type to construct and is generally smaller in size than other box types. It is forgiving of woofer parameter variations, making it a good choice when tight woofer parameter tolerances cannot be maintained. Closed boxes have a gradual 2nd order or 12 dB/octave low-frequency cutoff rate as depicted in the first graph below.

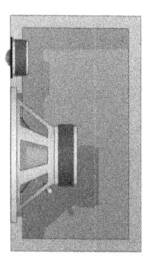  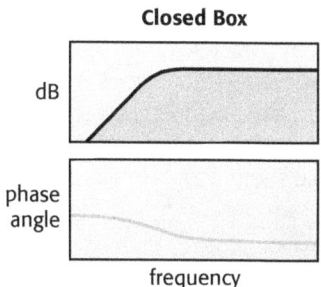

Since it has the most gradual cutoff rate, a closed box also has the least phase shift of the box designs—usually no more than 180° at the low frequency limit of its response (see the second graph above). This means that the low frequency sound waves created by the speaker will lag no more than half a wavelength and it also means that it will have the best low-frequency transient response. This is one reason why many audio purists believe that a closed box can provide the most accurate sound reproduction.

The 12 dB/octave cutoff rate of a closed box makes it a good match for use inside an automobile when a flat response is desired because most cars, trucks and vans have an acoustic response that rises at a rate of 12 dB/octave beginning at about 50 Hz. Notice what can happen when a closed box is chosen which perfectly compliments the acoustic response of a vehicle:

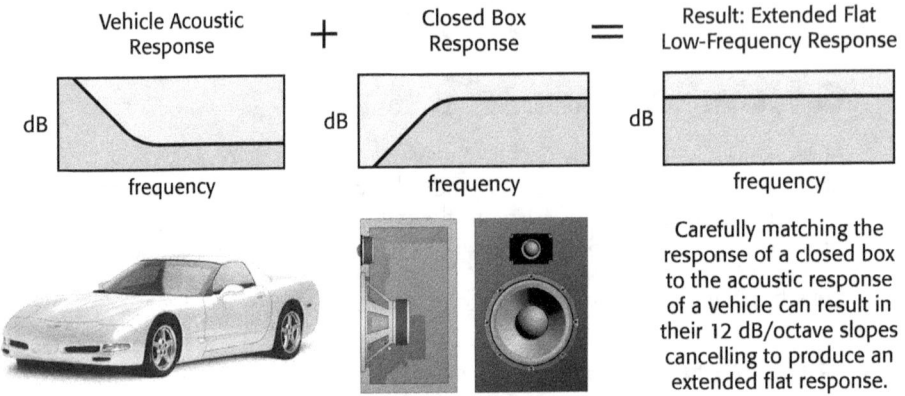

Carefully matching the response of a closed box to the acoustic response of a vehicle can result in their 12 dB/octave slopes cancelling to produce an extended flat response.

In this example, a closed box was chosen whose response begins to decrease at 12 dB/octave at the same time that the vehicle's acoustic response begins to increase at 12 dB/octave. The result is a very flat amplitude response. Of course, the excursion and thermal power handling limits of the woofer will also have a strong influence on the response at varying power levels.

Does a closed box handle more or less amplifier power than a comparable box with a vent or passive radiator? Does it have less or more nonlinear distortion? Is it more or less efficient? Does a closed box speaker cost more or less? These are important questions but they are sometimes difficult to answer. For example, some closed box characteristics enable it to handle more power while others reduce its power handling and which ones dominate will vary from one design to the next. The rest of this section will attempt to answer these questions.

Here are two characteristics that give a closed box a power handling advantage: First, without a vent or super-compliant passive radiator, a woofer in a closed box does not become "unloaded" below the box resonance frequency so it can handle more power at low frequencies before reaching its excursion limit. Second, a woofer in a closed box will have a higher electrical impedance at its cutoff frequency than a box with a vent or passive radiator so it can handle more power before the voice coil reaches its thermal limit.

However, the "unloading" problem described in the first point is often overcome by adding a subsonic high-pass filter to a vented or passive radiator box. When this is done, the gradual 12 dB/octave cutoff rate of a closed box causes it to handle less power than the higher cutoff rate of a comparable vented or passive radiator box. The gradual closed box cutoff rate causes the woofer's diaphragm to move greater distances at low frequencies so it will reach its excursion limit at a lower power level than a box with a vent or passive radiator. This requirement for a larger excursion means that closed box woofers must have a larger Xmax rating than a woofer that is used with a vent or passive radiator.

Two of the things which can increase the nonlinear distortion of a woofer are excessive excursion and changes in voice coil temperature. Since the gradual cutoff rate of a closed box can lead to greater excursion, it can also reduce the power handling and produce higher nonlinear distortion because the magnet system of the woofer will exert less control over the voice coil when fewer coils are within its gap. Fortunately, a closed box has an advantage with the second point, changes in voice coil temperature. The voice coil resistance (Re) increases when its temperature increases and visa versa. This in turn affects many characteristics of the woofer such as its electrical Q (Qes), motor strength (BL) and efficiency (ηo). Since the electrical impedance of a closed box is usually higher at its cutoff frequency than a comparable box with a vent or passive radiator, the voice coil of its woofer will not become as hot. With less heat, there is less change in temperature and the woofer of the closed box will suffer less dynamic nonlinearity.

To produce a similar low-frequency cutoff point (F3) as a box with a vent or passive radiator, a closed box woofer must have a lower free air resonance (Fs). This means that the woofer must have more moving mass (Mms) and this results in two disadvantages. First, the increased moving mass usually causes a closed box to be less efficient than other comparable box designs. Second, the increased moving mass also limits the upper frequency (midrange) response of the woofer, making it more likely that a full-range closed box speaker will need to use a midrange driver in addition to a tweeter, making it a more expensive three-way system.

Vented Boxes

A "vented box" has a single chamber with one or more openings that allow air to flow in and out of the box (see the illustration below). These openings are called "vents" and they are modest in size, compared to the overall size of the box, so that they will not affect the box's ability to trap the unwanted sound waves that emanate from the rear of the woofer and so they will not affect the compliance of the air inside the box.

A vent is usually constructed with a tube (also called a "duct" or "port") that is mounted through a round hole in the wall of the box. The vent tube dimensions are calculated to cause the box to resonate at a desired frequency (Fb) and the vent creates its own sound waves at this frequency. The box resonant frequency is often set close to the free air resonance of the driver (Fs). Two major advantages of a vented box are its ability to extend the bass response and its rapid 4th order or 24 dB/octave low-frequency cutoff rate as depicted in the first graph below.

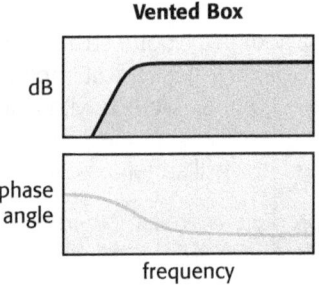

A vented box can have a phase shift as high as 360°—but 270° is more typical at the low frequency limit of its response (see the second graph above). This means that the low frequency sound waves created by the speaker may lag by as much as one full wavelength but usually by no more than three-quarters of a wavelength. With double the phase shift of a closed box, vented boxes reproduce low-frequency transient signals less accurately than a comparable closed box.

How does a vented box work?

The air inside the vent acts like a piston and vibrates in response to the movement of the woofer's diaphragm. The smaller the vent, the higher the air velocity and visa versa. Generally, the air velocity through the vent should be kept relatively low by using a vent of sufficient size. This avoids the problem of air turbulence noise in the vent. However the benefit of a large vent can be overshadowed by vent losses (QLv) because they increase when the vent size increases—a major source of vent losses is the viscosity friction of air as it moves through the vent. See the "Vent Coloration" section which follows for more information about air turbulence noise and other vent-related problems like "pipe" resonance.

Some of the vent's sound waves lag behind those of the woofer, creating a phase shift. At and above the resonance frequency of the woofer, the vent's sound waves have the same phase as those of the woofer so they reinforce each other. At the box resonance frequency, the vent also damps the woofer so that its diaphragm moves very little while the air velocity in the vent reaches a maximum.

Below the box resonance frequency, the phase of the vent's sound waves quickly shifts 180 degrees so that they are out of phase with the woofer's sound waves. This is both good and bad. It is good because the sound waves of both the woofer and vent begin to interfere and cancel each other below the box resonance. This produces a rapid 24 dB/octave low-frequency cutoff rate which can protect the woofer from excessive excursion. It is bad because it "unloads" the woofer which can make it susceptible to excessive excursion at ultra-low frequencies in spite of the rapid cutoff rate mentioned previously. Fortunately, the "unloading" problem is easy to overcome. All it takes is a low-frequency (usually subsonic) high-pass filter to protect the woofer from sound waves below its cutoff point.

Without a subsonic high-pass filter to protect it from "unloading", the power handling of a vented box is not as good as a comparable closed box. However, if this disadvantage is overcome with the filter mentioned above, it can be turned into an advantage. This is because the woofer in a vented box will require less excursion than a comparable woofer in a closed box so the vented box will generally have less nonlinear distortion from this source.

Another effect that results from the vent's damping action on the woofer is a significant reduction in the electrical impedance in the vicinity of the box resonance frequency. This reduced impedance has a bad side-effect. It causes the voice coil temperature to increase because the voice coil will now draw more current from the amplifier and as its temperature increases so does its resistance (Re). This in turn affects many characteristics of the woofer such as its electrical Q (Qes), motor strength (BL) and efficiency (ηo). The result: nonlinear distortion increases dynamically with changes in temperature.

A woofer that is designed for a vented box also tends to have a better midrange response than a woofer that is designed for a comparable closed box because the former can have a higher free air resonance (Fs). This means that the vented box woofer will have less moving mass (Mms) and be more efficient. It also means that a full-range vented box will more easily fit the requirements of a two-way system where the only other driver is a tweeter. When all of these factors are taken into consideration—less excursion (Xmax), higher resonance (Fs), less moving mass (Mms) and higher efficiency—it becomes evident that a woofer that is designed for a vented box will usually be less expensive than a woofer that is designed for a closed box. This combination of a less expensive two-way system design with a less expensive woofer makes vented box designs a good choice when cost is important.

Vent "Coloration"

Ideally, when accurate sound reproduction is desired, the woofer and vent in a vented box should reproduce the original audio signal as faithfully as possible. Unfortunately, an otherwise well-designed vent can sometimes alter the low-frequency response. This is commonly referred to as vent "coloration" and it has several sources. First, if the cross-sectional area of the vent is too small, then the air velocity through the vent may become too large and produce air turbulence noise such as a whistling or whooshing sound. As a general rule, the air velocity through the vent should be less than 10% of the speed of sound and some designers prefer that it never exceed 5%. Overcoming this problem requires that the total vent area be increased by using a larger vent diameter or by using more than one vent. However, as the total vent area increases, so too must the vent length and it is possible that the vent(s) will be too long to fit in the box.

Second, all vents have their own "pipe" resonance. As the size of a vent increases, these "pipe" resonance peaks move to lower frequencies and become stronger. One way to minimize this problem is to avoid excessively large vents. Fortunately this is seldom a problem for a standard vented box because the "pipe" resonance peaks are usually masked by the upper frequency response of the woofer. It is a more serious concern for bandpass boxes.

Third, reflected internal upper-frequency sound waves and standing waves that originally emanated from the backside of the woofer can sometimes escape through a vent. Remember, that one of the functions of the box is to trap the unwanted sound waves that emanate from the backside of the woofer. Fortunately, this problem can often be overcome with the use of damping material or "fill" inside the box. It can also be overcome by angling the internal walls in such a way as to prevent reflections from exiting the vent.

Vented Box with Active 2nd Order High-Pass Equalization Filter

BassBox Pro provides a variation of the vented box. This variation includes an external active 2nd order high-pass equalization filter. The filter usually has a Q of about 2 to produce a boost of about 6 dB followed by a 12 dB/octave cutoff rate. This filter can further extend the bass response of a vented box by as much as an octave below the woofer's free air resonance (Fs). It also increases the total cutoff rate of the speaker to a very steep 6th order or 36 dB/octave as shown in the third graph on the next page.

Here's how it works. First, a vented box with a maximally flat response is retuned, moving the box resonance (Fb) to a lower point. This is usually accomplished by lengthening its vent and it causes the low-frequency response to both extend and sag as shown in the second graph above. Second, an active 2nd order high-pass equalization filter is placed ahead of the amplifier input so that the amplified audio signal will boost the sagging bass response of the speaker and add a faster cutoff rate below the filter frequency as shown in the first graph. The third graph shows the results: an extension of the bass response and a 36 dB/octave cutoff rate.

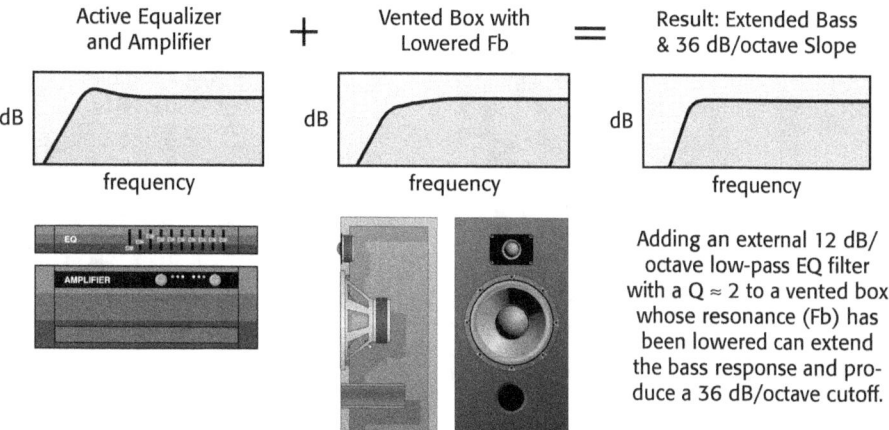

Adding an external 12 dB/octave low-pass EQ filter with a Q ≈ 2 to a vented box whose resonance (Fb) has been lowered can extend the bass response and produce a 36 dB/octave cutoff.

When designed properly, this type of vented box can handle more power because of the protection afforded by the steeper cutoff rate. The external filter also protects the woofer from the "unloading" problem mentioned earlier. However, in some cases this advantage can be overshadowed by increased nonlinear distortion due to higher excursion in the passband near the filter resonance frequency and increased voice coil temperature.

Vented boxes with active 2nd order high-pass equalization filters like the ones calculated in BassBox Pro are sometimes referred to as having a "B_6" alignment. This heralds back to the 6th order Butterworth alignment described by A. N. Thiele. However, BassBox Pro allows the user to control the filter parameters (Fx, Qx) and accepts a wider variety of woofers, resulting in response shapes that may not resemble a Butterworth shape.

Bandpass Boxes

A "bandpass box" is a very specialized form of a vented box. It is one of the most complicated types of boxes to construct because it completely encloses the woofer within a multi-chambered box so that sound waves can only emanate from one or more vents (see the illustration on the next page). *Note: It is possible to substitute passive radiators for the vents. However, BassBox Pro can only model a bandpass box that uses vents.*

As with regular vented boxes, bandpass boxes are vulnerable to vent "coloration", woofer "unloading" and higher voice coil temperatures than a comparable closed box. The vent coloration vulnerability is underscored by the fact that all of the sound waves created by a bandpass box emanate from its vents. Sources of vent coloration include: air turbulence noise, vent "pipe" resonance and reflected internal upper-frequency sound waves and standing waves that sometimes escape through a vent.

Most bandpass boxes have two internal chambers but boxes with more than one woofer can also have three chambers. The two outer chambers of a triple-chamber bandpass box are identical so the box still functions like a two-chamber box. An example of each is shown below.

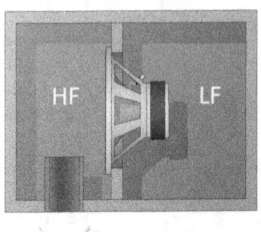
Double-Chamber
Bandpass Box

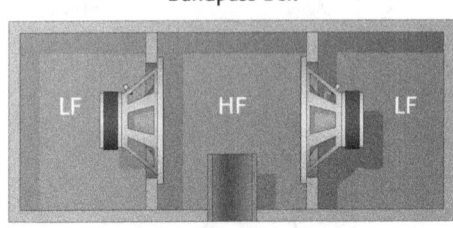
Triple-Chamber
Bandpass Box

HF Chamber controls the Upper Frequency Response
LF Chamber controls the Lower Frequency Response

In filter design, a band-pass filter is created by combining a high-pass filter with a low-pass filter. This is what a bandpass box does. One chamber acts like a high-pass acoustic filter to limit the lower frequency response—this is the sealed chamber in the illustration above. The second chamber acts like a low-pass acoustic filter to limit the upper frequency response—this is the vented chamber in the illustration above. The passband tends to be relatively small as shown in the graph below. For this reason, bandpass boxes are used almost exclusively as subwoofers and so they seldom have midrange drivers or tweeters.

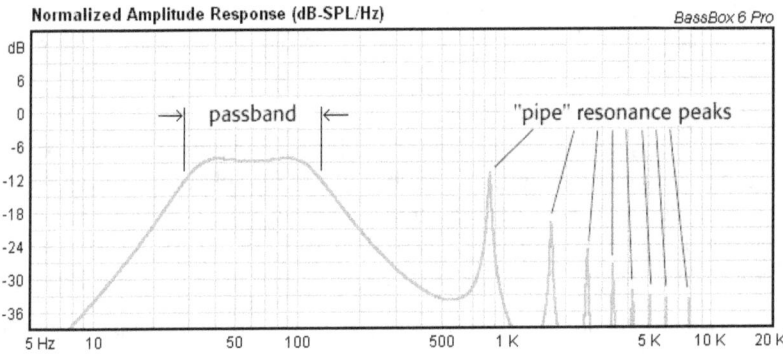

The low-pass quality of a bandpass box can be both good and bad. It can be good because it can eliminate the need for an external low-pass filter or the low-pass section in a crossover network. It can be bad because it can expose some rather "nasty" resonance peaks above the upper cutoff frequency (see the above illustration). These resonance peaks are created by the "pipe" resonance of the vents and they may or may not be a problem depending on their size. They can be minimized with either the addition of an appropriate low-pass filter or by keeping the vents to a modest size. However, if the vents are too small then the air velocity through them may be too high and produce another problem—air turbulence noise. Often, the vent resonance peaks are so narrow that they are very difficult to hear and measure. It is when they become large that they become a problem.

Note: So far, we have talked about one chamber acting like a high-pass acoustic filter to control the lower frequency response and another chamber acting like a low-pass acoustic filter to control the upper frequency response. This is a broad generalization which we will continue to use because it helps explain their predominant function. However, the frequencies of the high-pass and low-pass functions are spaced closely enough to "couple" the filters. This means that the chambers interact and so no chamber can really be viewed by itself because each one has an effect on the other. This is an important concept when you begin to experiment with a bandpass box design because changes to any one chamber will affect the overall response.

Another unique feature of a bandpass box is its ability to control efficiency. The system efficiency can be increased by narrowing the passband and/or raising the upper and lower cutoff frequencies. This makes it possible for a bandpass box to achieve much higher sound pressure levels than other box types. On the other hand, the system efficiency will decrease when the passband is widened and/or the lower and upper cutoff frequencies are lowered.

This concludes our brief overview of bandpass boxes. BassBox Pro models three different varieties. All three bandpass box types, single-tuned, parallel double-tuned and series double-tuned can each have either two chambers or, if more than one driver is used, three chambers. They are described next.

Single-Tuned Bandpass Box
A single-tuned bandpass box is the easiest bandpass box to design. It has a completely sealed chamber on one side of the driver to control the lower frequency response and a vented chamber on the other side to control the upper frequency response. Because only one chamber is vented it is said to be "single-tuned". Single-tuned bandpass boxes have a symmetrical 2nd order or 12 dB/octave cutoff rate for both the lower frequency and upper frequency response. This is the same cutoff rate as a closed box and a sample is shown in the first graph below.

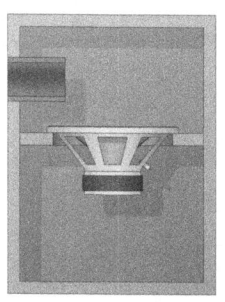

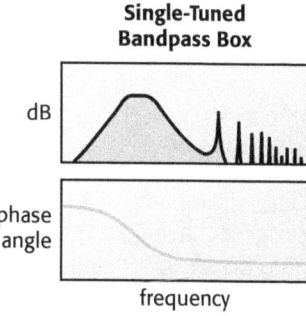

A single-tuned bandpass box has a phase shift that is similar to the maximum phase shift of a vented box—usually no more than 360° at the low frequency limit of its response (see the second graph above). This means that the low frequency sound waves created by the speaker can lag by as much as one full wavelength.

Like a closed box, a single-tuned bandpass box is a good match for use inside an automobile because it also has a 12 dB/octave low-frequency cutoff rate. A very flat and extended bass response is possible if this cutoff rate is carefully matched with the 12 dB/octave low-frequency rise that begins at about 50 Hz in most cars, trucks and vans.

Finally, to the best of our knowledge, single-tuned bandpass boxes are free of any patent restrictions. Double-tuned bandpass boxes are not.

Parallel Double-Tuned Bandpass Box
A parallel double-tuned bandpass box differs from a single-tuned bandpass box in that all of its chambers are vented. Because the chambers on each side of the driver are tuned to different frequencies the design is "double-tuned". Because all chambers vent to the outside, the tuning is "parallel". Parallel double-tuned bandpass boxes have an asymmetrical response with a 4th order or 24 dB/octave cutoff rate for the lower frequency response and a 2nd order or 12 dB/octave cutoff rate for the upper frequency response. A sample is shown in the first graph below.

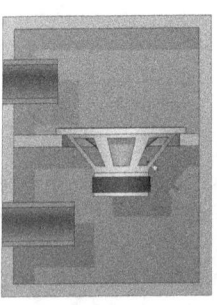
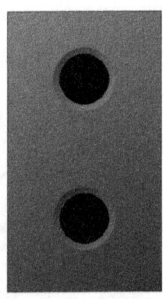
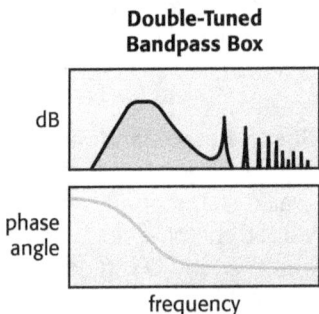

Double-tuned bandpass boxes (both parallel and series) have the greatest phase shift of all the box types that are modeled by BassBox Pro and therefore they have the worst low-frequency transient response. It can be as high as 540° at the low frequency limit of their response but a phase shift of 450° is more typical (see the second graph above). This means that the low frequency sound waves created by the speaker can lag by as much as one and a half wavelengths but usually by no more than one and a quarter wavelengths.

A significant advantage of a parallel double-tuned bandpass box over a single-tuned bandpass box having a similar passband and bandwidth is an increase in efficiency by as much as 6 dB. Unfortunately there is also a significant disadvantage: double-tuned bandpass boxes are more prone to vent "pipe" resonance problems.

Series Double-Tuned Bandpass Box

A series double-tuned bandpass box is similar to a parallel double-tuned bandpass box except that the low-frequency vent is enclosed inside the box. The vent is mounted in the wall that separates the chambers. This is the same inner wall to which the woofer is mounted. Because the lower frequency vent feeds into the upper frequency chamber, the only way for sound waves to leave the box is through the upper frequency vent, placing the lower frequency vent in "series" with the upper frequency vent. Series double-tuned bandpass boxes have an asymmetrical response that is very similar to a parallel double-tuned bandpass box (4th order lower frequency cutoff rate and 2nd order upper frequency cutoff rate). A sample is shown in the first graph below.

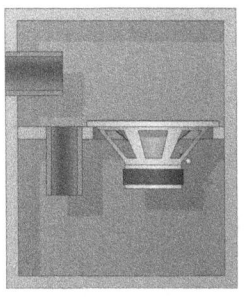

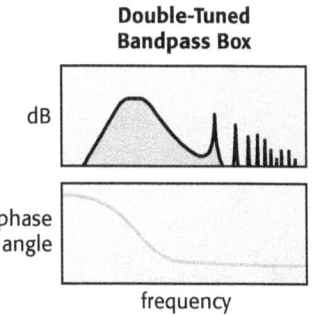

Series double-tuned bandpass boxes have the same phase shift (and transient response) as their parallel double-tuned siblings—450° to 540° at the low frequency limit of their response (see the second graph above). This means that the low frequency sound waves created by the speaker can lag by as much as one and a half wavelengths but usually by no more than one and a quarter wavelengths.

Since the lower frequency sound waves are forced to pass through the upper frequency vent, series double-tuned bandpass boxes are generally considered to be less desirable than their parallel counterparts. However, they can be useful when parallel vents are not physically possible because of design constraints.

The efficiency of a series double-tuned bandpass box is between that of a parallel double-tuned bandpass box and a single-tuned bandpass box. It is also more prone to vent "pipe" resonance problems like a parallel double-tuned bandpass box.

Passive Radiator Boxes

Like a closed box, a "passive radiator box" is a completely sealed box with only one chamber. However, the similarity to a closed box abruptly ends there because a passive radiator box also has one or more suspended diaphragms that vibrate in response to the movement of the woofer's own diaphragm. These former diaphragms are called "passive radiators" or "drone cones" because they radiate sound waves without a motor. In fact, you can convert a woofer into a passive radiator by removing its magnet assembly.

Unlike a vent, a passive radiator has a compliance and excursion limit. However, if the compliance and excursion limits are high enough, then the passive radiator will function very similarly to a vent. *Note: Usually a passive radiator should be capable of displacing 2½ times the air volume that the woofer can displace. If the passive radiator is the same diameter as the woofer, it will need to have an excursion limit that is 2.5 times greater.*

Its functional similarity to a vented box means that a passive radiator box has many of the same advantages such as an extended bass response and a rapid cutoff rate as depicted in the first graph below. A distinguishing characteristic of a passive radiator box is the presence of a notch or dip in the response at the resonance frequency of the passive radiator itself. This produces a more steep initial cutoff rate than a vented box and, unfortunately, a diminished transient response in the vicinity of the passive radiator resonance. *Note: The resonance frequency of a passive radiator is typically set at least an octave below the cutoff frequency of the box.*

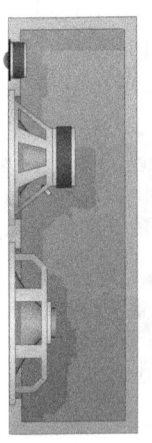

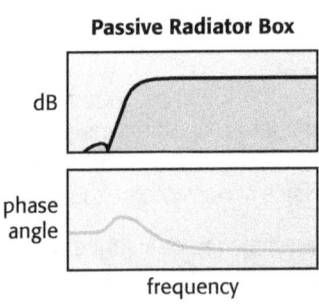

A passive radiator box exhibits a unique phase shift. It begins to increase at low frequencies very similar to a vented box. However, the phase shift suddenly decreases in the vicinity of the resonance of the passive radiator, causing the overall phase shift to decrease and assume a rise that is similar to a closed box (see the second graph above). Theoretically, a total phase shift of 450° is possible, but it is rare to see a phase shift that is greater than

270°. This means that the sound waves created by the speaker will usually lag by no more than three-quarters of a wavelength. However, the steep shift in phase at the resonance of the passive radiator reveals the diminished transient response mentioned earlier.

How does a passive radiator box work?
The passive radiator diaphragm vibrates in response to the movement of the woofer's diaphragm. However, some of the sound waves generated by the passive radiator lag behind those of the woofer, creating a phase shift. At and above the resonance frequency of the woofer, the passive radiator's sound waves have the same phase as those of the woofer and so they reinforce each other. At the box resonance frequency, the passive radiator also damps the woofer so that the woofer's diaphragm moves very little while the passive radiator's diaphragm reaches a maximum excursion.

This damping action on the woofer causes a significant reduction in the electrical impedance in the vicinity of the box resonance frequency. This reduced impedance has an undesirable side-effect. It causes the voice coil temperature to increase because the voice coil will now draw more current from the amplifier, and as its temperature increases so does its resistance (Re). This in turn affects many characteristics of the woofer such as its electrical Q (Qes), motor strength (BL) and efficiency (ηo). The result: nonlinear distortion increases dynamically with changes in temperature.

Below the box resonance frequency, the phase shift of the passive radiator's sound waves quickly shift 180 degrees so that they are out of phase with the woofer's sound waves. This produces the rapid 4th–5th order (24–30 dB/octave) low-frequency cutoff rate which can protect the woofer from excessive excursion. However, at the resonance frequency of the passive radiator, itself, the phase shift often creates a strong cancellation that produces a notch in the response as shown in the first graph of the preceding illustration. And somewhat like a vented box, the phase shift can begin to "unload" the woofer, making it susceptible to excessive excursion at ultra-low frequencies. Fortunately, the compliance and excursion limits of the passive radiator's suspension never allow the woofer to be fully "unloaded" and so this problem is less of a concern than it would be for a comparable vented box.

How does a passive radiator box compare to a vented box?
So far, we've seen that there are many similarities between a passive radiator box and a vented box. They both extend the bass response in a similar fashion and they both have a similar rapid low-frequency cutoff rate. And we have seen a few differences. A passive radiator box never fully "unloads" the woofer at ultra-low frequencies like a vented box and a passive radiator box has a distinguishing notch at the resonance frequency of the passive radiator which can diminish the transient response in this region. What other similarities and differences exist?

A passive radiator box can often achieve a slightly higher low-frequency output than a comparable vented box for two reasons. First, the area of a passive radiator is usually much larger than a vent and so it generates a lower air velocity. Second, a passive radiator does not have to overcome air viscosity friction like a vent. Both of these reasons result in fewer losses for a passive radiator.

A passive radiator avoids many of the problems associated with a vent. For example, a passive radiator may fit a box into which a vent will not fit because the required vent is too long. Also, a passive radiator avoids the vent "colorations" that can result from the noise of air turbulence through a vent, the "pipe" resonances of a vent and the reflected internal upper-frequency sound waves and standing waves that sometimes escape through a vent.

On the other hand, a passive radiator box is generally more difficult to design and construct than a vented box. Often one of the greatest difficulties is finding a suitable passive radiator. When one is found, its parameters must often be measured by the designer since few manufacturers provide full specifications for them.

Choosing a Box

Now that we've completed our review of each box type let's discuss their selection for different applications. Unfortunately there is no "absolute" technique for choosing a box type—any box type can be successfully used for most any application. Often it comes down to a matter of personal preference (or bias). Sometimes a box type must be selected based upon the components that are available. So the guidelines presented here are just that—guidelines—they are not absolute rules. Feel free to experiment with your owns ideas.

The flowchart on the next page is provided to help you make a selection. It is just one of the many possible approaches that can be employed in this decision-making process.

Guide: 3 How to Choose a Box

Selecting a Box Type

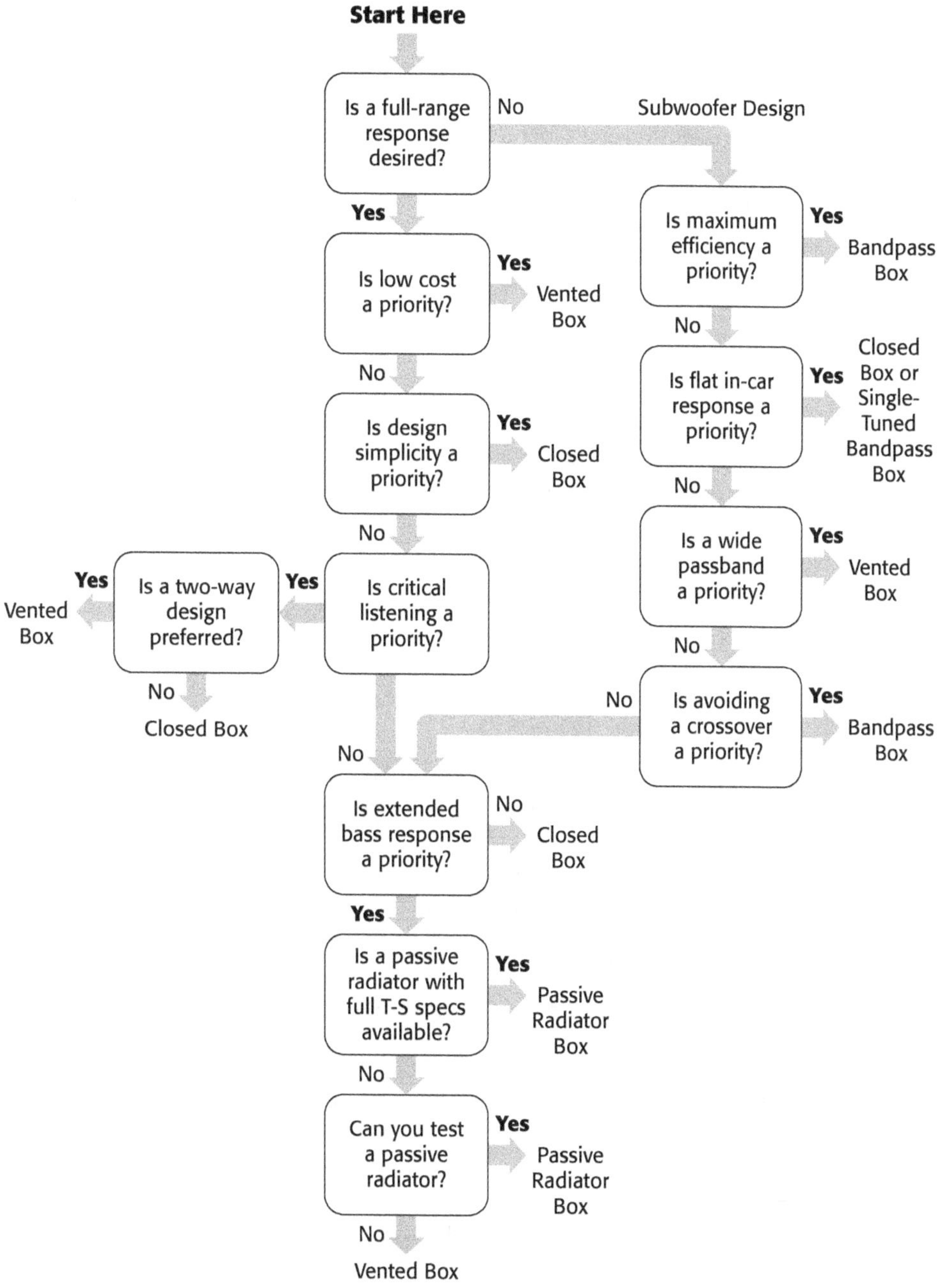

The preceding flowchart focused on how the speaker will be used. It probably will not help if part of the speaker already exists. For example, how do you choose a box for an existing driver? Or, how do you select a driver for an existing box? The following list of desired woofer characteristics for different box types should help you make these kinds of decisions.

Typical Characteristics of a **Closed Box Woofer**
- Qts greater than 0.3.
- Low resonance (Fs). High moving mass (Mms) and compliance (Vas, Cms).
- Large Xmax (small woofers: 2-4 mm; large woofers: 5-8 mm).
- Moderately strong motor (BL).
- EBP less than or equal to 50.

Typical Characteristics of a **Vented Box Woofer**
- Qts from 0.2 to 0.5.
- Moderate resonance (Fs) and moving mass (Mms).
- High compliance (Vas, Cms) is not necessary but can be beneficial.
- Moderate Xmax (small woofers: < 2 mm; large woofers: < 5 mm).
- Strong motor (BL).
- EBP greater than or equal to 100.
- Avoid woofers with excessive unknown losses.

Typical Characteristics of a **Single-Tuned Bandpass Box Woofer**
- Qts greater than 0.3.
- Low resonance (Fs). High moving mass (Mms) and compliance (Vas, Cms).
- Large Xmax (small woofers: 2-4 mm; large woofers: 5-8 mm).
- Moderately strong motor (BL).
- Compatible with a wide range of EBP values.

Typical Characteristics of a **Double-Tuned Bandpass Box Woofer**
- Qts greater than 0.4.
- Moderate resonance (Fs) and moving mass (Mms).
- High compliance (Vas, Cms) is not necessary but can be beneficial.
- Moderate Xmax (small woofers: < 2 mm; large woofers: < 5 mm).
- Strong motor (BL).

Typical Characteristics of a **Passive Radiator Box Woofer**
- Qts from 0.2 to 0.35.
- Moderate resonance (Fs) and moving mass (Mms).
- High compliance (Vas, Cms) is not necessary but can be beneficial.
- Moderate Xmax (small woofers: < 2 mm; large woofers: < 5 mm).
- Strong motor (BL).
- EBP greater than or equal to 100.
- Avoid woofers with excessive unknown losses.

Notes:
EBP is the efficiency bandwidth product and it is calculated by dividing Fs by Qes. If Qes is unknown, EBP can be estimated with: Fs / Qts. BassBox Pro provides an EBP indicator on the "Parameters" tab of the Driver Properties window and on the Driver Locator.

Although vented, double-tuned bandpass and passive radiator boxes do not usually require a woofer with a large Xmax, it won't hurt to have a large Xmax. In fact, a generous Xmax is always good because it reduces a source of nonlinear distortion.

Most vented and passive radiator boxes have a leakage loss parameter (QL) that is around 7. An excessively "lossy" driver can reduce this and should be avoided unless you know how to adjust QL accordingly. Sources of loss through a woofer are a porous dust cap or "lossy" surround.

When should you add an active 2nd order high-pass EQ filter to a vented box?

An external 2nd order high-pass filter is seldom added to a vented box because of the requirement for a separate low-frequency amplifier channel and the scarcety of suitable off-the-shelf active filters. Their most common application has been in the pro audio industry where bi-amp and tri-amp systems are common.

When should an external active 2nd order high-pass equalization filter be considered? The answer to this depends on two things: 1) Can the woofer handle a possible increase in excursion near its cutoff frequency? 2) Can the audio system accommodate the insertion of the required active filter between the preamplifier and power amplifier? If the answer to both questions is "yes" then you might want to consider it. It provides a way to extend the bass response without increasing the box size and it provides the subsonic protection that will prevent the driver from being "unloaded".

Which type of bandpass box should you choose?

If a bandpass box is selected, which one should you choose? First, let's examine some of the advantages of a single-tuned bandpass box. The single-tuned version is free from patent restrictions. The single-tuned version is the easiest to design and construct. With its 2nd-order cutoff rate, the single-tuned version is a good match for use inside an automobile. The single-tuned version has less phase shift and therefore potentially better low-frequency transient response.

Now let's examine some of the advantages of the parallel and series double-tuned bandpass boxes. Both double-tuned versions are more efficient (the parallel double-tuned is the most efficient). Both double-tuned versions provide a 4th-order cutoff rate which can increase displacement-limited power handling and can reduce some nonlinear distortion.

Summary

Selecting a box type is often a matter of "tradeoffs". For example, it may be desirable to accept the larger phase shift and reduced low-frequency transient response of a vented box in exchange for its extended low-frequency response over a closed box. At other times the opposite may be true. Understanding the strengths and weaknesses of closed, vented, bandpass and passive radiator boxes is the first step in making these kinds of decisions.

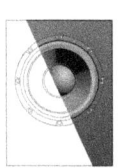 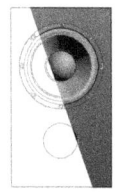 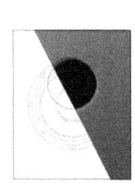 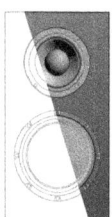

4 Constructing a Box

The purpose of the speaker enclosure or box is to trap unwanted sound emanating from the back of the driver and to add acoustic resistance so the driver can work more efficiently at low frequencies. For high-fidelity sound reproduction, the loudspeaker should not alter ("color") the sound but should faithfully reproduce the original signal. With these concepts in mind, this chapter will introduce some basic tips to help with box construction.

Avoiding Standing Waves

Most boxes are roughly shoebox or rectangular in shape. Technically speaking, this is a square prism. Unfortunately, these box shapes have a built-in flaw: box shapes with two or more sets of parallel sides will be prone to a phenomenon called "standing" waves. Standing waves are sound waves that reflect back and forth between two parallel sides inside the box. The frequency of a standing wave will have a wavelength equal to twice the distance between the two sides. For example, if two parallel walls are 24 inches (61 cm) apart, the standing wave that will develop between them will have a wavelength of 48 inches (122 cm). This corresponds to approximately 283 Hz. You can use the wavelength calculator in BassBox Pro to convert from the wavelength to the frequency (see page 327).

A cube is generally a bad box shape since all of its parallel sides are equidistant. This means that all the standings waves inside the box will have the same frequency and will sum together, making them louder. This reduces the fidelity of the loudspeaker. But all is not lost if you really like the cube shape because there are ways to overcome standing wave problems. Here are a few suggestions:

- Add damping to the interior of the box. Bear in mind, that most of the commonly-used absorption materials have poor absorption coefficients at low frequencies. If your box is large, you should probably calculate the frequencies of the standing waves and check the absorption coefficients at those frequencies for the damping material you plan to use. Also, see the Damping section in Chapter 4 of the *BassBox Pro Reference* later in this manual.

- Add one or more angled walls to the interior of the box. In this way, the outside of the box still has parallel walls, but the inside does not. These walls must be rigid and either have openings or be short in one dimension so that air can flow around them without any restrictions. Without parallel walls, the standing wave problem can be lessened.

- The frequencies of the standing waves can be spread apart so they do not sum together. This is accomplished with the "optimum square prism" box shape in BassBox Pro. It uses a 1.62 : 1 : 0.62 ratio between its three dimensions (height, width, depth) to spread apart the frequencies of the standing waves.
- Pick a box shape without parallel sides such as a four-sided pyramid.

Diffraction

The diffraction caused by some box shapes can have a very significant effect on fidelity. For example, a cube can introduce a ±5 dB variation to the amplitude response of a loudspeaker (depending upon the area of the mounting surface). A shape that produces very smooth amplitude response is a sphere. Unfortunately a sphere is not a very practical shape to use for many designs. Here are a few suggestions to help minimize diffraction:

- Locate the drivers so they are not the same distance from any two or more edges.
- Mount all drivers flush in the front panel.
- Round the edges and corners of the box. They must be rounded rather severely to be effective.

Materials

It has been suggested that the best loudspeaker box would be constructed with 12 inch thick lead. Others have constructed boxes with cement. Why such dense materials? Because a box with flimsy construction will vibrate or flex while the driver pumps air. And this will add distortion (coloration) and absorb power that would otherwise produce sound. Wood is not a very good material and yet it is the most common material from which boxes are made. Fortunately, good design and attention to detail can overcome the flexibility of wood and make a wooden box work very well.

What kind of wood is best? The most dense and least flexible. However, such hard woods are often very expensive. Plywood is not a good choice because it often has voids hidden in its laminations. Particle board (also know as "pressed board" or "chip board") is a fair inexpensive choice; however, medium density fiberboard (MDF) is considered one of the best low-cost materials.

In addition to reducing standing waves, adding damping material to the interior of the box can also reduce the unwanted resonances of wooden walls. For more information see the Damping section in Chapter 4 of the *BassBox Pro Reference* later in this manual.

Construction

Several construction tips follow. Of these, the two most important ones are airtight construction and rigid construction.

- A well constructed airtight box will have each joint glued and caulked. Wires that are run through a side will be carefully caulked where they pass through a wall. Obviously, a vented or bandpass box will not be airtight because it has a vent but all other aspects of the enclosure should be airtight just as with a closed box design. Make sure the driver is mounted to the box with an airtight seal. Use caulk if the driver does not have a suitable seal for this purpose.

- The walls of the box should not flex at all when the piston of the driver moves. Braces or cleats are usually fastened to the walls to make them more rigid and to divide the walls into smaller areas, each with their own resonant frequency. Braces can be added in any number of ways. One method is shown below:

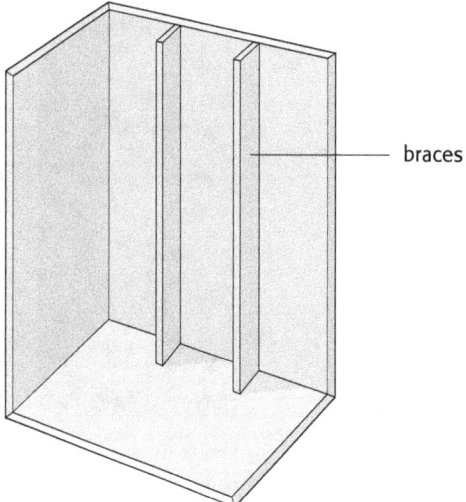

When you rap on the sides of a well constructed box with your knuckles you should not hear any "hollowness."

- Use an appropriate amount of damping material for the design. A small closed box often works well when stuffed 50% full (or more) with generous amounts of damping material because it makes the box "appear" to be larger. This helps to extend the low-frequency response a little. A medium or large closed box often has adequate volume without the need for lots of damping material. In this case, the damping material is used primarily to suppress unwanted standing waves and box resonances. Simply cover each surface inside the box with damping material. Vented, bandpass and passive radiator boxes are more sensitive to changes in the apparent box volume and usually have damping material added only to one of each pair of parallel sides. The goal being to absorb box resonances without affecting the apparent volume of the box. However, in actual practice, modest increases in damping material seldom cause vented or passive radiator designs any harm with one important caution: too much damping material can diminish the vent output of a vented box. The most important part of the box to damp are those surfaces which are directly behind or near to the driver.

- With multi-way loudspeakers, mount the drivers in a vertical line array for the best horizontal dispersion. And mount the drivers off-center so that diffraction will be minimized. An example three-way design is shown below:

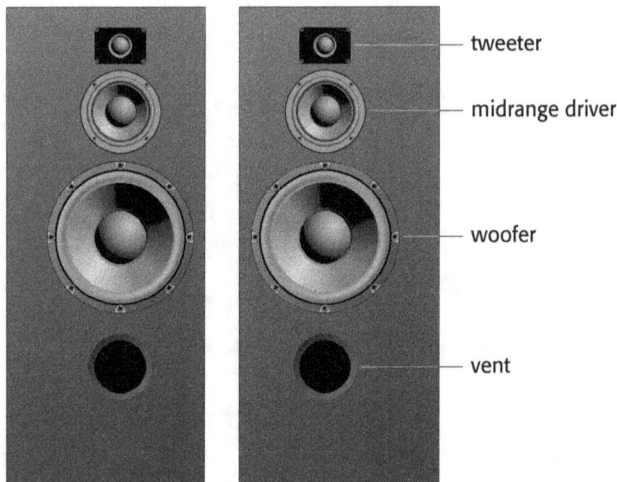

Lobing can usually be minimized by locating drivers with a shared crossover frequency as close together as possible. Try to keep the distance between their centers less than the wavelength of the crossover frequency. The tweeter should usually be located on top, closest to ear level since most direction cues come from the mid and higher frequencies. Driver placement can be greatly affected by the phase response of the crossover network. However, this leads into a subject that is beyond the scope of this manual.

- If a vented or bandpass box design is used, position the vent so that nothing obstructs the air flow in or out of the vent. A "rule of thumb" is to keep the end of the vent which is inside the box at least one diameter away from any internal walls.

- If present, vents should be located away from other structures, including box side walls. For round vents, keep each end at least one diameter away. For rectangular vents, keep each end the following distance away: square root of the height x width of the vent (one side of a square vent).

Slotted vents are not modeled by the program. They employ box walls as vent walls (usually, box walls are used for three of the vent sides and the fourth vent side is created by a ledge mounted behind the front wall—the ledge controls the vent length, Lv). Slot vents can be unpredictable to design because they usually appear longer than they really are, depending on the box wall geometry, since a larger mass of air is involved. They also tend to have higher air turbulence and higher friction as the air flows in and out. The rectangular vent with one flush end will provide a good starting point for a slot vent design. Then you will need to build a test box, measure the box's resonance (Fb) and shorten the vent (Lv) until the desired Fb is achieved.

Vent Placement Tips for Bandpass Boxes

Sometimes a chamber in a bandpass box is too small to accommodate the required vent length. Here are two ways to deal with this problem:

- Pass the vent through the adjoining chamber as shown in the left illustration below. Be careful that no air leaks in or out of the chamber through which the vent passes.

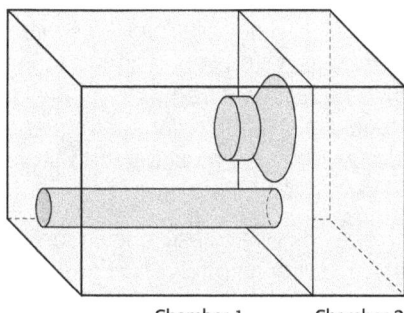
Chamber 1   Chamber 2

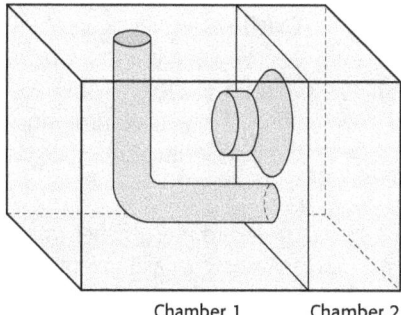
Chamber 1   Chamber 2

In some cases the vent may be too long or short to do this and you can try using an elbow to direct the vent to a different side of the box as shown in the right illustration above. *Note: If an elbow is used, be prepared to make adjustments to the duct length in order to achieve the desired resonance.*

Guide: 4 Constructing a Box

- Another way to handle a vent that won't fit inside a chamber is to allow the vent to project beyond the wall of the box as shown below. This is often a convenient way to direct the sound from an enclosure in the trunk of an automobile up to the rear deck.

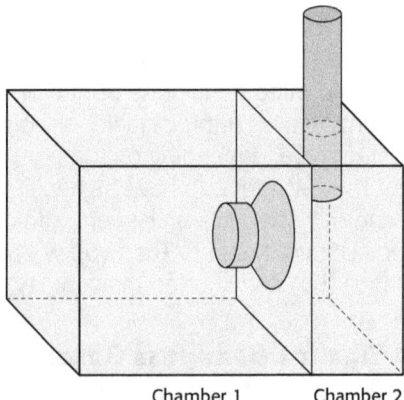

Chamber 1   Chamber 2

Note: Some of these examples may require you to manually adjust the vent length in order to achieve the desired resonance. Remember that BassBox Pro assumes that each end of the vent will be located away from any obstructions that would affect the air flow.

Caution: Very long vent lengths may produce an audible "pipe" resonance. Try to avoid excessive vent lengths. One way to reduce the required vent length is to increase the internal volume of the chamber. Remember that the resonant frequency of the chamber is a function of both its volume and the vent dimensions. If the resonant frequency is held constant, the smaller the chamber volume, the longer the vent must be and visa versa.

There are many other important considerations when designing a loudspeaker system such as the signal alignment between drivers, avoiding harmful reflections, crossover network design, high-frequency and mid-frequency driver selection, etc. Unfortunately these subjects are beyond the scope of this manual. For more information please refer to Appendix E for a list of suggested reading (pages 356-357).

Sample Designs

Several sample designs are presented in this section to illustrate the use of BassBox Pro. Imperfect designs where intentionally chosen to illustrate how to identify common problems like excessive cone excursion. Beginners are advised to begin with Chapter 1 because it is the most detailed, is written with the new box designer in mind, and lays a foundation for the vented, bandpass and passive radiator design samples that follow.

Chapters

1. Closed Box Example 71
2. Vented Box Example 99
3. Bandpass Box Example 117
4. Passive Radiator Box Example 139

SAMPLE DESIGNS

Sample 1: Closed Box Example BassBox 6 Pro

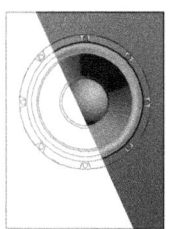

1 Closed Box Example

This chapter shows how to use BassBox Pro to design a closed box. The Design Wizard will not be used so that the operation of BassBox Pro can be examined in greater detail. This closed box example is intentionally more detailed than the others so that it will help first-time users of the program who are also novice box designers progress through an entire design from start to finish in a step-by-step fashion.

Don't let its simplicity fool you—a closed box is one of the most versatile and best sounding boxes around and this example will show how to design one for an automotive vehicle.

Jeep Wrangler

Box Properties

Our goal is to design a subwoofer for a Jeep Wrangler which will be mounted behind the rear seat. The front of the box will need to be slanted so that it will fit nicely in the available space. Since the space dictates the maximum available box size and shape, we will begin with the box dimensions and then find a suitable driver. Let's begin:

- Run BassBox Pro.
- When the title window appears, click the "Open Design Window" button shown at right.

Select the "Open Design Window" button.

BassBox 6 Pro User Manual © D.E.Harris

BassBox 6 Pro

Sample 1: Closed Box Example

1

- Select "New Design" from the File menu to begin a new design as shown below. This can also be accomplished from the keyboard by pressing [Ctrl]+[N].

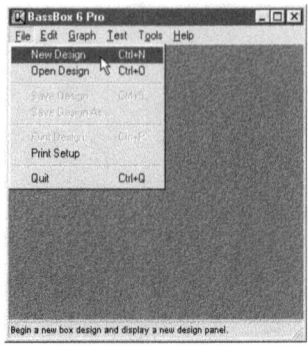

- Click on the "Box" button on the new design panel (shown at right) to open the Box Properties window. This can also be accomplished from the keyboard by pressing [Ctrl]+[B].

Select the "Box" button.

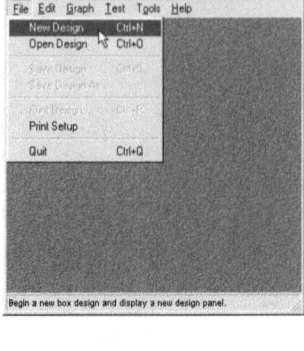

Select the "Box Design" tab.   Select the box shape.   Select external dimensions.

- Select the "Box Design" tab. Then select the "Prism, slanted front" box shape with the "Shape" drop-down list as shown at right.
- Set the dimensions to "External" in the upper right corner of the "Box Design" tab. The external dimensions are used because we must fit the box into an existing space.

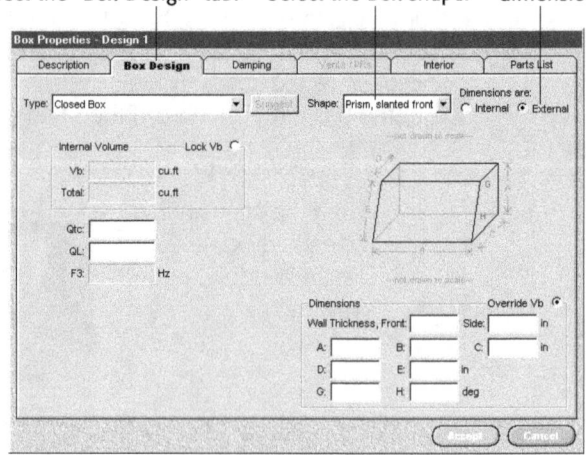

72 BassBox 6 Pro User Manual © D.E.Harris

Sample 1: Closed Box Example

BassBox 6 Pro

- Enter the box dimensions. *Note: Before entering the dimensions, check to see if the desired units are selected (see the illustration below). If not, click on one of the unit labels to change them before entering the dimensions. The box dimension unit labels are synchronized so they will change in unison when just one of them is clicked. They will advance from "mm" (millimeters) to "cm" (centimeters) to "m" (meters) to "in" (inches) with each single mouse click—double-clicks will be ignored. In our example we chose inches.*

  Begin by entering a wall thickness of 0.75 inches (19 mm) for both the "Front" and "Side" because we will be constructing the box entirely with ¾-inch marine plywood. We chose marine plywood because the Jeep is a convertible and may accidentally get wet. Dimension "B" is 34 inches (864 mm) because this is the width behind the rear seat. Dimension "C" is 12 inches (305 mm) because this is the bottom depth behind the rear seat along the floor. We chose a box height of 10 inches (254 mm) for dimension "A". At this height the top depth will also be 10 inches (254 mm) and so we entered this for dimension "D". After entering the above dimensions (B, C, A and D) BassBox Pro now has enough information to fully model the box size and shape. So, the remaining dimensions will be automatically calculated for you (dimension "E" = 10.2 inches and angles "G" = 101.3° and "H" = 78.69°). See the example below:

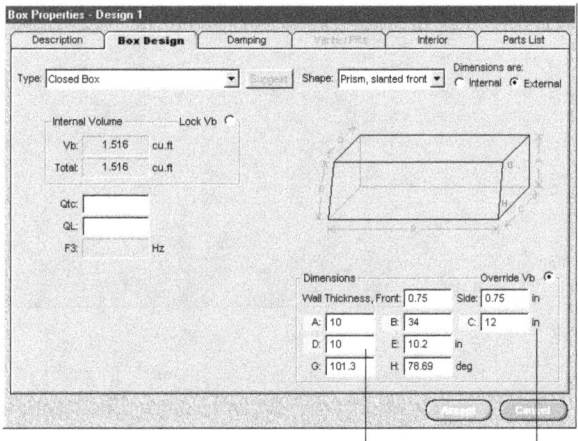

Enter dimensions.   Click on a unit label to change units.

Notice that the label background color of the dimensions we entered (A, B, C and D) is highlighted with a lighter color. This is done to identify which dimensions were entered and are therefore protected or "locked" so they will not be changed by the program. If desired, you can unlock them later by clicking on their labels.

Also note that "Override Vb" is turned on. This control should be turned on whenever you want the program to calculate the net internal box volume (Vb) from the dimensions you entered. If you look on the left side of the window you will see that this box has a net internal volume of 1.516 cubic feet (42.94 liters). *Note: If desired, you can change the volume units by clicking on one of the volume unit labels. The box volume unit labels are synchronized so they will change in unison when just one of them is clicked. They will advance from "cu.cm" (cubic centimeters) to "liters" to "cu.m" (cubic meters) to "cu.ft" (cubic feet) to "cu.in" (cubic inches) with each single mouse click—double-clicks will be ignored.*

- Next, click on the "Damping" tab at the top of the Box Properties window so we can set the amount of "fill" or "stuffing" that will be used inside the box. Make sure that the "Use Classical box calculations" control is turned off and then click on the "Heavy" setting for the "Amount of Acoustic Absorption" as shown below.

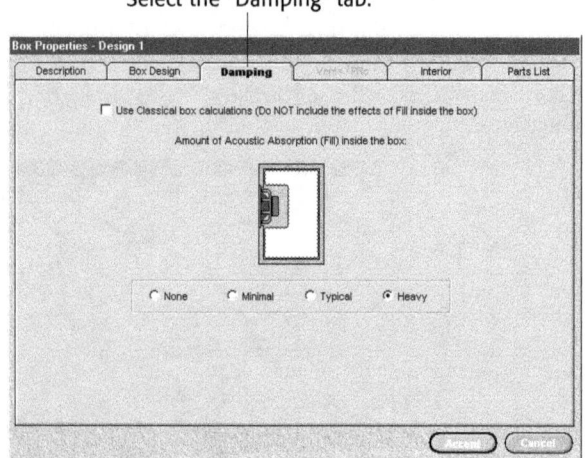

The reason we plan to use a "heavy" amount of fill is because we want to maximize the box's damping. By increasing the damping we make the box function like a bigger box and this provides smoother bass response and better low-frequency transient response. You can use a variety of different fill materials such as fiberglass, long-fiber synthetics (Acousta Stuf, Dacron, polyester, etc.), acoustic foam or long-fiber wool. The material you select should absorb sound, have a low density and a high specific heat. *Note: Having a high specific heat means that it will take a lot of heat to raise the temperature of the material. Therefore the fill material will be relatively insensitive to the heat generated inside the box by the woofer.*

We plan to use acoustic-grade fiberglass because of its excellent sound absorption, good availability and low cost. Since our box will be sealed we don't need to be concerned about stray glass fibers being blown out a vent where they may be inhaled by a passenger.

- If desired, you can now close the Box Properties window by clicking on its "Accept" button.

Driver Properties

Next let's find a suitable driver. We would like to mount the drivers in the top of the box so they fire upward past the back of the seat and past the tailgate. Since the top of our box is only 10 inches (254 mm) deep, it doesn't allow enough room for a large woofer. So we plan to use four 8-inch (203 mm) woofers as shown in the drawing below. This should enable us to move enough air for a very "musical" and "rock-solid" bass response.

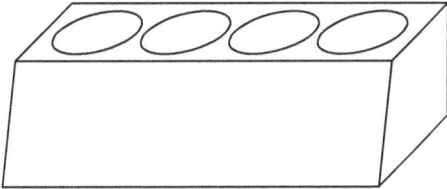

Another goal for our subwoofer is to achieve a smooth low-frequency bass extension. To do this we want the box to have a 50 Hz cutoff frequency (F3) and have a smooth response with a Qtc between 0.7 and 1.0. This should mesh well with the in-car bass rise that begins at the same frequency.

Finding the right driver is largely a matter of trial and error. Here are a few guidelines:

- The Fs of the drivers must be lower than the desired F3 of the box. In this case we want the box to go to 50 Hz so each driver's Fs must be lower than 50 Hz.

- The Qts of the drivers must be lower than the desired Qtc of the box. In this case we want the box Qtc to be between 0.7 and 1.0 and so the driver Qts must be a little lower than that. However, an extremely low value should be avoided since this is a closed box.

- The sum of the Vas values of all the drivers should be larger than the box volume but it should not be too large or the Qtc of the box may be pushed higher than the upper limit of 1.0. Since there will be four drivers, the individual Vas of each driver should be greater than 0.38 cubic feet (10.7 liters). This is found by dividing the box volume by the number of drivers.

We found three drivers (XZ-801, XZ-802, XZ-803) that we think might work and so we will enter their parameters and see what kind of response they produce. They all have polypropylene cones and dust caps and rubber surrounds so that they will be resistant to water damage (remember that our Jeep Wrangler is a convertible). In order to remain neutral regarding manufacturer and model preference we will use a fictitious driver manufacturer named "Acme" and fictitious model names. The driver parameters are listed below:

Model:	XZ-801	XZ-802	XZ-803
Parameters			
Fs (Hz):	34	30	34
Qms:	5	6.3	5
Vas (liters):	15	50	20
Mms (g):	95	36.6	71
Xmax (mm):	3.5	2.8	4
Sd (sq.cm):	214	214	214
Qes:	0.435	0.725	0.68
Re (ohms):	6.3	6.5	6.8
Le (mH):	0.8	0.5	0.7
Z (ohms):	8	8	8
BL (Tm):	17.1	7.9	12.3
Pe (watts, continuous):	75	60	75
Qts:	0.4	0.65	0.6
1-W SPL (dB):	84	85	85
Dimensions (inches)			
Outside diameter:	8	8	8
Mounting hole diameter:	7.5	7.5	7.5
Cutout diameter:	7	7	7
Magnet diameter:	5	5	5
Magnet depth:	1.75	1.75	1.75
Front depth:	0.25	0.25	0.25
Mounting depth:	4.75	4.75	4.75

Notice that each driver has an 8 ohm impedance. We don't want to go any lower than this because we want to be able to wire them all in parallel and not have the net impedance go lower than 2 ohms. Here are the steps to enter and evaluate each driver:

- Click on the "Driver" button on the design panel in the main window as shown at right. This will cause the Driver Properties window to open. It can also be opened from the keyboard by pressing [Ctrl]+[D].

Select the "Driver" button.

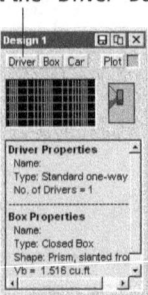

Sample 1: Closed Box Example BassBox 6 Pro

- Before we enter the driver parameters, let's enter a description for the driver. If it is not already selected, click on the "Description" tab in the upper left corner of the Driver Properties window. Enter "XZ-801" for the model name and "Acme" for the company name as shown at right:

  Select the "Description" tab.

  By entering the model name of the woofer, it will help us avoid confusion after all three woofer models have been entered. The Driver Type should be set to "Standard one-way driver".

  Additional description lines are available for specific parts of the driver. They are available so that you can provide a detailed description. If you want to enter miscellaneous notes about the driver, you can do so in the "Notes" box in the upper right corner. We will skip most of the "Description" tab since much of it is not essential for this design.

- Next, let's configure the woofer. Click on the "Configuration" tab at the top of the Driver Properties window as shown at right:

  Select the "Configuration" tab.

  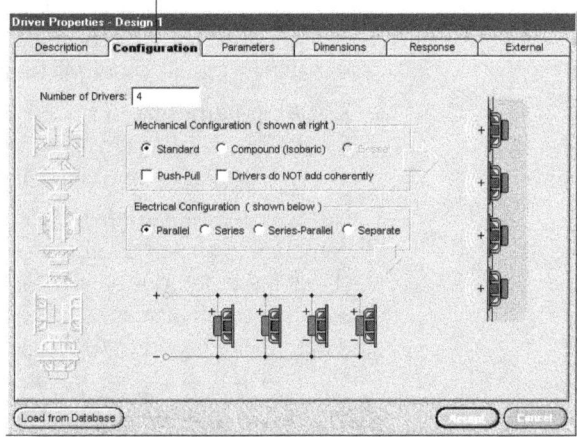

  Enter "4" into the "Number of Drivers" box. We plan to mount all of the drivers so that they face the same direction and we plan to wire them in parallel, so the "Mechanical Configuration" should be set to "Standard" and the "Electrical Configuration" should be set to "Parallel". Since all of the drivers will be mounted and wired the same way, the "Push-Pull" option should not be checked. The "Drivers do NOT add coherently" option should not be checked also

BassBox 6 Pro User Manual © D.E.Harris **77**

since the drivers will be mounted close together and will produce only low frequencies allowing their sound waves to sum coherently.

- At last we are ready to enter the woofer parameters. Click on the "Parameters" tab. Next, click on the "Expert Mode" control so that all of the parameters are displayed as shown below:

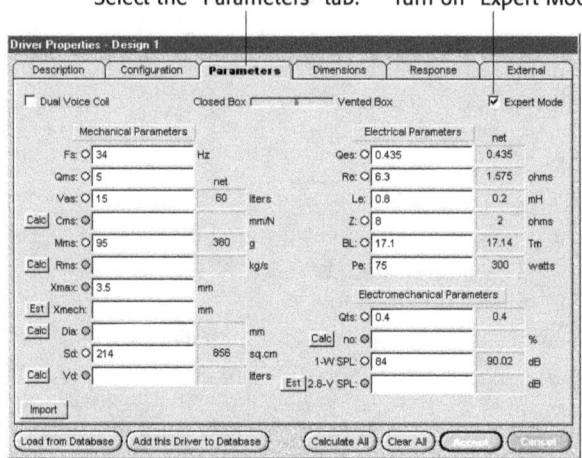

Enter the parameters for XZ-801 as shown above. Important: Be careful to set the units for each parameter <u>before</u> entering its value in the input box. To change a unit, click on the unit label, itself. For example, click on "liters" to change the units for Vas to "cu.ft" (cubic feet). Each single mouse click will advance the units to the next available setting. Double-clicks will be ignored.

XZ-801 parameters: Fs = 34 Hz, Qms = 5, Vas = 15 liters, Mms = 95 g, Xmax = 3.5 mm, Sd = 214 sq.cm, Qes = 0.435, Re = 6.3 ohms, Le = 0.8 mH, Z = 8 ohms, BL = 17.1 Tm, Pe = 75 watts, Qts = 0.4, 1-W SPL = 84 dB.

Notice in the "Parameters" tab that a "net" value is displayed beside many of the parameters to show how the program will internally adjust them to account for the fact that there are four parallel-wired woofers. Notice also that many of the parameters have green indicators beside them to show that they have successfully passed the automatic "expert mode" tolerance test.

The EBP indicator at the top of the "Parameters" tab is in the middle between "Closed Box" and "Vented Box". It shows that the XZ-801 is equally well suited for use in either a closed box design or a vented box design.

If you want, you can click on individual "Calc" or "Est" buttons or the "Calculate All" button at the bottom of the Driver Properties window to calculate most of the missing parameters. However, this is not necessary because sufficient parameters have been entered for this woofer.

- The last woofer information to enter is its dimensions. Click on the "Dimensions" tab. Select a "Round" outer shape and a "Cone" piston type. Enter the dimensions as shown below:

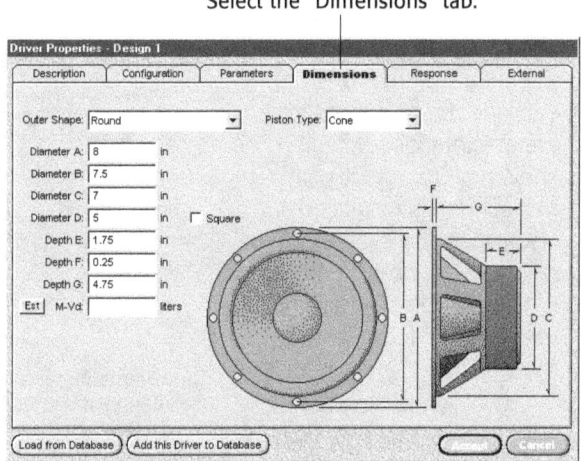

Select the "Dimensions" tab.

XZ-801 dimensions: Outside diameter (A) = 8 inches, mounting hole diameter (B) = 7.5 inches, cutout diameter (C) = 7 inches, magnet diameter (D) = 5 inches, magnet depth (E) = 1.75 inches, front depth (F) = 0.25 inches, mounting depth (G) = 4.75 inches.

The only item we did not enter was the mounting displacement volume (M-Vd). This is the volume displaced in the box by a single woofer. Not many manufacturers provide this value but it can be estimated from the previous dimensions. You can click on the "Est" button to calculate it now. But this isn't necessary because the program will estimate it internally whenever it is needed just as long as enough dimensions are known.

Why enter the woofer dimensions? Answer: So the program can calculate the net internal volume (Vb) inside the box. It does this by subtracting the volume displaced in the box by the woofers from the total box volume. In the next step we'll refine this volume calculation by setting the driver mounting.

- Let's switch back to the Box Properties window and specify how the woofers are mounted so that BassBox Pro can calculate the woofer displacement volume accurately. Close the Driver Properties window by clicking on its "Accept" button then click on the "Box" button of the design panel (or press [Ctrl]+[B]) to reopen the Box Prop-

BassBox 6 Pro

Sample 1: Closed Box Example

1

erties window. Select the "Interior" tab as shown at right:

We plan to flush-mount the woofers to the box for a professional "look" so select the "Flush" driver mounting. The mounting selection tells the program how much of each woofer protrudes into the box interior so it will know how much internal box volume is displaced. It also allows the program to calculate the portion of the mounting hole that contributes to the interior box volume.

Select the "Box Design" tab and let's see how the program adjusts the net internal volume (Vb) now that the woofer dimensions and mounting setting have both been entered.

Remember that Vb previously equaled 1.516 cubic feet (42.94 liters). Notice that it now equals 1.36 cubic feet (38.5 liters)—a 10% reduction. This is because each of the four woofers displaces approximately 0.039 cubic feet (1.1 liters) inside the box after taking into account the mounting hole size and wall thickness.

Select "Flush" driver mounting.   Select the "Interior" tab.

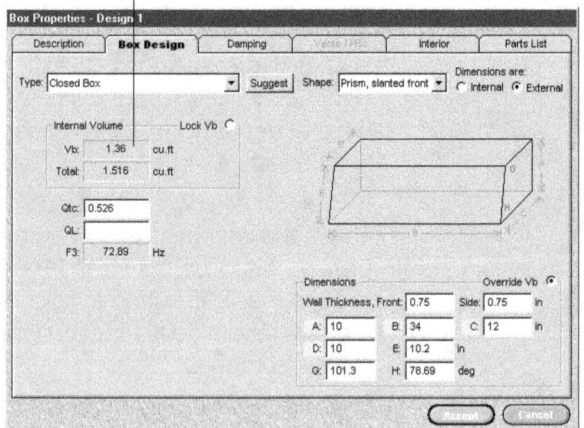

The volume displaced in the box by the woofers has been subtracted from the total box volume to determine the net internal volume (Vb).

BassBox 6 Pro User Manual © D.E.Harris

Sample 1: Closed Box Example

- Click the "Accept" button at the bottom of the Driver Properties window to close it and let's return to the main window. Notice that the mini graph for the design has now been plotted. It is automatically plotted as soon as enough driver and box information is present.

    From now on, changes to the driver or box parameters will cause this mini graph to be automatically updated to provide immediate feedback.

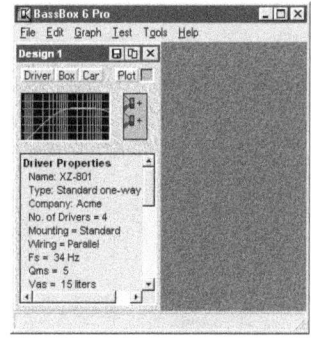

- Next, let's enter the information for the second and third woofer models. The easiest way to do this is to duplicate the first design twice and then replace the woofer information in the second and third designs with the new woofer information. This will also enable us to easily do a side-by-side comparison. Begin by clicking on the design copy button twice (shown at right):

    Clicking on the copy button twice should create Design 2 and Design 3. Both are exact duplicates of Design 1 with the same woofer and box information as shown below.

Select the copy button.

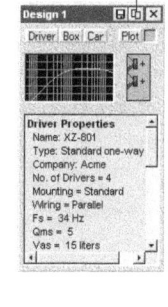

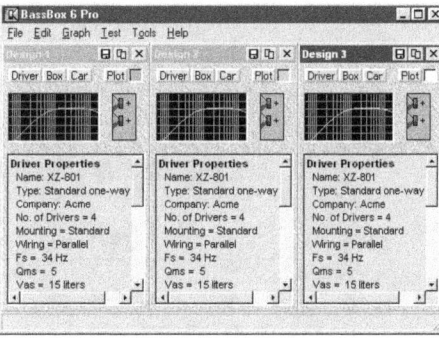

- Click on the "Driver" button of the Design 2 panel (or press [Ctrl]+[D]) to open the Driver Properties window and select the "Parameters" tab. *Note: The Driver Properties window displays the information for the currently selected design. You can quickly switch to a different design by clicking on its design panel in the main window. For example, clicking anywhere on the Design 3 panel in the main window will cause the Driver Properties window to immediately display the information for Design 3. In this case, we clicked on the "Driver" button of the Design 2 panel so De-*

sign 2 was preselected when the Driver Properties window opened.

- Use the "Clear All" button on the bottom of the window and then enter the parameters for woofer XZ-802 into Design 2 as shown below:

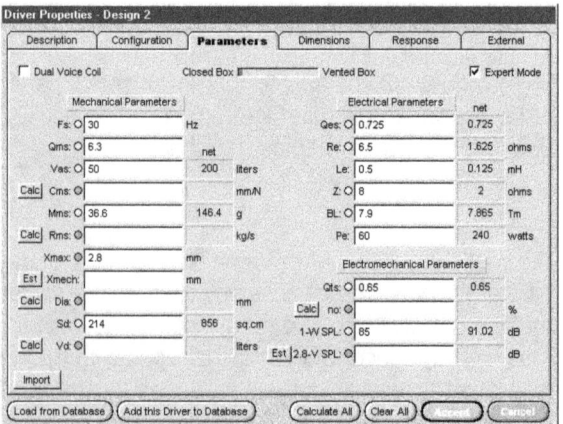

XZ-802 parameters: Fs = 30 Hz, Qms = 6.3, Vas = 50 liters, Mms = 36.6 g, Xmax = 2.8 mm, Sd = 214 sq.cm, Qes = 0.725, Re = 6.5 ohms, Le = 0.5 mH, Z = 8 ohms, BL = 7.9 Tm, Pe = 60 watts, Qts = 0.65, 1-W SPL = 85 dB.

Notice that the EBP indicator at the top of the "Parameters" tab shows that the XZ-802 is more suited for a closed box design rather than a vented box design. This is good since our box will be closed.

- Switch to the "Description" tab and change the driver model name for Design 2 to "XZ-802".

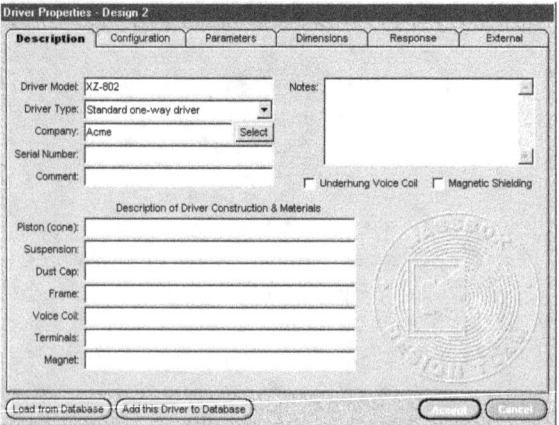

Sample 1: Closed Box Example   BassBox 6 Pro

- Now let's do the same thing for Design 3. Switch to Design 3 by clicking anywhere on the Design 3 panel in the main window. You may need to move the Driver Properties window temporarily out of the way if the main window is covered by it.

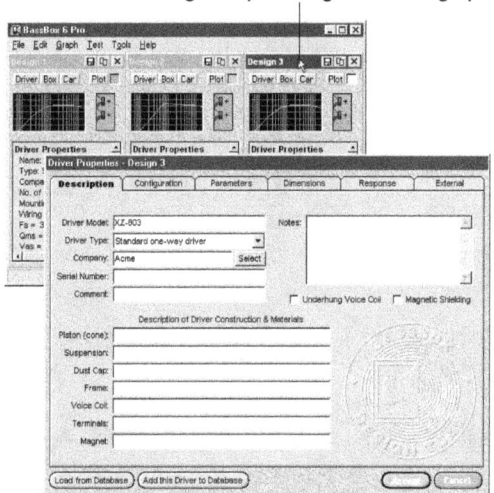

Select Design 3 by clicking on its design panel.

Note: The main window always stays behind the other BassBox Pro windows.

- Since the "Description" tab of the Driver Properties window is still selected, let's begin by entering "XZ-803" for the driver model name of Design 3 as shown above.
- Next, select the "Parameters" tab, click the "Clear All" button on the bottom of the window and enter the parameters for woofer XZ-803 as shown below.

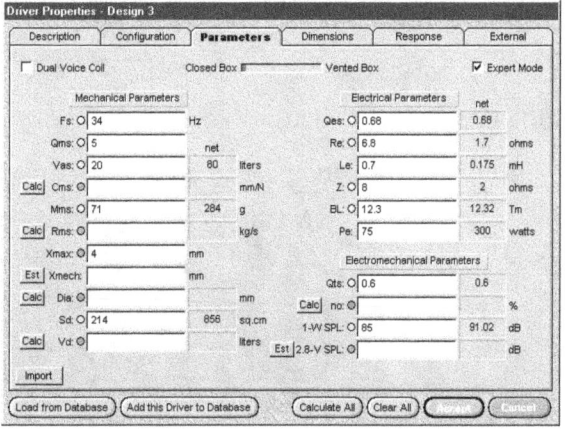

BassBox 6 Pro User Manual © D.E.Harris

XZ-803 parameters: Fs = 34 Hz, Qms = 5, Vas = 20 liters, Mms = 71 g, Xmax = 4 mm, Sd = 214 sq.cm, Qes = 0.68, Re = 6.8 ohms, Le = 0.7 mH, Z = 8 ohms, BL = 12.3 Tm, Pe = 75 watts, Qts = 0.6, 1-W SPL = 85 dB.

The EBP indicator at the top of the "Parameters" tab shows that the XZ-803, like the previous XZ-802, is well suited for a closed rather than a vented box design.

- All three woofers should now be entered. We don't need to change the woofer configurations on the "Configuration" tab or the dimensions on the "Dimensions" tab because all three woofers have the same configuration and dimensions. Click on the "Accept" button to close the Driver Properties window.

Evaluating the Performance

Its now time to compare the three woofer models to see which one provides the best in-car response with our box. Follow these steps:

- We will use the single-window graph mode. If it is not already selected, select the "Display Mode > Single Window" command from the Graph menu as shown below:

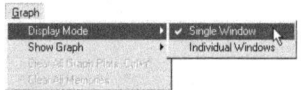

- Now let's open the graph window. Select the "Show Graph > Amplitude—Normalized" command from the Graph menu as shown below (or press Ctrl+F1).

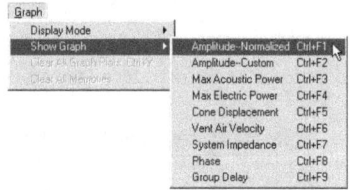

The graph window should open as shown below:

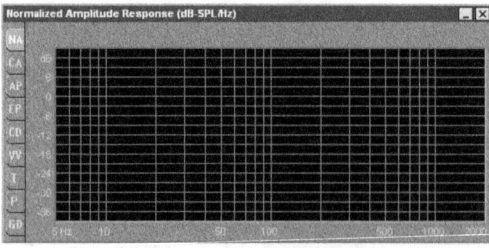

Sample 1: Closed Box Example **BassBox 6 Pro**

Notice the tabs along the left edge of the graph window. Tab "NA" is selected because we asked the program to show the **Normalized Amplitude Response** graph. Later we will use these tabs to examine the other graphs. The Normalized Amplitude Response graph will allow us to compare the free field amplitude response of each design.

- Next, click on the "Plot" button of each design as shown below to plot it in the graph. This can also be done from the keyboard with the shortcuts [Ctrl]+[1], [Ctrl]+[2] and [Ctrl]+[3].

Use the "Plot" buttons to plot each design in the graph.

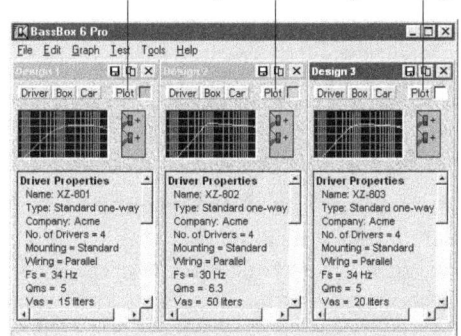

The graph should now look like the example below:

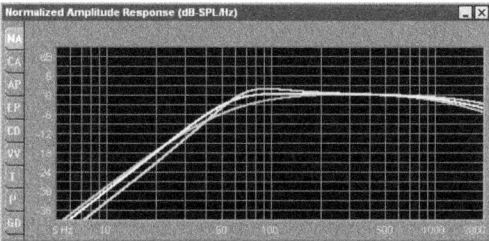

Since this graph is "normalized", it does not show which design is louder or more efficient. The response is "free field" because it does not yet include the effects of the in-car acoustic response. Notice that Design 1 (medium grey plot line) has the lowest level at the "knee" of the response and Design 2 (light grey plot line) has the highest level. *Note: The "knee" is the region where the low-frequency response begins to drop off.*

Turn on the Overlay feature.

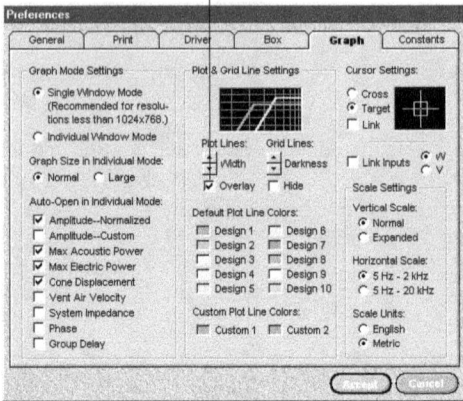

If the plot lines of all three designs are not overlaid as shown on the previous page, the overlay feature must be turned on before proceeding. To do this, select the "Preferences" command from the Edit menu of the main window. Select the "Graph" tab and turn on the "Overlay" feature as shown at right and click on the "Accept" button. Then replot the designs so they are overlaid and can be easily compared as shown on the previous page.

- The total Q of each box (Qtc) determines the response shape. The –3 dB frequency (F3) determines the cutoff point. Let's return to the "Box Design" tab of the Box Properties window to see what the Qtc and F3 values of each design are. Click on the "Box" button of Design 1. The Qtc and F3 values of Design 1 are shown below:

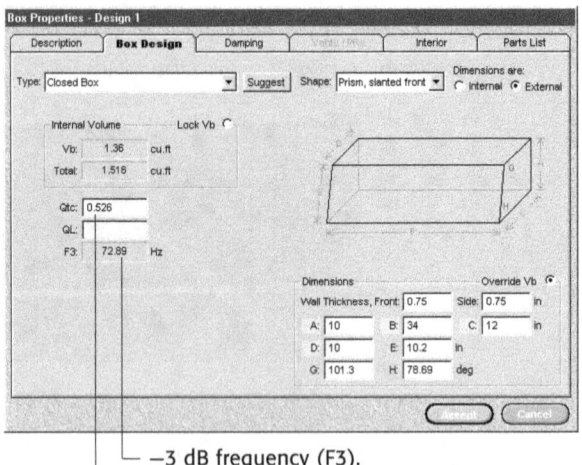

—–3 dB frequency (F3).

The total Q of the Box (Qtc).

You can switch the "Box Design" tab from one design to the next without closing the Box Properties window by clicking on each design panel back in the main win-

dow. If you do, you'll see that each design has the following Qtc and F3 values:
Design 1 (XZ-801) Qtc = 0.526, F3 = 72.89 Hz.
Design 2 (XZ-802) Qtc = 1.088, F3 = 52.84 Hz.
Design 3 (XZ-803) Qtc = 0.793, F3 = 50.98 Hz.

Remember our original design goal. We want a subwoofer with a Qtc between 0.7 and 1.0 and an F3 of 50 Hz. With this in mind, the XZ-802 woofers of Design 2 or the XZ-803 woofers of Design 3 may satisfy our requirements. How do we decide which is best? Let's add the in-car acoustic response to the graphs and see what it reveals.

Note: So far we have ignored the box leakage loss parameter (QL). This is because it is automatically estimated for us internally by BassBox Pro based on the box type and box volume.

Click the "Close" button to close the Box Properties window.

- To add the in-car response we must configure its settings with the Car Acoustic Properties window. Click on the "Car" button of the Design 1 panel to open it. It can also be opened from the keyboard by pressing [Ctrl]+[A].

Select the "Car" button.

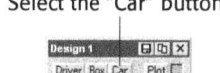
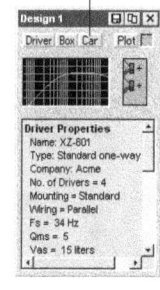

If the design panels have a "Room" button instead of a "Car" button then you will need to change the acoustic environment setting of BassBox Pro before proceeding. To do this, select the "Preferences" command from the Edit menu of the main window. Then select the Automotive acoustic environment as shown below and click on the "Accept" button.

Select the Automotive acoustical environment.

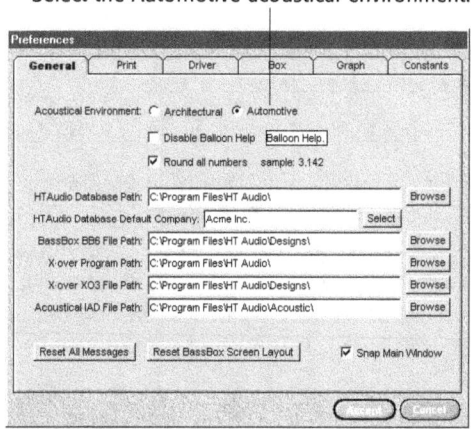

The Car Acoustic Properties window is shown below.

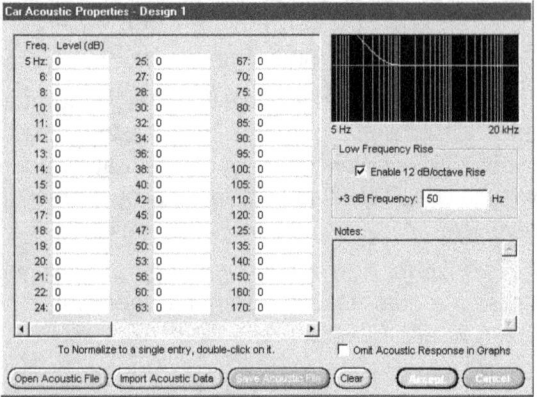

Since we do not have an acoustical measurement of our Jeep Wrangler, we can ignore the individual level settings on the left side of this window. Instead, we will estimate the in-car response, so let's focus on the right side. Begin by turning on the "Enable 12 dB/octave Rise" control. Set the "+3 dB Frequency" to 50 Hz if it does not already have this value. Turn off the "Omit Acoustic Response in Graphs" control. Now notice how the Car Acoustic Properties graph shows the 12 dB/octave low-frequency boost to simulate the bass response inside the car. Also notice how the mini graph in the Design 1 panel of the main window has replotted to show the effects of the in-car response.

- Next, let's repeat the above steps for Design 2 and 3. With the Car Acoustic Properties window still open, go back to the main window and click on the Design 2 panel. This will switch the Car Acoustic Properties window to Design 2 so that you can turn on its "Enable 12 dB/octave Rise" option and set the "+3 dB Frequency" to 50 Hz. Then do the same thing for Design 3.

- Click on the "Accept" button and close the Car Acoustic Properties window.

- Now let's clear the main graph and replot the designs with the in-car response. Several ways are provided to clear the graphs in order to make it convenient. From the keyboard press (Ctrl)+(Y) or select "Clear ALL Graph Plots" from the Graph menu of the main window (shown below) or select "Clear > ALL Graphs" from the graph popup menu or use (Shift)+"Clear" on the Graph Properties window.

Sample 1: Closed Box Example **BassBox 6 Pro**

- Next, click on the "Plot" buttons of each design ([Ctrl]+[1], [Ctrl]+[2], [Ctrl]+[3]) to plot them in the graph window. Your results should match the illustration below:

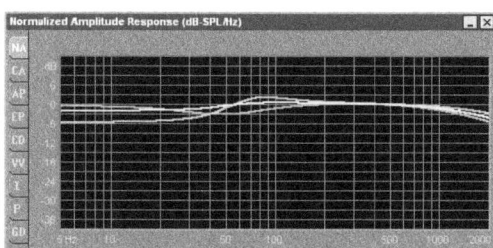

Notice that the bass response has risen so that it is now almost flat. A flat response is our goal for the most accurate bass reproduction.

It is obvious from the graph that Design 3 with the XZ-803 woofers (white plot line) achieves the most flat response and so we will use the XZ-803s in our design. But before we conclude this design, let's examine the other preference graphs and evaluate the design from additional perspectives.

The **Custom Amplitude Response** graph is selected with the "CA" graph tab ([Ctrl]+[F2]). It is very similar to the Normalized Amplitude Response graph except that it shows the sound pressure level (SPL) that the speaker can produce on-axis at 1 meter with a specified input power. We used 300 watts because it is the total maximum power for all four XZ-803 woofers (75 watts each).

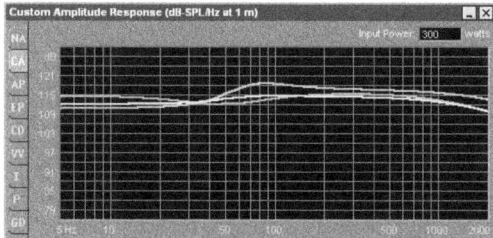

Notice that Design 3 with the XZ-803 woofers (white plot line) achieves a sound pressure level of approximately 115 dB for much of its passband. This is the total sound pressure level from all four woofers. It should be plenty loud enough for our Jeep—especially since the 115 dB level is based on a continuous power rating and the speaker should be able to pass even higher peak levels. Design 2 (light grey plot line) achieves a higher sound pressure level but it is not uniform.

The **Maximum Acoustic Power** graph is selected with the "AP" graph tab (Ctrl+F3). It shows how much acoustic power the speaker can produce before reaching its steady-state displacement limit or its thermal limit. Acoustic power is sort of like the sound pressure level because they both are a measure of the amount of energy being radiated from a speaker. Acoustic power is the total power radiated by the speaker in all directions. The sound pressure level changes with direction and distance from the speaker. For this reason the acoustic power provides a good way to compare the total output of different speakers—especially if the speaker will be used in a "live" or reverberant space.

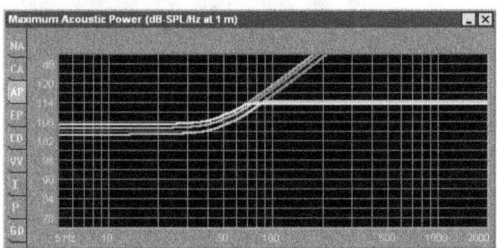

At lower frequencies the steady-state displacement limit of the woofer's diaphragm and voice coil will usually dominate the acoustic power limit. It is calculated from the peak displacement limit (Xmax). At higher frequencies the mid-band thermal limit of the voice coil will usually dominate. The thermal limit may be a steady-state or a peak level depending on the nature of driver's power rating (Pe). In the graph above, the curved portion of the plot lines on the left half of the graph represent the displacement limit. Notice also that these lines are allowed to continue to the top of the graph with a darker shade to show what the maximum acoustic power would be if there were no thermal limit. The flat region of the plot lines on the right half of the graph represents the thermal limit.

This graph shows us that the maximum loudness of Design 3 will begin to drop below 70 Hz until it is about 6 dB lower. Will the speaker actually be 6 dB quieter at 30 Hz than at 80 Hz? Probably not because the displacement limits in this graph represent a "steady-state" or RMS level rather than a peak level. Only the peak motion of the driver diaphragms will reach the Xmax limit at the levels shown in this graph. However, it is probably possible for the excursion to safely exceed Xmax until the mechanical limit (Xmech) is reached. Nonlinear distortion will begin to increase when Xmax is exceeded because, with fewer coils of wire in the gap of the magnet, the motor strength will diminish. So the difference in level between 30 Hz and 80 Hz will be no more than 6 dB and probably less.

The **Maximum Electric Input Power** graph is selected with the "EP" graph tab (Ctrl+F4). It is derived from the previous Maximum Acoustic Power graph and it shows how much amplifier power the speaker can handle before reaching the first of either its displacement or thermal limits.

Sample 1: Closed Box Example

BassBox 6 Pro

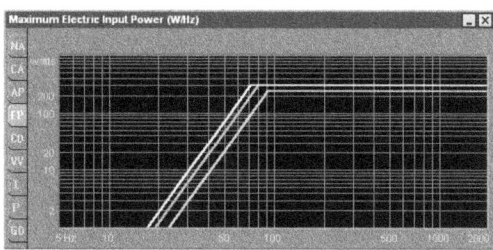

Notice that Design 3 with the XZ-803 woofers (white plot line) handles more power than the other designs. It has the same thermal limit as Design 1 with the XZ-801 woofers (medium grey plot line) but it has the best displacement limit because it has the highest Xmax.

This graph shows us that Design 3 may not be able to handle full RMS power (300 watts) from the amplifier below 70 Hz because of the diaphragm displacement limits. However, as with the previous Maximum Acoustic Power graph, the displacement limit represented here is the steady-state limit based on a peak excursion at Xmax. The driver diaphragms will be able to move a little farther, although with increasing distortion levels. It would be helpful to know what the mechanical excursion limit of the drivers is because it would tell us how far the diaphragms can move before reaching their physical limits or before risking damage. Fortunately, this isn't as bad as it looks because the low-frequency 12 dB/octave rise of the in-car acoustic response pulls the bass response up as shown in the preceding Maximum Acoustic Power graph so that the displacement-limited level is within 6 dB of the thermal-limited level.

The important thing to learn from this graph is that the speaker may not be able to handle a full 300 watts from our amplifier at very low frequencies so we may want to consider adding a low-frequency high-pass filter to protect it.

The **Cone Displacement** graph is selected with the "CD" graph tab (Ctrl+F5). It shows how far the diaphragm and voice coil of each woofer will travel at the specified input power level. We used 300 watts because it is the total maximum power level for all four XZ-803 woofers (75 watts each).

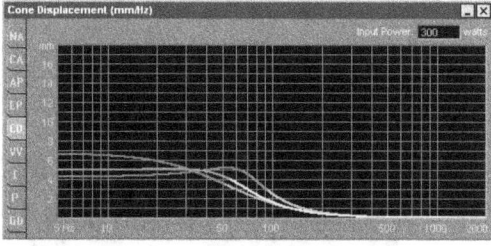

Notice that each plot line changes intensity when it rises above the Xmax level of the woofer. For example, the Xmax of XZ-803 is 4 mm (Design 3) so the white plot line changes to a darker shade at about 55 Hz when it exceeds this level.

Some may think that there is a discrepancy between the Cone Displacement graph and the Maximum Acoustic Power and Maximum Electric Input Power graphs. Why does the Cone Displacement graph show that the XZ-803 woofers exceed Xmax at 55 Hz while the Maximum Acoustic Power and Maximum Electric Input Power graphs show that displacement limiting of the same woofers begins at approximately 70 Hz? The answer has two parts: First, the Cone Displacement graph shows the peak cone excursion so that you will know exactly when the diaphragm will reach or exceed Xmax. However, the Maximum Acoustic Power and Maximum Electric Input Power graphs both show a lower "steady-state" or RMS excursion limit which more closely represents the way we perceive loudness. Second, the Cone Displacement graph assumes a constant input voltage while the Maximum Acoustic Power and Maximum Electric Input Power graphs assume that the peak excursion is constantly at Xmax and so the input voltage will be forced to continually change with frequency.

Ideally, the motion of the diaphragm should not exceed Xmax (one way from its resting position). If it does, nonlinear distortion will increase as mentioned previously. It could be argued that this is the weakest aspect of all three designs. However, several factors can reduce the seriousness of this problem with Design 3. First, we do not plan to drive the speaker continuously at full power. In fact, we won't even drive it continuously at half power. And yet, at 150 watts, the 1 meter SPL will still be 112 dB and the peak excursion of each woofer will not exceed Xmax. Second, the mechanical excursion limit (the distance the diaphragm can travel before reaching the limits of the suspension or before the voice coil former strikes the back plate) should be larger than Xmax. This allows the diaphragm to move farther than Xmax without risk of damage. Third, the nonlinear distortion that increases when the excursion exceeds Xmax is subtle and is not audible to most listeners until it becomes severe.

The **System Impedance** graph is selected with the "I" graph tab ([Ctrl]+[F7]). It shows the net impedance of the speaker. In each of our designs, we wired four 8 ohm speakers in parallel. This was specified on the "Configuration" tab of the Driver Properties window and it results in a net impedance of 2 ohms for each design as shown in the graph below:

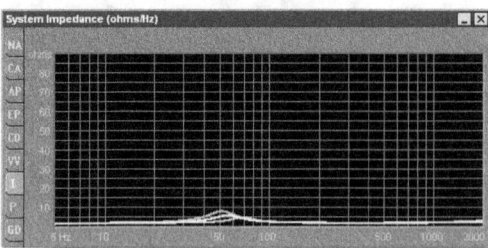

Sample 1: Closed Box Example

Notice that each plot line has a peak in the middle of the graph. This represents the resonance of the drivers in the box. For example, the peak at 56.2 Hz in the white plot line is the resonance of the XZ-803 drivers in the box. *Note: The resonance of a driver is always at a higher frequency in a box than in free air (Fs).* It can sometimes be difficult to determine the precise location of the peak. **Tip:** The phase of the impedance should pass through zero degrees at the peak.

The System Impedance graph is useful for a couple of reasons. If the speaker will be connected directly to an amplifier, it shows the load that the amplifier must "drive". In this case the amplifier must be capable of driving a 2 ohm load. If the speaker will include a passive crossover network, the graph shows the impedance that the crossover network will "see". This is very important when designing the crossover network because it usually wants to see a flat impedance.

The **Phase Response** graph is selected with the "P" graph tab ([Ctrl]+[F8]). It shows how much the sound waves which emanate from the speaker will lag behind the input signal that feeds the speaker. This delay is expressed as a phase angle in degrees and it is literally the difference between the phase of the input signal and the phase of the output signal.

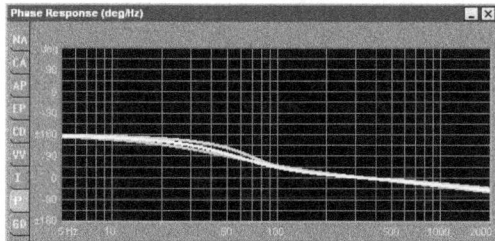

Notice that the low frequency sound waves of all three designs never lag by more than 180°. This means that the output signal will not lag behind the input by more than a half wavelength. Remember that a full wavelength requires 360° of phase rotation. *Note: The reason the phase response drops below 0° above approximately 330 Hz is because of the inductive reactance of each woofer's voice coil.*

Ideally the phase response should be a perfectly flat line with 0° (zero degrees) of phase shift. Fortunately, a gradual change in phase angle as shown above will not significantly harm the fidelity. However, sharp and sudden changes in phase angle are a significant problem and should be avoided when possible.

The **Group Delay** graph is selected with the "GD" graph tab ([Ctrl]+[F9]). It is very similar to the Phase Response graph because it also shows how much the sound waves which emanate from the speaker will lag behind the input signal that feeds the speaker. However, this graph expresses this information as a delay in milliseconds and it is derived from the slope of the phase response.

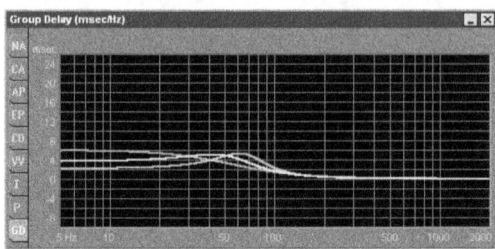

Ideally, there should be no delay. In practice, a gradual change as shown above will not significantly harm fidelity. However, sharp and sudden changes in group delay are a significant problem and should be avoided when possible.

Note: The **Vent Air Velocity** graph (graph tab "VV") was omitted because this is a closed box design and therefore it has no vents.

External Series Resistance

There is one more item to consider: external series resistance. Let's return to the Driver Properties window and select the "External" tab. The "External Resistance" section is located on the left side of the window as shown below. *Note: Because we will be using an active crossover network with this subwoofer, we will skip the "External Network" section on the right side of this tab because our subwoofer will not need a passive network.*

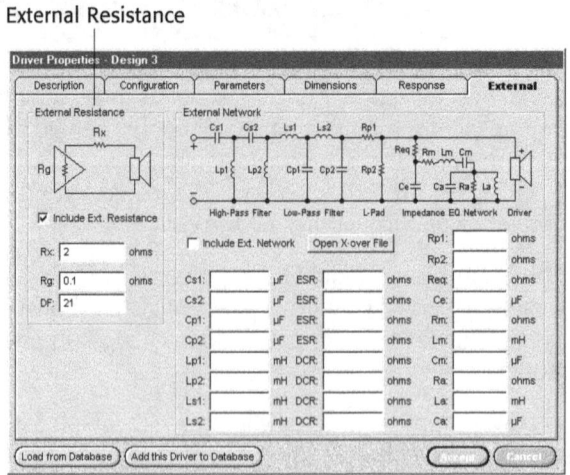

There are three external resistance parameters. Rx is the series resistance of the speaker cables and connectors that connect the speaker to the amplifier. If a passive crossover network is present, Rx does not include the series resistance of the network. Since a speaker cable uses two conductors, Rx is the total resistance of both. A value of 2 ohms is shown above—a very high value for our design. Ideally, Rx should be less than one quarter of the net DC resistance of the speaker. Since the net DC resistance of our speaker is only 1.7 ohms, Rx should ideally be no higher than 0.425 ohms. There are three ways to keep Rx low:

- Use speaker cables that are made with a large wire gauge having low resistance.
- Locate the amplifier close to the speaker and keep the speaker cable length short.
- Use high-quality low-resistance connectors that make a good electrical and mechanical connection to the speaker cables.

Rg is the source resistance of the amplifier. Most modern amplifiers have a very low source resistance and so this parameter is seldom a problem. The 0.1 ohm value shown above is actually a bit high. DF is the damping factor and it can be calculated from Rg if the speaker impedance (Z) is known. DF is provided as an alternative in case Rg is unknown, so you can enter DF instead.

Admittedly, we used an Rx value that was very high and an Rg value that was mildly high to illustrate how they can affect the response of a speaker. The graph below shows the custom amplitude response for Design 3 with and without external resistance:

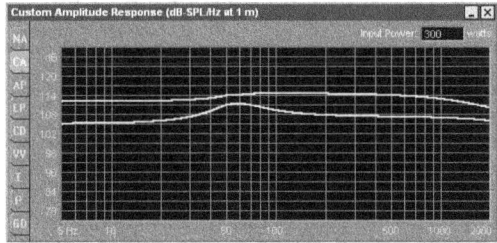

The top plot line is the response of Design 3 with no external series resistance. The bottom plot line shows the response of Design 3 with the external series resistance (Rx = 2 ohms, Rg = 0.1 ohms). Notice how the external resistance causes an overall 6 dB loss in level and a pronounced ripple. Obviously external series resistance can have a dramatic effect on the design. Why is this?

The answer is found in the electrical Q (Qes). Its net value is increased from 0.68 to 1.52. This in turn affects the following parameters: Qts increases from 0.6 to 1.166 and BL decreases from 12.3 to 8.237 Tm (Tesla meters). This is displayed on the "Parameters" tab of the Driver Properties window.

In our actual design, Rx and Rg did not play a significant role because their values were very low. However, this exercise illustrates just how dramatically a seemingly small increase in external resistance can result in large consequences—especially if the net impedance of the speaker is, itself, low.

Conclusion

The design is now complete. We began by designing a box to fit the space behind the rear seat of a Jeep Wrangler and decided to place four 8-inch woofers in it. Next, we examined three different woofer models and selected the Acme XZ-803 because it produced the smoothest in-car response. Last, we examined the role that external series resistance can play in the design. Where does that leave us? It's time to build the box and BassBox Pro can help by providing a parts list with dimensions and drawings. This is provided on the "Parts List" tab of the Box Properties window.

Click on the "Box" button of the Design 3 panel in the main window (or press [Ctrl]+[B]). Then select the "Parts List" tab as shown below:

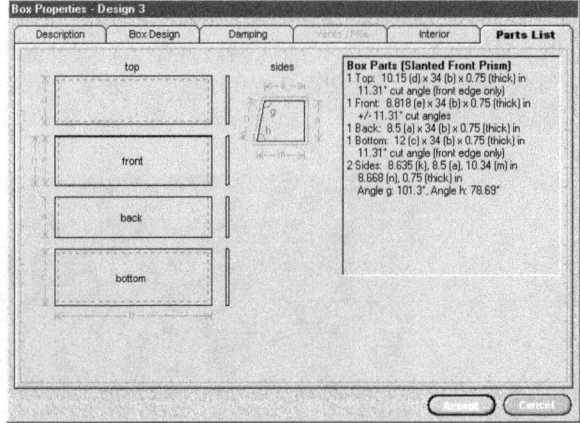

Notice that a drawing of each part is provided on the left. (Notice also that the dimension labels are different here than on the "Box Design" tab because of the different number of dimensions. To help you remember this, the dimension labels use lower case characters here and upper case characters on the "Box Design" tab.) A list of the parts, their dimensions and, if appropriate, cut angles are provided on the right. Both the drawings and the list can be included in a printout of the design.

Sample 1: Closed Box Example   **BassBox 6 Pro**

The last step is to save the design so that we can open it again later to review or edit it. Select the "Save Design As…" command from the File menu as shown below.

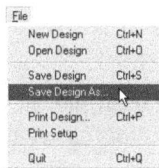

Then enter a file name. We used a descriptive name that will make it easy for us to recognize later on. For example, we used the name "Jeep Wrangler 4-driver subwoofer.bb6" as shown below. All BassBox design files must end with the file name extension ".bb6" and it will be automatically added if omitted.

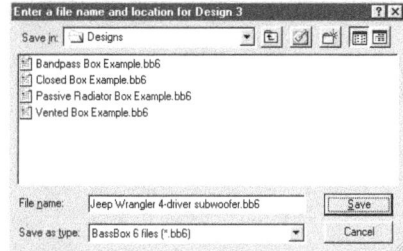

At this point, only the selected design (Design 3) is saved. If desired, we can also save Designs 1 and 2 in case we want to review them again later. Or, we can simply close Designs 1 and 2 without saving them.

The finished box should look something like the illustration below. We chose to add a strong wire grill (not shown) over each driver so they will not be harmed if something is dropped on top of the speaker.

Subwoofer boxes for automotive use are often carpeted to match the interior of the vehicle. Instead, we decided to stain and waterproof our box with a clear acrylic coating. The carpet would have been a mess if it got wet—remember our Jeep is a convertible. For additional information on box construction we suggest that you review Chapter 4 of the previous *Box Designer's Guide*.

This concludes the closed box example. We hope that it has been helpful. Please review the other examples for additional information about box design.

Sample 2: Vented Box Example BassBox 6 Pro

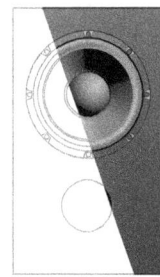

2 Vented Box Example

This chapter shows how to use BassBox Pro to design a vented box. Vented boxes are very popular because they are well suited to two-way speakers and because of their ability to extend the bass response. This example will focus on the second characteristic to design two identical high-performance subwoofers for our home theater.

New speaker designers are advised to review Chapter 1 ("Closed Box Example") if they have not yet done so. Chapter 1 includes details that will help the beginner get "up to speed" with BassBox Pro a little faster.

Driver Properties

Many designers already have a driver model in mind when they begin a speaker design—that approach will be used here. For a long time we have eagerly desired to use a 15-inch (381 mm) dual-voice coil Acme SledgeBeast 6000 woofer to create a powerful subwoofer that will add a vital element of tactile realism when we watch "disaster" movies in our home theater. In this example we will use these legendary (and fictitious) drivers to design a couple of awesome subwoofers that will augment our home theater's existing speaker system below 90 Hz. Our goal is to achieve a solid F3 of 20 Hz! Let's begin:

- Run BassBox Pro, go to the main design window and create a new design ([Ctrl]+[N]).
- Let's begin by entering the driver information first, since our driver is already selected. Click on the "Driver" button of the new design panel ([Ctrl]+[D]) to open the Driver Properties window. Select the "Parameters" tab and turn on the "expert" mode if it is not already turned on. Since the SledgeBeast 6000 has two voice coils, turn on the "Dual Voice Coil" option.

Sample 2: Vented Box Example

Enter the following driver parameters (remember to set the units of each parameter before entering its value):

Model:	SledgeBeast 6000		
Fs (Hz):	17		
Qms:	9		
Vas (liters):	245		
Cms (mm/N):	0.254		
Mms (g):	345.1		
Rms (kg/s):	4.095		
Xmax (mm):	16		
Dia (mm):	323.9		
Sd (sq.cm):	824		
Vd (liters):	1.32		
VC Wiring:	Separate	Parallel	Series
Qes:	0.8	0.4	0.4
Re (ohms):	5.64	2.82	11.28
Le (mH):	2.1	1.05	4.2
Z (ohms):	8	4	16
BL (Tm):	16.12	16.12	32.24
Pe (watts, continuous):	350	600	600
Qts:	0.74	0.38	0.39
ηo (%):	0.145	0.29	0.29
1-W SPL (dB):	83.82	86.83	86.83
2.8-V SPL (dB):	85.33	91.35	85.33

As you can see, the SledgeBeast 6000 has excellent specs for a subwoofer with its generous Xmax of 0.63 inches (16 mm) and a low Fs of 17 Hz. *Note: Although the EBP indicator (shown below) suggests that this woofer is more suitable for a closed box, it will still work well in a vented box. Many woofers with an Fs this low will also have a low EBP. Remember that the EBP indicator is only a general guideline—not a rule.*

The "Parameters" tab will look similar to the one on the next page after the parameters are entered correctly.

Now we must decide how to wire the voice coils. Our choices are explained below:

Separate (individual) The voice coils are used independently. Each one is driven by a different amplifier channel. This method makes it possible to drive a single subwoofer with a stereo signal without a summing amplifier because one voice coil can be driven with the "left" signal and the other voice coil with the "right" signal.

Parallel The voice coils are wired in parallel and are both driven from a single amplifier channel. The net impedance (Z) and DC resistance (Re) will be half

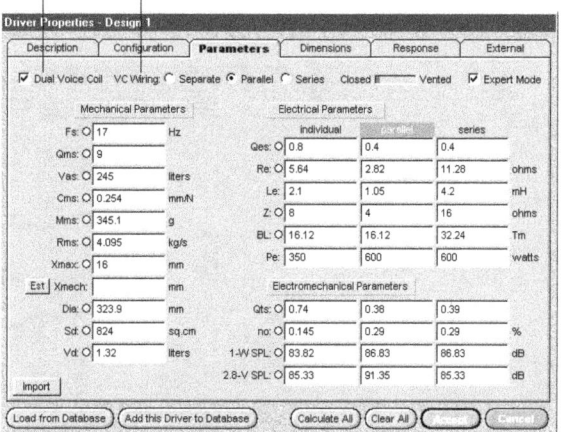

Turn on the "Dual Voice Coil" option.   Select the voice coil wiring method.

that of a single voice coil. This method offers the advantage that if one voice coil "opens", the other can continue to function while drawing half of the amplifier power. However, the lower impedance that results from parallel wiring may make it unwise to combine the woofer in parallel with other similarly wired woofers in a multi-woofer design because the overall impedance may be too low.

Series The voice coils are wired end-to-end in series and are both driven from a single amplifier channel. The net impedance (Z) and DC resistance (Re) will be double that of a single voice coil. This method has the disadvantage that if one voice coil "opens", they both stop working. Its higher impedance will also draw less power from the amplifier. However, the high impedance can be used to advantage if the woofer will be combined in parallel with other similarly wired woofers in a multi-woofer design because it prevents the overall impedance from being too low.

We plan to use only one woofer per box so we will wire the voice coils in parallel because the resulting 4-ohm load will make it easier to draw full power from our amplifier. Select the "Parallel" setting for the VC Wiring option.

- Select the "External" tab. We will enter the external resistance information in the "External Resistance" section.

The subwoofers will be located close to the amplifier so the speaker cables are not very long. Low-resistance cables made with large-gauge "heirloom-quality rare earth copper" will be used. We shorted one end of the cable and measured its resistance to be 0.25 ohms. We entered this for Rx. Our amplifier has a damping factor of 1000 for frequencies at and below 100 Hz and so we entered 1000 for DF. The program then calculated Rg to be 0.004 ohms as shown below. Turn on the "Include Ext. Resistance" checkbox.

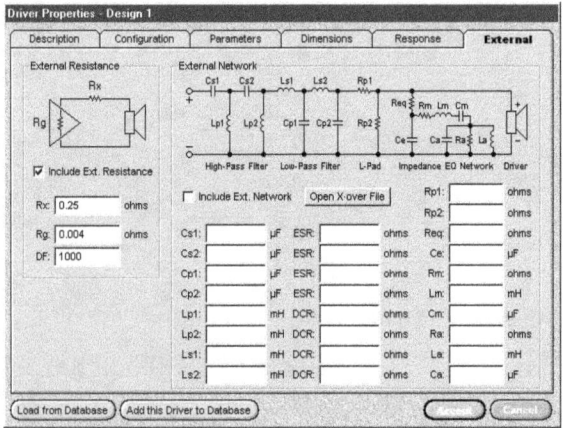

We do not plan to include an external network in this design so the "External Network" portion of this tab can be ignored.

- Select the "Description" tab and enter "SledgeBeast 6000" for the driver model name and "Acme" for the company name. The driver type should be set to "Standard one-way driver".
- Select the "Dimensions" tab. Select a "Round" outer shape and a "Cone" piston type. Enter the dimensions listed below:

    Dimensions (inches)
    Outside diameter (A):   15.25
    Mounting hole diameter (B):   14.8
    Cutout diameter (C):   14
    Magnet diameter (D):   5.375
    Front depth (F):   0.75
    Mounting depth (G):   7.125
    M-Vd (liters):   2.38

All of the driver information should now be entered. Click the "Accept" button and close the Driver Properties window.

Box Properties

Fortunately for this design, there is no upper limit on how large the box can be. We will focus entirely on performance and construct the best box design. Our primary goal will be to achieve a smooth, strong response down to 20 Hz. This will undoubtedly require a large box and one or more large vents. Let's begin:

- Open the Box Properties window by clicking on the "Box" button in the design panel of the main window or use the keyboard shortcut [Ctrl]+[B].

- Select the "Damping" tab and select "None" for the amount of acoustic absorption. The "Use Classical box calculations" option will need to be turned off for you to do this. The reason we will not be adding fill to the box is because it would take a lot to have much effect at frequencies below 100 Hz and this might decrease the vent output. Remember that our subwoofer will be used to produce frequencies up to only 90 Hz.

    Note: It is usually best to set the damping setting before designing a box since the damping setting will affect the box calculations.

- Select the "Box Design" tab and set the box type to "Vented Box". Let's ask BassBox Pro to suggest a box and see how close it comes to producing an F3 of 20 Hz. Click on the "Suggest" button.

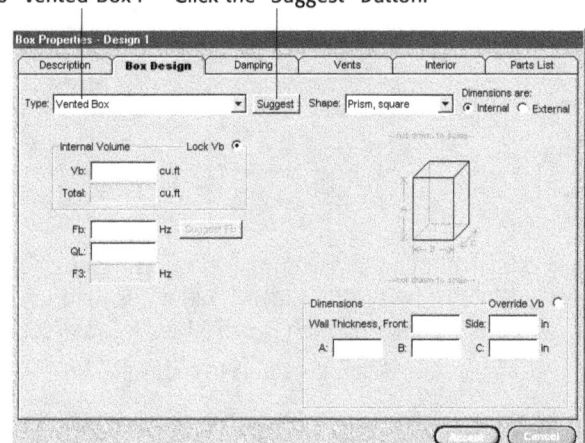

BassBox 6 Pro

Sample 2: Vented Box Example

The Design Priority window shown at right will open. *Note: The Design Priority window is only available for closed box and standard vented box designs.*

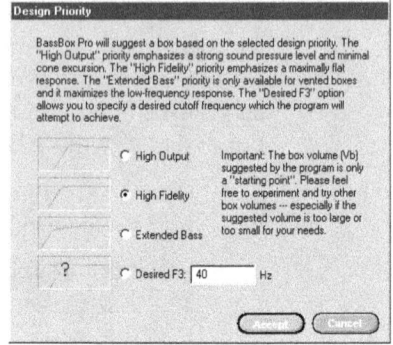

Select the "High Fidelity" option because we want a flat response. After clicking the "Accept" button you should see similar results as shown below:

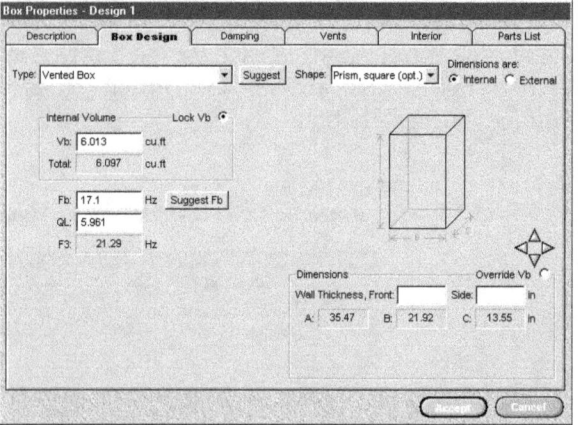

Notice above that BassBox Pro recommends a net internal box volume (Vb) of 6.013 cubic feet (170.3 liters) and a box resonance (Fb) of 17.1 Hz. This produces an F3 of 21.29 Hz which is very close to our goal.

Click on the "Accept" button to close the Box Properties window.

- Next, let's examine the response of the speaker. We will use the combination graph window so select the "Display Mode > Single Window" command from the Graph menu. Then select the "Show Graph > Amplitude—Normalized" command from the Graph menu (or press [Ctrl]+[F1]) to open the graph window. Finally, click on the "Plot" button of the design panel ([Ctrl]+[1]) back in the main window and the **normalized amplitude response** should look like the example on the next page:

Sample 2: Vented Box Example

BassBox 6 Pro

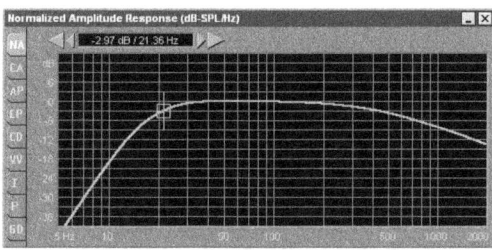

We enabled the cursor with the "Cursor > Show" command in the Graph popup menu ([Ctrl]+[U]) and verified that F3 is located a little above 21 Hz. *Note: The graph popup menu is summoned by clicking on the graph with the right mouse button (also called "right-clicking").*

Notice that the "knee" of the response (the region where the bass response begins to drop) is quite rounded. Normally the shape of the knee should be a bit sharper than this if the response is "maximally flat". With this in mind, let's see if we can do better by manually adjusting the box volume and tuning frequency ourselves.

- Reopen the Box Properties window ([Ctrl]+[B]) and let's return to the "Box Design" tab. Position the Box Properties window on your Windows desktop so that you can see both it and the mini preview graph for this design in the main window.

  Next, let's experiment by making small changes in the box volume (Vb) and box resonance (Fb) and observe their effect on F3 and the mini preview graph. We found that by making Vb = 6.5 cubic feet (184.1 liters) and Fb = 20 Hz that we could achieve both of our goals. Enter these values into Vb and Fb and F3 moves to 20.07 Hz and the knee of the response sharpens as desired.

  Click on the "Accept" button to close the Box Properties window again and let's return to the graph.

- Begin by changing the plot color. This is done by clicking on the plot color indicator on the design panel. It will advance to the next color as shown at right.

Click on the plot color indicator to toggle the color.

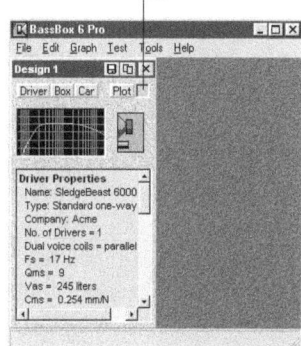

BassBox 6 Pro

Sample 2: Vented Box Example

Finally, click on the "Plot" button (Ctrl+1) and the graph should look like the one at right:

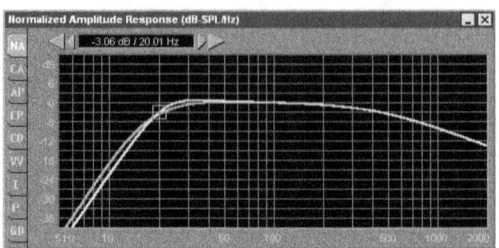

The white plot line is a very satisfactory response and shows that the SledgeBeast 6000 will produce the response we desire. We will explore other aspects of its performance soon. Next we must calculate the vent dimensions.

- Reopen the Box Properties window (Ctrl+B) and select the "Vents" tab. We plan to use PVC pipe to construct the vent, so select "Round" for the vent cross section shape and "One Flush End" for the vent end type. Next let's ask BassBox Pro to recommend a vent. This is done by clicking on the "Suggest Minimum Vent Area for Xmax" button as shown below:

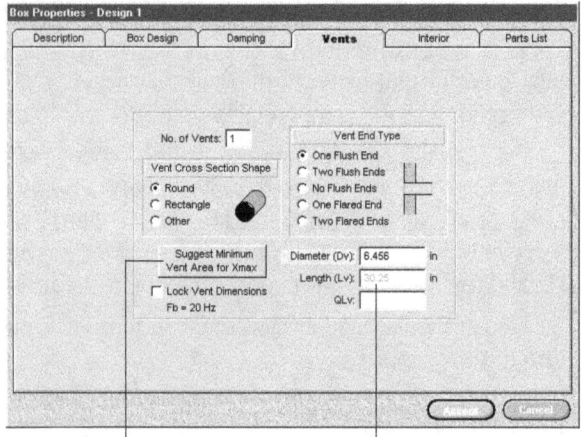

Click on the "Suggest Minimum Vent Area for Xmax" button.

The vent length (Lv) is red when it may be too long to fit in the box.

BassBox Pro estimates that a vent inside diameter of 6.456 inches (164 mm) is the minimum size that will keep the vent air velocity low and avoid vent turbulence noise. This is a "worst case" calculation that assumes that the woofer is being driven to the full limit of its linear excursion (Xmax). Because the vent diameter is large and because the desired box resonance frequency is low (20 Hz), the vent must be quite long. In fact, the vent length (Lv) must be 30.25 inches (768.3 mm). The program realizes that this length is probably too long to fit inside the box and so it warns us by making the color of the Lv value red. *Note: Although we haven't entered the box dimensions yet, the program temporarily estimates them using the optimum square*

Sample 2: Vented Box Example   **BassBox 6 Pro**

prism ratio. For more information about the optimum square prism box shape, see page 227 in the BassBox Pro Reference portion of this manual.

The total cross-sectional area of the vent(s) and the volume of air swept by the moving diaphragm of the woofer determine the velocity of the air in the vent. This gives us only two ways to **reduce the air velocity** in the vent:

1. Increase the total vent area. This can be done by increasing the number of vents or by increasing the diameter of the vent(s).
2. Decrease the amplifier power driving the speaker. This will decrease the excursion of the woofer and lower the volume of air swept by its diaphragm.

The length of the vent is controlled by the combination of the desired box resonance (Fb), the box internal volume (Vb) and the total vent cross-sectional area. This gives us three ways to **shorten the vent length** (Lv):

1. Decrease the total vent cross-sectional area. This can be done by reducing the diameter (Dv) of the vent(s) or, if there are more than one, reducing the number of vents.
2. Increase the internal box volume (Vb).
3. Raise the desired box resonance (Fb) to a higher frequency.

Do we really need a vent with a diameter as large as 6.456 inches? Probably not because we do not plan to "ride the ragged edge" and drive the subwoofer to Xmax—at least not very often. A "Recommend Vent Cross Section Size for Nominal Music Playback" table is included on page 234 of the Reference portion of this manual. It states that a 6 inch (152 mm) vent would probably be sufficient for a typical 15 inch woofer and we may even be able to get by with a bit less.

The area of a 6.456 inch diameter vent is 32.74 square inches (211.2 sq.cm). We would like to try using two 4 inch (101.6 mm) vents because 4-inch PVC pipe is readily available. The area of one 4 inch diameter vent is 12.57 square inches and so the total area for two vents is 25.14 square inches (162.1 sq.cm). Let's try this and see how it works. Enter "2" for the "No. of Vents" then enter 4 inches for the vent diameter (Dv) as shown at right:

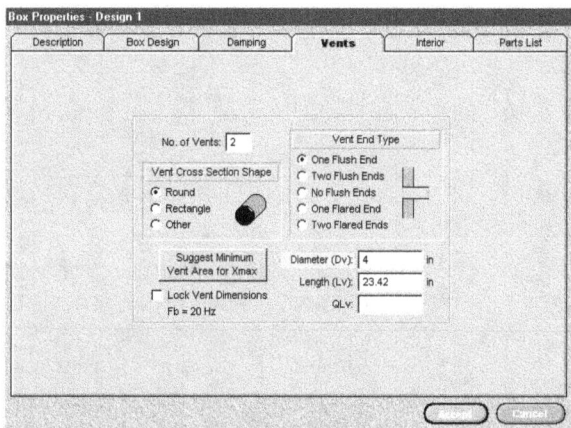

BassBox Pro still calculates a long vent but this time it is a more usable length of 23.42 inches (594.9 mm). It is also short enough to fit inside the box and the color of the Lv value is now black rather than red.

We will ignore the vent losses (QLv) because the program will estimate them internally. *Note: This length may still seem long and it may result in higher vent losses due to the viscosity friction of air moving through two vents of this length. However, this is the unavoidable by-product of tuning the box to such a low frequency.*

We will revisit the vent again when we examine the Vent Air Velocity graph.

- Select the "Box Design" tab and let's work on the box dimensions. The default box shape for new single-chamber boxes is the optimum square prism. It forces the dimensions to fit a certain ratio in order to minimize the problem of "standing waves" in the box. We do not need to be concerned with standing waves because our subwoofer will operate only as high as 90 Hz with a wavelength of 151 inches (3835 mm) and this is many times longer than the largest box dimension we will use.

We want full control over the box dimensions so we will switch to the square prism shape. Set the box shape to "prism, square" as shown below:

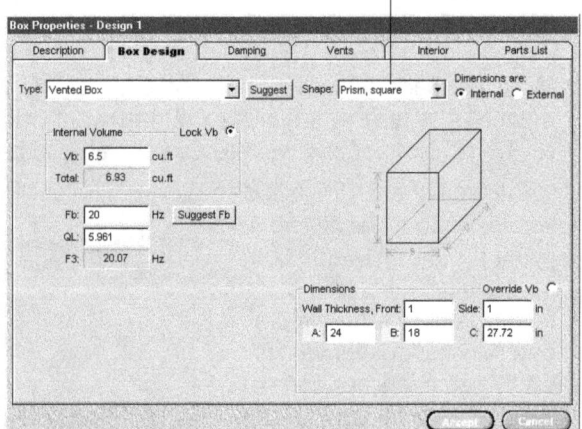

Select the "square prism" box shape.

We will use "Internal" dimensions because it will help us keep an eye on the internal depth to see how close it is to the vent length. If necessary, set the dimensions to "Internal" now. Because our box is big we want to construct it with a strong rigid material. MDF (medium density fiberboard) is a popular construction material for boxes. We will use 1 inch MDF so enter 1 inch (25.4 mm) for the front wall thickness and side wall thickness. We don't want the front of the box to look too huge so let's try

Sample 2: Vented Box Example   BassBox 6 Pro

entering a height (A) of 24 inches (609.6 mm) and a width (B) of 18 inches (457.2 mm). Since the internal volume (Vb) is locked, BassBox Pro will calculate the final dimension. It calculates the depth (C) to be 27.72 inches (704.1 mm). The three-dimensional box drawing should look like the one in the preceding illustration.

Notice that the background of the "A" and "B" labels are highlighted with a lighter shade to show that these dimensions were entered and are therefore protected (locked) and can only be changed by us. In contrast, dimension "C" can be changed by the program whenever it sees that the box volume has changed. *Note: You can change which dimensions are locked and which are unlocked by clicking on their labels.*

- Next let's enter the driver mounting method and vent wall thickness. This will enable BassBox Pro to property account for the volume that the driver and vents displace inside the box. Select the "Interior" tab.

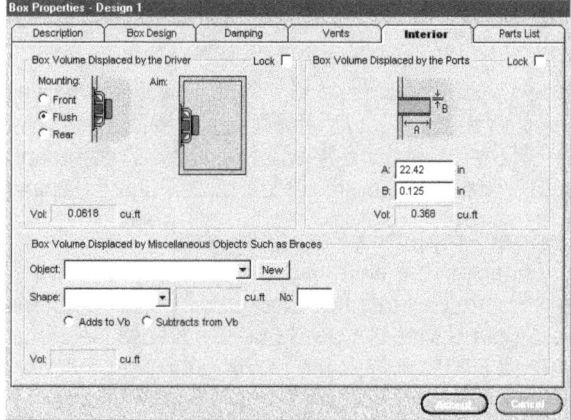

We plan to flush-mount the driver in the box so set the driver mounting to "Flush" as shown above in the "Box Volume Displaced by the Driver" section in the upper left portion of the "Interior" tab. BassBox Pro estimates that the driver will displace 0.0618 cubic feet (1.8 liters) inside the box.

The PVC pipe we plan to use for the vents has a wall thickness of 1/8 inch (3.2 mm) so enter 0.125 inch for dimension "B" in the "Box Volume Displaced by the Ports" section in the upper right portion of the "Interior" tab. This is the default value for "B". Notice also that dimension "A", the depth that the vent protrudes into the box, is already calculated for us by subtracting the box front wall thickness from the vent length (Lv). BassBox Pro calculates that our two vents will displace a total of 0.368 cubic feet (10.4 liters) inside the box.

BassBox 6 Pro

Sample 2: Vented Box Example

We plan to stiffen the box walls with braces. If we wanted to, we could enter the sizes and quantity of braces in the lower portion of the "Interior" tab so that the volume displaced by the braces would also be subtracted from the net internal box volume (Vb). However, this usually isn't necessary unless the braces will occupy a sizeable portion of the box interior. (This is also true for the driver and vents—however, the program automatically accounts for them.) We have chosen to ignore the brace displacement volume because we expect it to be relatively small.

We are finished with the Box Properties window for the time being. Close it by clicking on its "Accept" button.

Evaluating the Performance

So far everything looks good. Our design produces the response we want. But how does it perform in other areas? How loud will the speaker go? Since we used a smaller total vent area than was suggested, will there be problems? The performance graphs will answer these questions. We'll skip the Normalized Amplitude Response graph since it was discussed previously.

Since the graphs contain two plot lines, let's clear them and get a fresh plot. If it is on, begin by turning off the cursor with the "Cursor > Hide" command of the graph popup menu ([Ctrl]+[H]). Then select the "Clear > All Graphs" command from the graph popup menu ([Ctrl]+[Y]). Finally, click on the "Plot" button of the design panel ([Ctrl]+[1]).

The **Custom Amplitude Response** graph is selected with the "CA" graph tab ([Ctrl]+[F2]). It is very similar to the Normalized Amplitude Response graph except that it shows the sound pressure level that the speaker will produce on axis at 1 meter (3.3 feet) with the specified input power. We used 600 watts because this is the maximum power level of the SledgeBeast 6000 and it is the maximum power that our amplifier will produce per channel into a 4 ohm load.

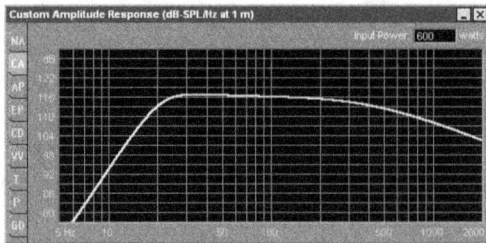

Notice the result: the speaker will produce 117 dB from about 28 to 51 Hz and continue with 116 dB up to 176 Hz. This should be very adequate for our home theater—especially since the power rating of both the woofer and the amplifier is a continuous RMS level and both can handle higher peak levels.

Sample 2: Vented Box Example **BassBox 6 Pro**

The **Maximum Acoustic Power** graph is selected with the "AP" graph tab ([Ctrl]+[F3]). It shows how much acoustic power the speaker can produce before reaching its steady-state displacement limit or its thermal limit.

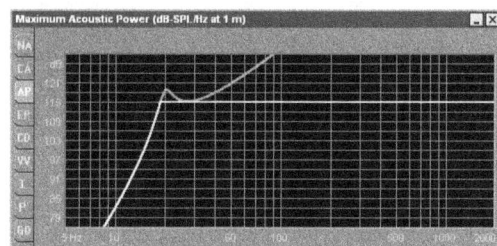

In our case, the speaker is thermal-limited down to 19 Hz. Below 19 Hz it is displacement-limited by Xmax and the maximum acoustic power drops off rapidly. It is uncommon for a woofer to go this low before reaching its steady-state excursion limit and it is a testament to its high Xmax rating.

The **Maximum Electric Input Power** graph is selected with the "EP" graph tab ([Ctrl]+[F4]). It is derived from the Maximum Acoustic Power graph and it shows how much amplifier power the speaker can handle before reaching the first of either its displacement or thermal limits.

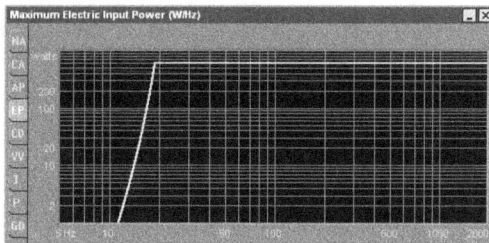

This graph shows us that the speaker should be able to handle a full 600 watts of power from the amplifier down to 19 Hz provided of course that the amplifier's output signal is not heavily distorted or "clipped". It would be wise to protect the driver below this point with a subsonic high-pass filter. This would also protect the woofer from becoming "unloaded" by the vent.

The **Cone Displacement** graph is selected with the "CD" graph tab ([Ctrl]+[F5]). It shows how far the diaphragm and voice coil must travel with the specified input power. We used 600 watts because this is the maximum power level of the SledgeBeast 6000.

BassBox 6 Pro User Manual © D.E.Harris   **111**

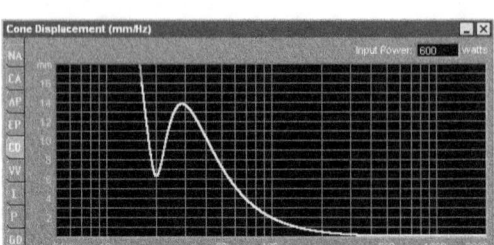

Notice that the plot line changes intensity when it rises above the Xmax level of 16 mm to show when the woofer will exceed its peak linear excursion limit. This happens at about 17 Hz and underscores the need to add a subsonic high-pass filter to protect the woofer as mentioned previously. This graph is very positive because the woofer stays within its excursion limit down to and a little below 20 Hz so it should have relatively low distortion.

The **Vent Air Velocity** graph is selected with the "VV" graph tab (Ctrl+F6). It shows how fast air will vibrate in the vent with the specified input power. We used 600 watts because this is the maximum power level of the SledgeBeast 6000.

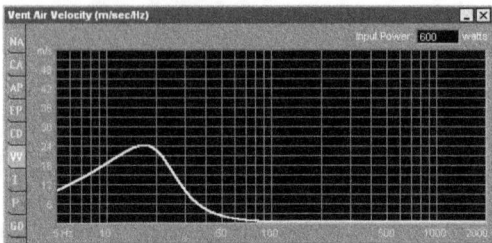

Earlier we decided to use a smaller total vent area than was recommended in order to reduce the vent length to a more manageable size. This caused concern because a reduction in total vent area causes the vent air velocity to increase. Fortunately, as shown in the above graph, we needn't have worried because the vent air velocity never rises above 35 meters per second (113 feet per second). If the plot line had risen above this level, it would have changed intensity like the plot line did on the Cone Displacement graph.

Note: 35 m/s is 10% of the velocity of sound in air. Ideally the vent air velocity should be kept below this level to avoid air turbulence noise. Some designers even prefer to keep the vent air velocity below 5% of the velocity of sound.

The **System Impedance** graph is selected with the "I" graph tab (Ctrl+F7). It shows the impedance of the speaker.

Sample 2: Vented Box Example **BassBox 6 Pro**

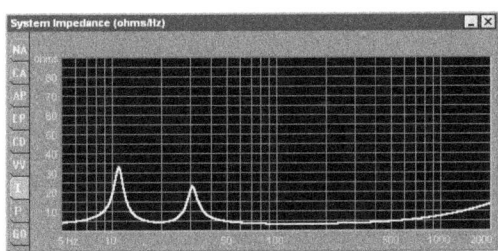

The first peak at 11.1 Hz is the resonance of the vents, themselves. The minima at 19.7 Hz is the system resonance of the box. The second peak at 30.4 Hz is the resonance of the driver in the box. *Note: The resonance of a driver is always at a higher frequency in a box than in free air (Fs).* It can sometimes be difficult to determine the precise location of a peak or minima. **Tip:** The phase of the impedance should pass through zero degrees at each peak and minima (unless the peaks are spaced closely together).

The System Impedance graph is useful for a couple of reasons. If the speaker will be connected directly to an amplifier, it shows the load that the amplifier must "drive". In this case the amplifier must be capable of driving a 4 ohm load. If the speaker will include a passive crossover network, the graph shows the impedance that the crossover network will "see". This is very important when designing the crossover network because it usually wants to see a flat impedance.

The **Phase Response** graph is selected with the "P" graph tab (Ctrl+F8). It shows how much the sound waves which emanate from the speaker will lag behind the input signal that feeds the speaker. This delay is expressed as a phase angle in degrees and it is literally the difference between the phase of the input signal and the phase of the output signal.

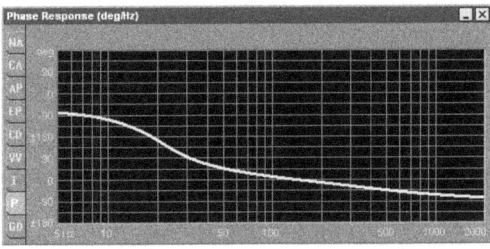

Notice that the low frequency sound waves never lag by more than 270°. This means that the output signal will not lag behind the input by more than three quarters of a wavelength. Remember that a full wavelength requires 360° of phase rotation. *Note: The reason the phase response drops below 0° above 140 Hz is because of the inductive reactance of the woofer's voice coil.*

Ideally the phase response should be a perfectly flat line with 0° (zero degrees) of phase shift. Fortunately, a gradual change in phase angle as shown above will not significantly harm the fidelity. However, sharp and sudden changes in phase angle are a significant problem and should be avoided when possible.

The **Group Delay** graph is selected with the "GD" graph tab ((Ctrl)+(F9)). It is very similar to the Phase Response graph because it also shows how much the sound waves which emanate from the speaker will lag behind the input signal that feeds the speaker. However, this graph expresses this information as a delay in milliseconds and it is derived from the slope of the phase response.

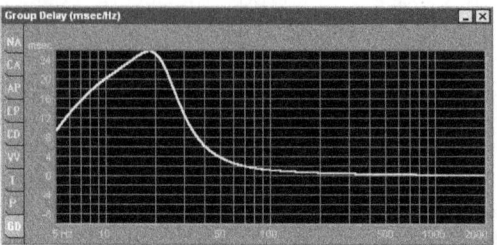

Ideally there should be no delay. In practice, a gradual change will not significantly harm fidelity. However, sharp and sudden changes in group delay are a significant problem and should be avoided when possible. Our design has significant group delay which can diminish the transient response. This is probably the weakest aspect of this design. Fortunately, the group delay does not begin to rise rapidly until about 40 Hz and at such low frequencies it is doubtful if most listeners will discern a problem. If group delay were a serious concern, we would need to switch to a closed box design. However, if we did this, we would sacrifice bass response because the lowest F3 we would get from a closed box design would be about 33 Hz instead of the 20 Hz we achieve with this vented box design.

When deciding which tradeoffs to make in a speaker design it is important to focus on the purpose of the speaker. Our primary purpose is increased "tactile realism" in our home theater. In other words, we want to "feel" the bass. A solid response down to 20 Hz will certainly enable us to experience this. If this subwoofer were being designed as a part of a monitor speaker for a recording studio or other critical listening environment which has good low-frequency acoustics, then we might place a greater priority on a flat group delay response.

Conclusion

The last step in our design is to review the parts list and, if desired, create a printout. Open the Box Properties window by clicking on the "Box" button on the design panel ((Ctrl)+(B)). Select the "Parts List" tab. The drawing of each part and a list of dimensions should appear

Sample 2: Vented Box Example **BassBox 6 Pro**

as shown at right:

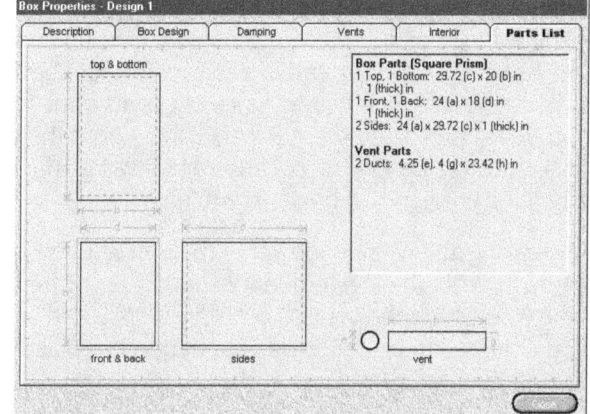

Notice that the vents are also included in the drawing and parts list.

Since this box is large its walls will tend to flex more easily than a smaller box. Ideally, sound should only emanate from the woofer and the vent—not the sides. This is one reason why we choose to use 1-inch thick MDF for all sides. In addition, we heavily braced the interior of the box to further stiffen its walls. See Chapter 4 of the *Box Designer's Guide* earlier in this manual for additional information about box construction.

We decided to mount our two vents on the front of the box underneath the woofer as shown in the illustration below.

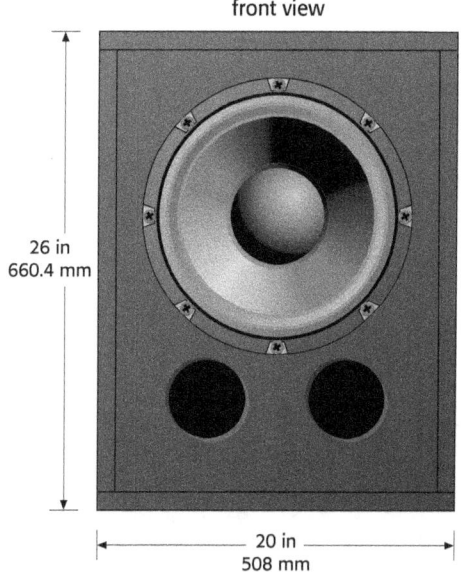

It is important to keep both ends of each vent away from walls and braces inside the box so that the air flow in and out of the vent is not affected. To accomplish this we moved them close to the woofer. Ideally, the end of each vent should be at least one diameter away from the closest surface or structure. Since our vents have an inside diameter of 4 inches (101.6 mm) we should keep them 4 inches away from the side walls and bottom. As you can see this proved to be difficult because there wasn't much room left over after the woofer mounting position had been determined.

Generally speaking, the woofer and vents could have been mounted anywhere—even the back side—because, at the low frequencies produced by this subwoofer, the sound emanating from the woofer and the vents is omnidirectional. And since the wavelengths are so long, we do not need to be concerned about phase problems if the vents are mounted on a different side than the woofer.

In summary, vented boxes are more complicated to design than closed boxes. Most of the design problems center on the vent. The vent needs to be large enough to avoid air turbulence noise. This leads to the most common problem: a vent that is too long to fit in the box. This frequently happens when trying to tune a small box to a low frequency. There will be situations where a vent just won't work because the box is too small. This can easily happen with compound multi-driver designs because an isobaric pair of drivers requires only half the box volume that a single driver would require. A passive radiator box is a good alternative when these problems become unsolvable.

This concludes the vented box example. We hope that it has been helpful. Please review the other examples for additional information about box design.

Sample 3: Bandpass Box Example

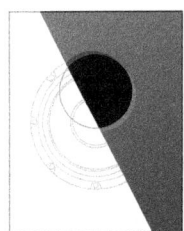

3 Bandpass Box Example

This chapter shows how to use BassBox Pro to design a bandpass box. Bandpass boxes have the unique feature that they produce sound in a relatively narrow band of frequencies. Since the box acts like an acoustic band-pass filter and controls both the low-frequency and high-frequency response of the driver, it is possible to avoid a crossover network. Bandpass boxes also provide the unique feature of allowing the designer to control efficiency. This example will design an advanced isobaric bandpass box for use as a subwoofer in an automotive vehicle. We want the upper limit (upper F3) of the response to be between 95 and 100 Hz so that we can avoid adding another filter to our crossover network. We want the lower limit (lower F3) of the response to be around 50 Hz so that it will blend well with the in-car bass response.

This example uses a single-tuned bandpass box because it is often better suited for use in a motor vehicle than a double-tuned bandpass box. The 2nd-order low-frequency cutoff rate of a single-tuned bandpass box blends nicely with the typical in-car acoustic response which boosts the bass at the same rate. The result is a smooth, extended low end.

If you are new to speaker box design and have not yet reviewed the earlier design examples, we recommend that you do so before studying this one. Begin with Chapter 1 because it includes extra details for beginners. Then continue with Chapter 2 because it adds additional details about box design and the use of BassBox Pro.

Driver Properties

This design begins with the driver. Selecting a suitable driver can take considerable time and often involves experimentation with several different models in one or more box designs before deciding which one is best. The following driver qualities are often desired for a single-tuned bandpass box (however, there are many exceptions):

- Qts greater than 0.3.
- Low resonance (Fs). High moving mass (Mms) and compliance (Vas, Cms).
- Large Xmax (small woofers: 2-4 mm; large woofers: 5-8 mm).
- Moderately strong motor (BL).

 **BassBox 6 Pro** *Sample 3: Bandpass Box Example*

We've found a driver that seems to satisfy most of these requirements. Its the 10-inch (254 mm) Acme BassBomb 100. We will use two of these fictitious drivers in an isobaric configuration. This will reduce the required box volume while increasing the power handling. Let's begin:

- Run BassBox Pro, go to the main design window and create a new design (Ctrl+N).
- Click on the "Driver" button of the new design panel (Ctrl+D) to open the Driver Properties window and select the "Parameters" tab. Turn on the "expert" mode if it is not already turned on. Enter the following driver parameters (remember to set the units of each parameter before entering its value):

Model:	BassBomb 100
Fs (Hz):	30
Qms:	8
Vas (liters):	99.11
Cms (mm/N):	0.586
Mms (g):	48
Rms (kg/s):	1.132
Xmax (mm):	8.35
Dia (mm):	209.6
Sd (sq.cm):	345
Vd (liters):	0.288
Qes:	0.412
Re (ohms):	6.4
Le (mH):	1.1
Z (ohms):	8
BL (Tm):	11.86
Pe (watts, continuous):	100
Qts:	0.392
ηo (%):	0.626
1-W SPL (dB):	90.17
2.8-V SPL (dB):	91.13

The "Parameters" tab should look similar to the one on the next page after the parameters are entered.

Sample 3: Bandpass Box Example   BassBox 6 Pro

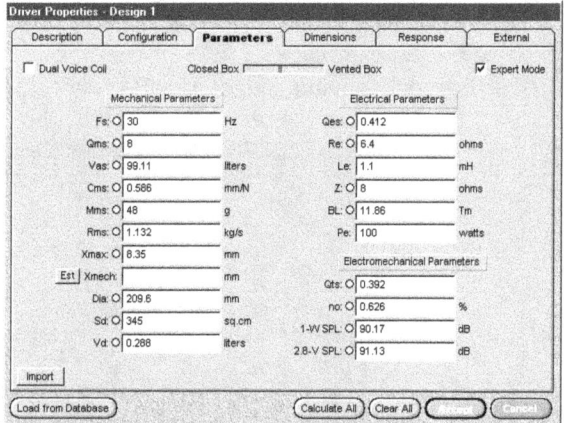

- Select the "Configuration" tab. We plan to use two BassBomb 100s in our subwoofer so enter "2" for the number of drivers. The two drivers will face each other on the interior baffle of the box in an isobaric configuration so select both "Compound" and "Push-Pull" for the mechanical configuration. We plan to wire the drivers in parallel so select "Parallel" for the electrical configuration. *Note: The "Parameters" tab will contain "net" values for most of the parameters now that this is a multi-driver design.*

  The "Configuration" tab should look like the one below:

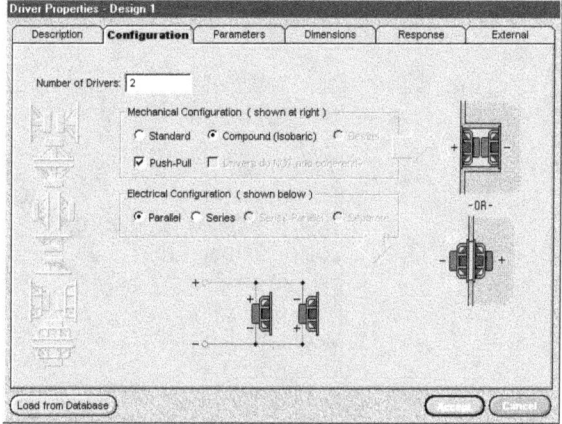

- Select the "External" tab. The subwoofer will be located close to the amplifier and so the speaker cable has a short length with a resistance of only 0.2 ohms. Enter 0.2 for Rx in the "External Resistance" section on the left side of the "External" tab. Our amplifier has a damping factor of 200 for low frequencies. Enter 200 for DF. The pro-

Sample 3: Bandpass Box Example

gram should calculate Rg to be 0.0201 ohms. Turn on the "Include Ext. Resistance" checkbox.

The "External" tab should look like the one below:

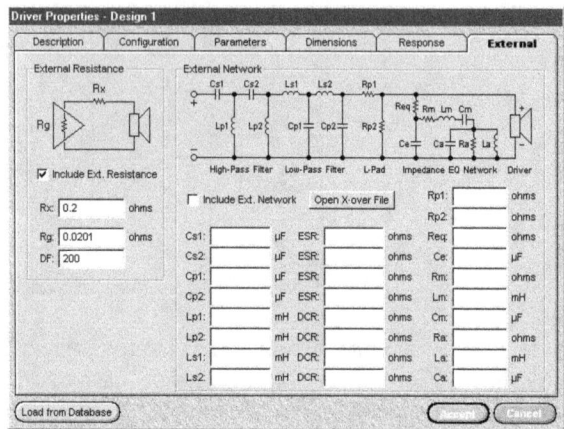

We will ignore the "External Network" section on the right because we do not plan to use a passive network with this subwoofer.

- Select the "Description" tab and enter "BassBomb 100" for the driver model name and "Acme" for the company name. The driver type should be set to "Standard one-way driver".

- Select the "Dimensions" tab. Select a "Round" outer shape and a "Cone" piston type. Enter the dimensions listed below:

    Dimensions (inches)
    Outside diameter (A):   10.125
    Mounting hole diameter (B): 9.8
    Cutout diameter (C):   9.2
    Magnet diameter (D):   5
    Front depth (F):   0.3
    Mounting depth (G):   5.125
    M-Vd (liters):   2.079

All of the driver information should now be entered. Select the "Accept" button and close the Driver Properties window.

Sample 3: Bandpass Box Example   **BassBox 6 Pro**

Box Properties

We would like to construct a cylindrical box with Sonotube®. Sonotubes are very strong, rigid tubes manufactured from spiral wound and laminated fiber layers by Sonoco. They are used in the construction industry as forms for the pouring of concrete columns and are available in a wide variety of diameters from 6 inches (152.4 mm) to 48 inches (1219.2 mm). We will use a 12 inch (304.8 mm) Sonotube for our enclosure. The illustration below shows what we have in mind:

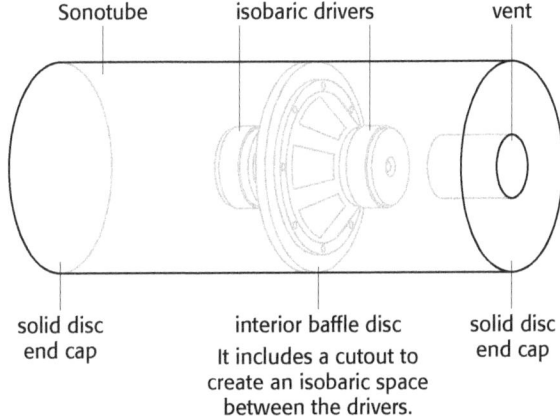

Let's begin the box design by first calculating the required chamber volumes and tuning frequency. Then we will select the box shape and determine its dimensions.

- Click on the "Box" button of the new design panel ([Ctrl]+[B]) to open the Box Properties window and then select the "Box Design" tab. Set the box type to "Bandpass Single-Tuned Box".

- Select the "Damping" tab. Choose "Heavy" for the amount of acoustic absorption for chamber 1 and "None" for chamber 2 as shown on the next page. The "Use Classical box calculations" option will need to be turned off for you to do this. The reason we want to use a lot of fill in chamber 1 is because it is a sealed chamber and we would like to minimize its size by adding as much damping as possible. Remember that chamber 1 is the lower frequency chamber and so it will usually be the largest chamber. The reason we do not want to add fill to chamber 2 is because it is vented and we do not want to attenuate the vent. Remember that chamber 2 is the upper frequency chamber and so it will tend to be small and may not have much room for a vent. We don't want to make this problem any worse by adding fill. Standing waves are not a concern because we expect the size of each chamber to be too small for them to develop in the frequency range of this speaker.

 **BassBox 6 Pro** *Sample 3: Bandpass Box Example*

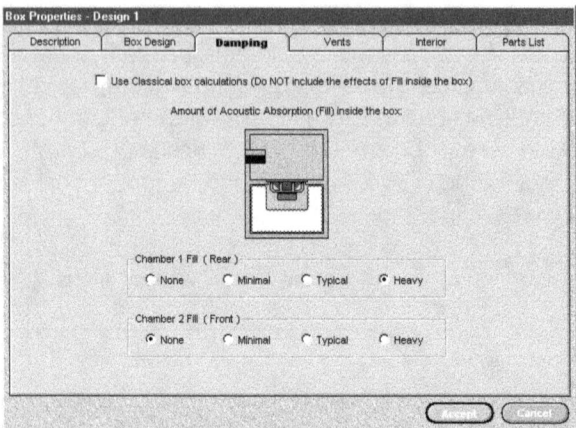

Note: It is usually best to set the damping setting before designing a box since the damping setting will affect the box calculations.

- Select the "Box Design" tab again and let's ask BassBox Pro to suggest a box and see how close it comes to producing a lower F3 of 50 Hz and an upper F3 of 95–100 Hz. Click on the "Suggest" button and you should see similar results as shown below:

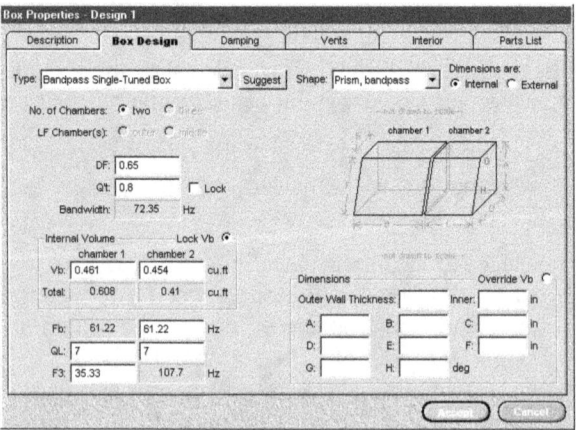

Notice in the example above that BassBox Pro recommends a net internal chamber volume (Vb) of 0.461 cubic feet (13.05 liters) for chamber 1 and 0.454 cubic feet (12.87 liters) for chamber 2. The program also recommends that chamber 2 be tuned (Fb) to 61.22 Hz. This combination produces a lower F3 of 35.33 Hz and an upper F3 of 107.7 Hz. This doesn't satisfy our design goals but its a good start.

Click the "Accept" button to close the Box Properties window.

Sample 3: Bandpass Box Example BassBox 6 Pro

- Let's see what the amplitude response looks like. We will use the combination graph window so select the "Display Mode > Single Window" command from the Graph menu. Then select the "Show Graph > Amplitude—Normalized" command from the Graph menu (or press Ctrl+F1) to open the graph window. Finally, click on the "Plot" button of the design panel (or press Ctrl+1) back in the main window and let's look at the response. The **normalized amplitude response** should look like the example below:

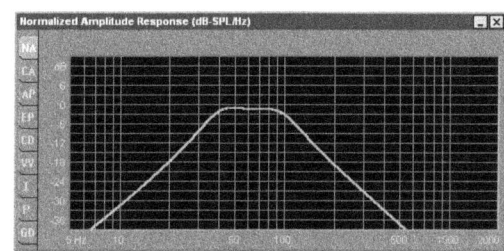

This response doesn't look bad. In fact, it might be an excellent subwoofer response to augment the bass of a small bookshelf audio system in a bedroom or dorm room.

- Since our subwoofer will be used in a car, we need to turn on the car acoustic response and see how it will sound in that environment. Click on the "Car" button of the design panel (Ctrl+A) to open the Car Acoustic Properties window (shown below). *Note: If the design panel has a "Room" button instead of a "Car" button, you will need to change the acoustic environment setting to "automotive" with the "General" tab of the Preferences window before proceeding.*

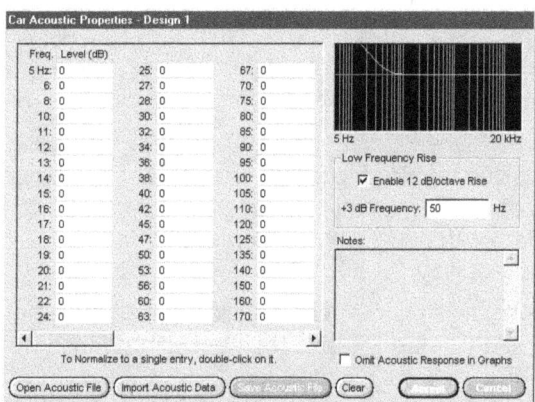

We will ignore the individual level settings on the left side of the Car Acoustic Properties window because we do not have an acoustical measurement of the car. Instead, we will estimate the in-car response with the "Low Frequency Rise" section on the right side of the window.

Turn on the "Enable 12 dB/octave Rise" control. Set the "+3 dB Frequency" to 50 Hz if it does not already have this value. Turn off the "Omit Acoustic Response in Graphs" control. Now notice how the Car Acoustic Properties graph shows the 12 dB/octave low-frequency boost to simulate the bass response inside the car. Also notice how the mini graph in the design panel of the main window has replotted to show the effects of the in-car response.

Click on the "Accept" button to close the Car Acoustic Properties window.

- Let's return to the graph and plot the new response. Begin by changing the main graph's plot color by clicking on the plot color indicator of the design panel in the main window. Then click on the "Plot" button ([Ctrl]+[1]) to replot the graph. The in-car response should appear as shown below with the white plot line:

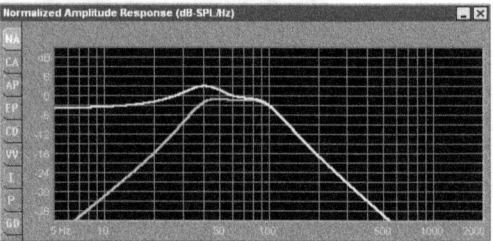

The in-car response is not as smooth as we would like, especially around 40 Hz where it peaks. This is because the lower F3 is at 35.33 Hz instead of the desired 50 Hz. Now its time to experiment with the box volumes and the resonance of chamber 2 and see if we can improve things. Ideally, we would like the bass response to be flat and the upper F3 to be between 95 and 100 Hz. We will continue to include the in-car response while we adjust the design.

- Reopen the Box Properties window ([Ctrl]+[B]) and return to the "Box Design" tab. Position the Box Properties window on your Windows desktop so that you can see both it and the mini preview graph for this design in the main window. Let's experiment with the chamber volumes and chamber 2 tuning and see if we can achieve our design goals.

We began our experimentation by adjusting the volume of chamber 1 and we noticed that making it smaller made things much worse, so we gradually increased it until it equaled 1.2 cubic feet (33.98 liters). This flattened the bass response very nicely. Next, we experimented with the volume and tuning of chamber 2. We were concerned about chamber 2 being too small for a vent so we decided to try to obtain an upper F3 of 95 Hz rather than 100 Hz since this would result in the largest volume. We found that by increasing its volume to 0.6 cubic feet (16.99 liters) and raising its resonance to 65 Hz that we could move the upper F3 down to 95.35 Hz. This seemed to produce a relatively flat response with only minor ripple. The final response is shown in the graph below:

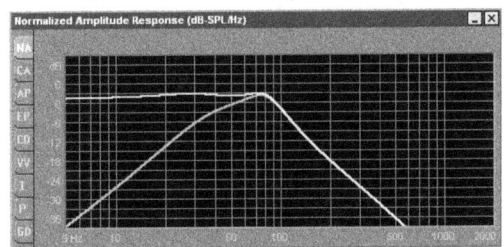

We included both the in-car response (white plot line) and the subwoofer response without it (medium grey plot line) in the graph to illustrate how the speaker response was tailored to fit the car's acoustic response. By itself, the speaker response doesn't look very good because it does not have a flat passband. However, when the car is added, the response is very smooth and, if its level is balanced properly with the rest of the car's audio system, should provide accurate sound reproduction.

- Next, let's enter the box dimensions so we can see how much room there is in chamber 2 for the vent. Select the "Box Design" tab and set the box shape to "Cylinder, bandpass". Set the dimensions to "Internal" as shown below:

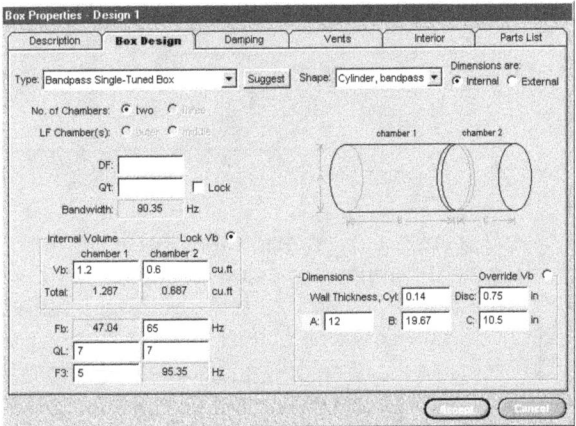

We will use a 12-inch (304.8 mm) Sonotube so enter 12 inches for the diameter (A). BassBox Pro will automatically calculate the length of each chamber. Chamber 1

should be 19.67 inches (499.6 mm) long (dimension B) and chamber 2 should be 10.5 inches (266.8 mm) long (dimension C). The 12-inch Sonotube has a 0.14 inch (3.56 mm) wall thickness so enter this for the cylinder wall thickness. We will use ¾-inch MDF (medium density fiberboard) for the ends and the baffle wall inside the box so enter 0.75 inches (19.05 mm) for the disc wall thickness.

We would like to mount the vent in the flat end along the radial axis of the enclosure but with an interior length of 10½ inches, chamber 2 will be a problem because almost 5½ inches of it will be occupied by one of the drivers and this will not leave much room for a vent. *Note: Once the vent dimensions are determined, the length of chamber 2 will automatically increase in proportion to the volume displaced by the vent. However, it may not increase enough.*

- Select the "Vents" tab and let's see just how big the vent needs to be. We plan to use PVC pipe to construct the vent, so select "Round" for the vent cross section shape and "One Flush End" for the vent end type. Next let's ask BassBox Pro to recommend a vent. This is done by clicking on the "Suggest Minimum Vent Area for Xmax" button shown below:

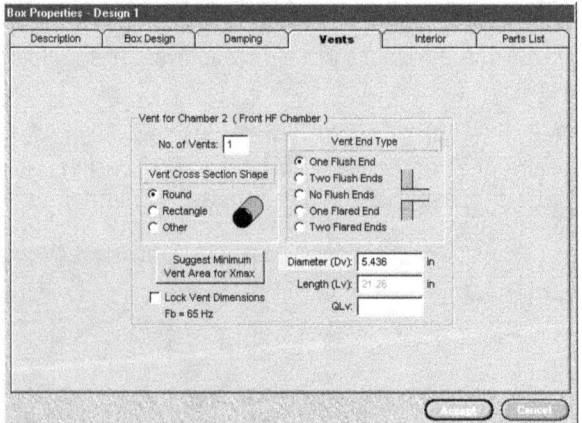

BassBox Pro estimates that a vent inside diameter (Dv) of 5.436 inches (138.1 mm) is the minimum size that will keep the vent air velocity low and avoid air turbulence noise. This is a "worst case" calculation that assumes that the woofers are being driven to the full limit of their linear excursion (Xmax). However, the suggested diameter is an estimate only and sometimes it produces a larger-than-needed value for the upper frequency chamber of a bandpass box. If a 5.436 inch diameter vent is used it will need to have a length (Lv) of 21.26 inches (539.9 mm). The program realizes that this length is probably too long to fit inside the box and so it warns us by making the color of the Lv value red.

Sample 3: Bandpass Box Example **BassBox 6 Pro**

Do we really need a vent with a diameter as large as 5.436 inches? Let's plot the vent air velocity and see. Select the **Vent Air Velocity** graph with the "VV" graph tab (Ctrl+F6) of the graph window. We entered an input power of 200 watts since this is the total maximum power that both BassBomb 100 drivers can handle (100 watts each). Finally, plot the response (Ctrl+1). You should see results similar to the illustration below:

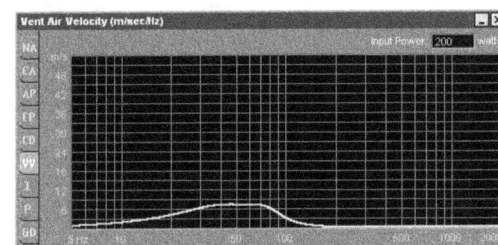

Generally, the vent air velocity should be less than 10% of the velocity of sound in air to prevent noise from air turbulence. This means we should try to keep the vent air velocity to 35 meters per second (113 ft/s) or less. Fortunately, the graph shows that the vent air velocity in the 5.436-inch diameter vent never rises any higher than 7.35 m/s and so we should be able to safely reduce the vent diameter.

- Let's return to the "Vents" tab of the Box Properties window and try both a 3 inch (76.2 mm) and a 2 inch (50.8 mm) vent diameter (Dv). This results in a 5.129 inch (130.3 mm) length for the 3 inch diameter and a 1.613 inch (40.96 mm) length for the 2 inch diameter. The 3-inch vent is shown below.

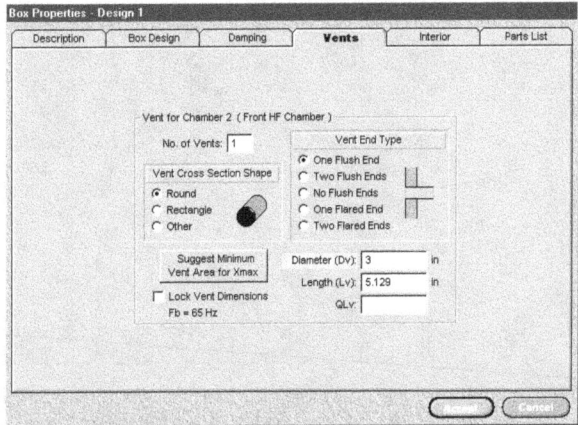

We will ignore the vent losses (QLv) because the program will estimate them internally.

The graph below shows the vent air velocity of all three vents:

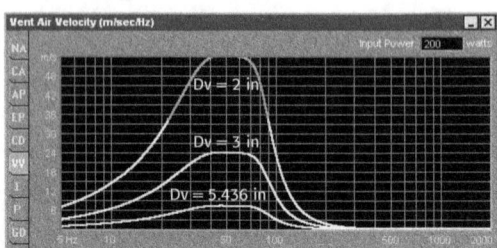

Notice that the air velocity of the 2-inch diameter vent rises to 54.3 m/s. Notice also that the intensity of the plot line darkens to warn you when it rises above 35 m/s. This vent may result in air turbulence noise when the speaker is driven to full excursion. It may not be a problem for nominal playback levels but we don't want to risk these problems since all of the sound waves generated by a bandpass box must emanate from its vent.

The vent air velocity of the 3-inch diameter vent only rises to 24.1 m/s so we will use it. With a length of 5.129 inches, it is still too long but we may be able to deal with this.

- Select the "Box Design" tab and let's return to the box dimensions and see how much the chamber 2 length has increased now that the vent dimensions are known.

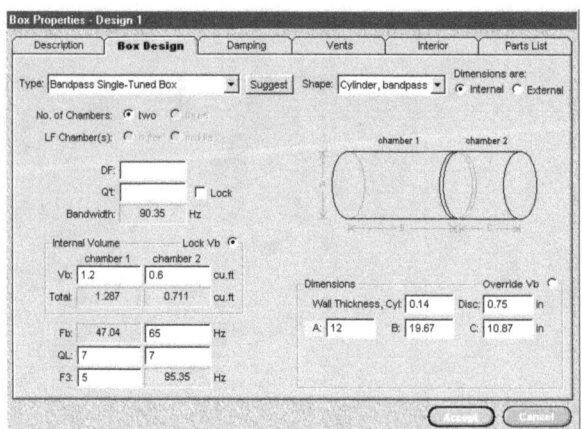

When the volume displaced by the 3-inch diameter vent is considered, the chamber 2 length (dimension C) increases to 10.87 inches (276.1 mm). This is still not long enough to accommodate the vent if it is mounted axially in the end. We'll explain: One of the two drivers protrudes 5.425 inches (137.8 mm) into chamber 2, leaving 5.445 inches (138.3 mm) of available space between the back of the driver's magnet and the end wall. The end of the vent should be at least one diameter away from the closest wall or structure. In this case, the closest structure is the magnet of the driver. Subtracting 3 inches from the available space leaves only 2.445 inches (62.1 mm) for the vent. Since the vent will pass through the 0.75 inch end wall, we can add this to the available length for a final length of 3.195 inches (81.2 mm). The bottom line: our 5.129 inch long vent is about 2 inches too long to fit inside the enclosure.

We "brainstormed" and came up with the three possible solutions shown below.

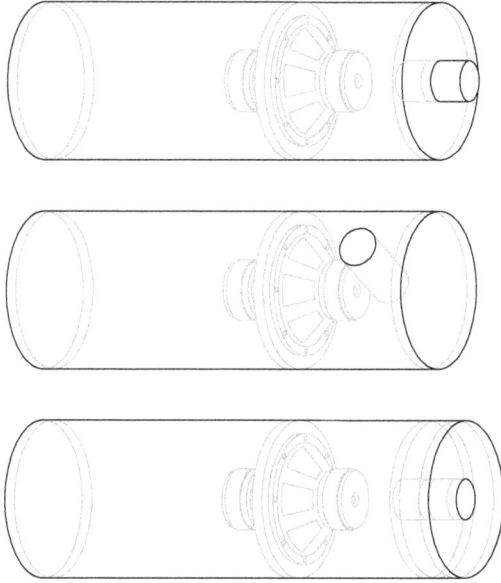

Our first idea was to simply allow the vent to extend beyond the enclosure. Although it would only need to extend about 2 inches outside of the enclosure, we didn't like this idea for aesthetic reasons. Our second idea was to abandon axial mounting and instead mount the vent to the cylinder wall. Unfortunately, this idea places the vent too close to either the end wall or the driver magnet—remember that each end of the vent should be at least one diameter away from the closest wall or structure so that the air flow is not affected. Our third idea expanded on the first. We lengthened the enclosure with a sealed, unused space to enclose the external portion of the vent. Notice that we added a new disc end cap so that the volume of chamber 2 would not change. We decided to use the third idea.

- Now we must adjust the box interior settings accordingly. Select the "Interior" tab and we will set the actual distance that the vent will protrude into chamber 2. We decided to allow the vent to protrude exactly 2 inches (50.8 mm) into chamber 2 so enter this value for dimension "A" of the "Box Volume Displaced by the Port" section in the upper right corner of the "Interior" tab as shown below.

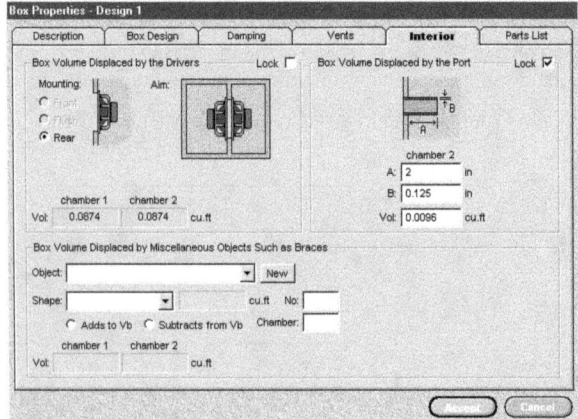

Dimension "B" should be set to 0.125 inches (3.2 mm) because we will construct the vent from PVC pipe with a 1/8 inch wall thickness. Notice that BassBox Pro calculates the volume displaced by the vent to be 0.0096 cubic feet (0.272 liters).

Lock the vent displacement volume by clicking on the "Lock" checkbox in the upper right corner. This will prevent BassBox Pro from changing the vent displacement volume (we can still change it manually if we desire later).

Notice that the driver mounting in the "Box Volume Displaced by the Drivers" section in the upper left portion of the "Interior" tab cannot be changed because the drivers are configured as an isobaric pair and are therefore assumed to be rear-mounted on the interior baffle. BassBox Pro estimates that each driver displaces approximately 0.0874 cubic feet (2.475 liters) in each chamber.

Finally, we will not enter any brace information at the bottom of the "Interior" tab. This is because the cylindrical shape of this enclosure and its relatively small MDF disc walls will be very rigid

- Let's return to the "Box Design" tab and see how BassBox Pro used the interior information to update the length of chamber 2.

Notice that dimension "C" has been reduced from 10.87 inches (276.1 mm) to 10.65 inches (270.5 mm) to reflect the fact that the vent protrudes only 2 inches

Sample 3: Bandpass Box Example

BassBox 6 Pro

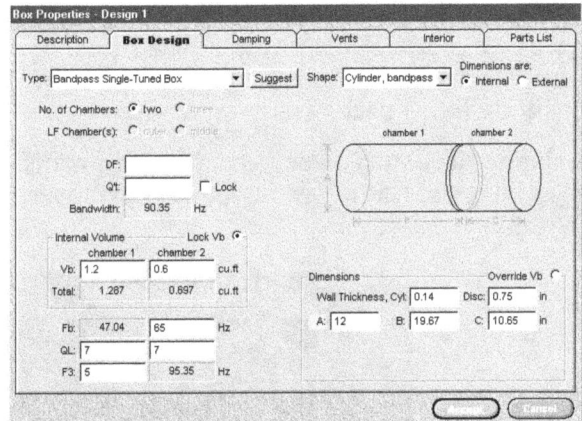

into chamber 2. Admittedly, this is a minor adjustment but we use this example to illustrate how BassBox Pro allows you to control these details of the design.

We are finished with the Box Properties window for the time being. Close it by clicking on its "Accept" button and we'll examine the performance of our design.

Evaluating the Performance

So far we have a decent in-car amplitude response and an acceptable vent air velocity. Now lets examine the remaining performance graphs and complete the evaluation of our design. We will skip the Vent Air Velocity graph since it was discussed previously. We will revisit the Normalized Amplitude Response graph at the end when we discuss vent "pipe" resonance. Before examining the remaining graphs clear them and create a fresh plot.

The **Custom Amplitude Response** graph is selected with the "CA" graph tab ([Ctrl]+[F2]). It is very similar to the Normalized Amplitude Response graph except that it shows the sound pressure level that the speaker will produce on axis at 1 meter (3.3 feet) with the specified input power. We used 200 watts because this is the total maximum power level for both BassBomb 100 drivers (100 watts each) and it is the maximum power that our amplifier will produce per channel into a 4 ohm load.

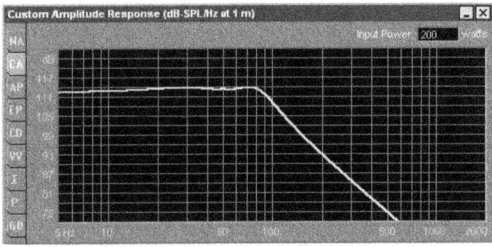

 *Sample 3: Bandpass Box Example*

Notice the result: the speaker should produce 114 dB for much of its passband. This is pretty good for a couple of 10-inch drivers and should be adequate for our car—especially since the power rating of both the woofers and the amplifier is a continuous RMS level and both can handle higher peak levels.

The **Maximum Acoustic Power** graph is selected with the "AP" graph tab (`Ctrl`+`F3`). It shows how much acoustic power the speaker can produce before reaching its steady-state displacement limit or its thermal limit.

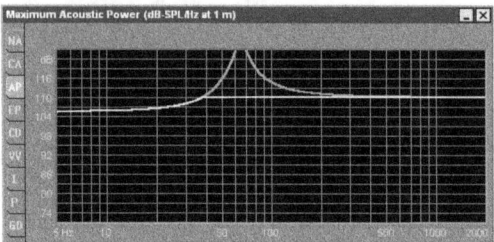

In our case, the speaker is thermal-limited down to 37 Hz. Below 37 Hz it is displacement-limited by Xmax and the maximum acoustic power drops about 4 dB. This is fairly good. We'll discuss the peak displacement limits again later with the Cone Displacement graph. By the way, the darkened portion of the plot line that extends to the top of the graph shows what the maximum acoustic power would be if there were no thermal limit.

The **Maximum Electric Input Power** graph is selected with the "EP" graph tab (`Ctrl`+`F4`). It is derived from the Maximum Acoustic Power graph and it shows how much amplifier power the speaker can handle before reaching the first of either its displacement or thermal limits.

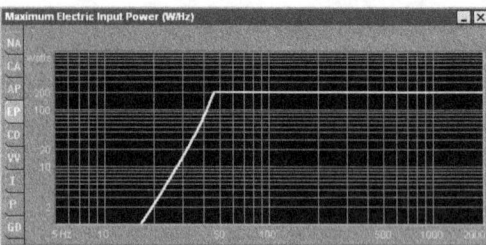

This graph shows us that the speaker should be able to handle a full 200 watts of power from the amplifier down to about 46 Hz provided, of course, that the amplifier's output signal is not heavily distorted or "clipped". It would be wise to protect the driver below this point with a subsonic high-pass filter. This would also protect the woofer from becoming "unloaded" by the vent. The reason that the displacement-limit is reached at 46 Hz here rather than 37 Hz as shown in the preceding Maximum Acoustic Power graph is because this graph does not include the effects of the in-car acoustic response like the Maximum Acoustic Power graph does.

The **Cone Displacement** graph is selected with the "CD" graph tab ([Ctrl]+[F5]). It shows how far the diaphragm and voice coil must travel with the specified input power. We used 200 watts because this is the total maximum power level for both BassBomb 100 woofers (100 watts each).

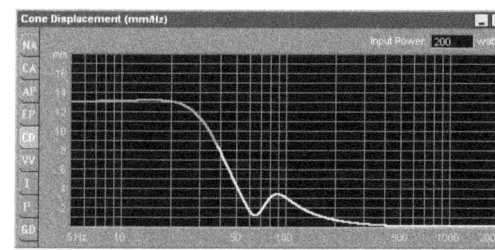

Notice that the plot line changes intensity when it rises above the Xmax level of 8.35 mm to show when the woofers will exceed their peak linear excursion limit. This happens at about 38 Hz and may underscore the need to add a subsonic high-pass filter to protect the woofers as mentioned previously.

Ideally, the peak cone excursion should not exceed Xmax (one-way from the cone's resting position). If it does, nonlinear distortion will increase. It could be argued that this is the weakest aspect of this design. However, several factors can reduce the seriousness of this problem. First, we do not plan to drive the speaker continuously at full power. In fact, we won't even drive it continuously at half power. And yet, at 100 watts, the 1 meter SPL will still be 111 dB and the excursion will not exceed Xmax until 28 Hz and never by more than 1 mm. Second, the mechanical excursion limit (the distance the diaphragm can travel before reaching the limits of the suspension or before the voice coil former strikes the back plate) would be larger than Xmax. This allows the diaphragm to move farther than Xmax without risk of damage. Third, the nonlinear distortion that increases when the excursion exceeds Xmax is subtle and is not audible to most listeners until it becomes severe.

The **System Impedance** graph is selected with the "I" graph tab ([Ctrl]+[F7]). It shows the impedance of the speaker.

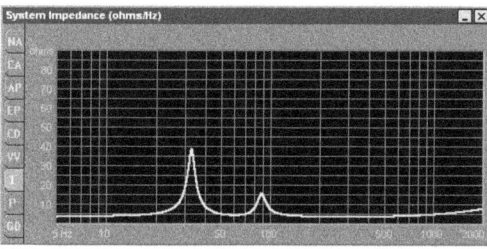

The first peak at 32.9 Hz is the resonance of the drivers in the enclosure. The minima at 64.1 Hz is the system resonance of the enclosure. The second peak at 87.7 Hz is the resonance of the vent itself. *Note: The resonance of a driver is always at a higher frequency in an enclosure than in free air (Fs).* It can sometimes be difficult to determine the precise location of a peak or minima. **Tip:** The phase of the impedance should pass through zero degrees at each peak and minima (unless the peaks are spaced closely together).

The System Impedance graph is useful for a couple of reasons. If the speaker will be connected directly to an amplifier, it shows the load that the amplifier must "drive". In this case the amplifier must be capable of driving a 4 ohm load. If the speaker will include a passive crossover network, the graph shows the impedance that the crossover network will "see". This is very important when designing the crossover network because it usually wants to see a flat impedance.

The **Phase Response** graph is selected with the "P" graph tab ([Ctrl]+[F8]). It shows how much the sound waves which emanate from the speaker will lag behind the input signal that feeds the speaker. This delay is expressed as a phase angle in degrees and it is literally the difference between the phase of the input signal and the phase of the output signal.

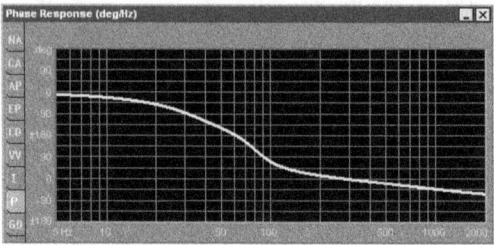

Notice that the low frequency sound waves never lag by more than 360°. This means that the output signal will not lag behind the input by more than one full wavelength. *Note: The reason the phase response drops below 0° above 290 Hz is because of the inductive reactance of the woofers' voice coils.*

Ideally the phase response should be a perfectly flat line with 0° (zero degrees) of phase shift. Fortunately, a gradual change in phase angle as shown above will not significantly harm the fidelity. However, sharp and sudden changes in phase angle are a significant problem and should be avoided when possible.

The **Group Delay** graph is selected with the "GD" graph tab ([Ctrl]+[F9]). It is very similar to the Phase Response graph because it also shows how much the sound waves which emanate from the speaker will lag behind the input signal that feeds the speaker. However, this graph expresses this information as a delay in milliseconds and it is derived from the slope of the phase response.

Sample 3: Bandpass Box Example   **BassBox 6 Pro**

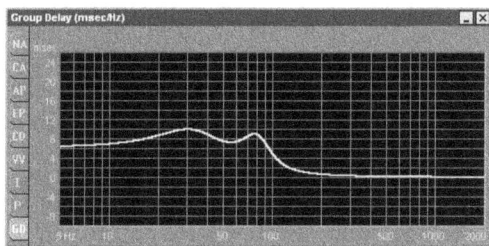

Ideally, there should be no delay. In practice, a gradual change will not significantly harm fidelity. However, sharp and sudden changes in group delay are a significant problem and should be avoided when possible. Our design has a somewhat "lumpy" group delay which can diminish the transient response a little. But overall, the group delay isn't terrible.

There is one more item to examine before we leave the performance evaluation. That item is the "pipe" resonance of the vent that results from the natural tendency of a vent to resonate and produce additional harmonic waves. To illustrate this let's return to the **Normalized Amplitude Response** graph (Ctrl+F1) and change its plot color. Then select the "Include > Vent Resonance Peaks" command from the graph popup menu (Ctrl+V) and replot the graph as shown below.

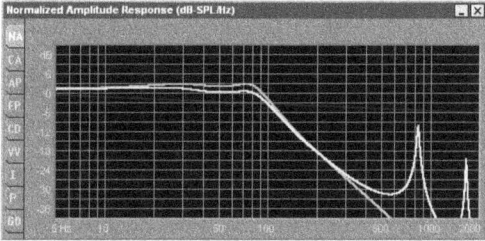

The medium grey plot line reflects the amplitude response without the vent resonance peaks. The white plot line includes the vent resonance peaks. Notice that two resonance peaks are visible. The tallest one is at 843 Hz and is 13 dB below the passband level. Both peaks appear to be low enough in level to be masked by the rest of the car's speaker system, so they can probably be safely ignored. Notice also that there is a 2 dB decrease in level near the upper knee of the response curve. If the resonance peaks were larger and higher in level, they may be audible and we would have to consider one of two possible remedies: 1) reduce the vent size or 2) add a low-pass filter to attenuate the peaks. See the "Vents" section of Chapter 4 of the *BassBox Pro Reference* portion of this manual for more information.

Conclusion

The last step in our design is to review the parts list and, if desired, create a printout. This is complicated by the fact that we decided to extend the length of our enclosure as explained below. Reopen the Box Properties window ([Ctrl]+[B]) and select the "Parts List" tab.

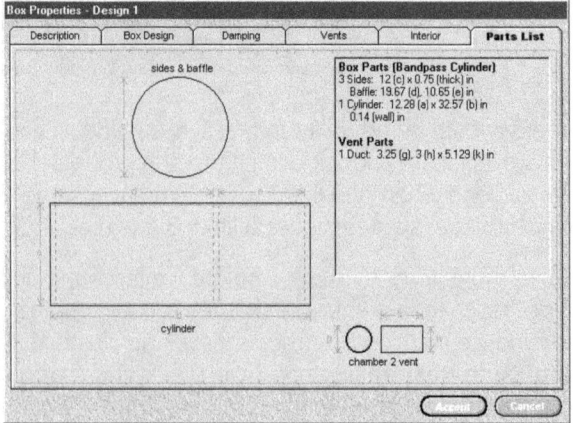

This parts list contains the correct internal dimensions for chambers 1 and 2. However, the overall length of the Sonotube (dimension "b") needs to be lengthened 2.379 inches so that it will cover the portion of the vent that will protrude beyond the end of chamber 2. So dimension "b" should really be 34.95 inches (887.7 mm). A 4th round side will also need to be added for the extended end of the Sonotube. This is shown in the illustration below.

This would be a good time to create a printout of the parts list and parts drawings. You can then note the changes on the printout.

A cutaway drawing of the final speaker is shown on the next page. *Note: We finished the enclosure by covering it with carpet so that it will match the interior of the car.*

Sample 3: Bandpass Box Example **BassBox 6 Pro**

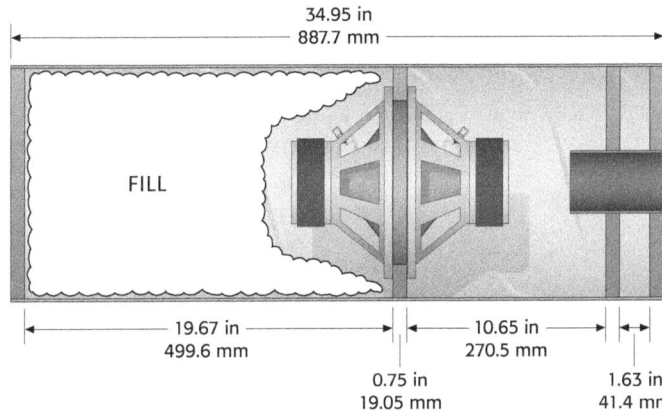

In summary, bandpass boxes are the most complicated type of box to design. Like regular vented boxes, most of the design problems of bandpass boxes center on their vents. The vents need to be large enough to avoid air turbulence noise. This leads to the most common problem: a vent that is too long to fit in the chamber. This frequently happens when trying to tune a small chamber to a low frequency. There will be situations where a vent just won't work because the chamber is too small. This can easily happen with compound multi-driver designs because an isobaric pair of drivers requires only half the box volume that a single driver would require.

This concludes the bandpass box example. This design has had its share of problems. This was not by accident—we intentionally chose a problematic design to illustrate the kinds of difficulties you may face and to show how BassBox Pro can help you identify them before box construction begins. Two points worth remembering are: 1) Although the decreased box volume of an isobaric design can be a big advantage for many box designs, it can also be a big disadvantage for a vented chamber in a bandpass box because the vent may not fit. 2) It sometimes takes a little creativity to fit a vent into a small chamber. Finally, we'd like to encourage you to review the other box design examples for additional information about box design.

BassBox 6 Pro

Sample 4: Passive Radiator Box Example

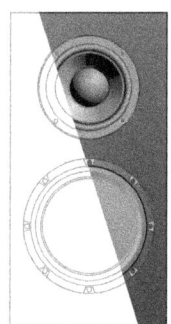

4 Passive Radiator Box Example

This chapter shows how to use BassBox Pro to design a passive radiator box. Boxes with passive radiators have many similarities to boxes with vents, including suitability to two-way speaker designs and an ability to extend the bass response. Unlike a vent, which is tuned by making its length longer or shorter, a passive radiator is tuned by adding or subtracting mass from its diaphragm. This makes it possible to tune small enclosures to lower frequencies than would otherwise be possible with a vent because the vent would be too long to fit in the box. And some designers simply prefer the sound of a passive radiator over a vent.

Often the greatest stumbling block to designing a passive radiator box is obtaining a suitable passive radiator. Very few are available on the market and, of the ones that are, few manufacturers provide complete specifications for them (a set of Thiele-Small parameters is required). This forces some designers to make their own and/or measure the parameters themselves. *Note: Almost any woofer with a nonporous dustcap can be turned into a passive radiator by removing its magnet.*

This chapter will focus on designing a full-range two-way speaker for our hi-fi system. We will use one of the 12-inch (304.8 mm) passive radiators from the fabled (and fictitious) Acme company.

If you are new to speaker box design and have not yet reviewed the earlier design examples, we recommend that you do so before studying this one. Begin with the closed box example in Chapter 1 because it includes extra details for beginners. Then continue with the vented box example in Chapter 2 and bandpass box example in Chapter 3 because they add additional details about box design and the use of BassBox Pro.

Driver Properties

In this example, our goal is to design a full-range two-way speaker. We have already selected a 1-inch (25.4 mm) soft dome tweeter and an 8-inch (203.2 mm) woofer. Both drivers have a similar sensitivity and a similar coverage pattern at the selected crossover frequency of 1500 Hz. We want to get the best bass response possible out of the woofer and this is where our box design will be focused. By "best bass response" we mean a maximally flat response.

BassBox Pro is not concerned about the tweeter—although we will consider its mounting dimensions when we determine the size of the box's front panel. The program's concern is limited to the parameters of the woofer and the passive radiator. Let's enter the woofer parameters:

- Run BassBox Pro, go to the main design window and create a new design ([Ctrl]+[N]).
- Click on the "Driver" button of the new design panel ([Ctrl]+[D]) to open the Driver Properties window. Select the "Parameters" tab and turn on the "expert" mode if it is not already turned on. Enter the following driver parameters (remember to set the units of each parameter before entering its value):

Model:	DRV 8
Fs (Hz):	32.5
Qms:	5
Vas (liters):	64.22
Cms (mm/N):	0.858
Mms (g):	28
Rms (kg/s):	1.135
Xmax (mm):	5.6
Sd (sq.cm):	231
Qes:	0.39
Re (ohms):	5.48
Le (mH):	1.15
Z (ohms):	8
BL (Tm):	8.9
Pe (watts, continuous):	100
Qts:	0.37
ηo (%):	0.54
1-W SPL (dB):	89.3

The "Parameters" tab should look like the one on the next page after the parameters are entered.

Sample 4: Passive Radiator Box Example BassBox 6 Pro

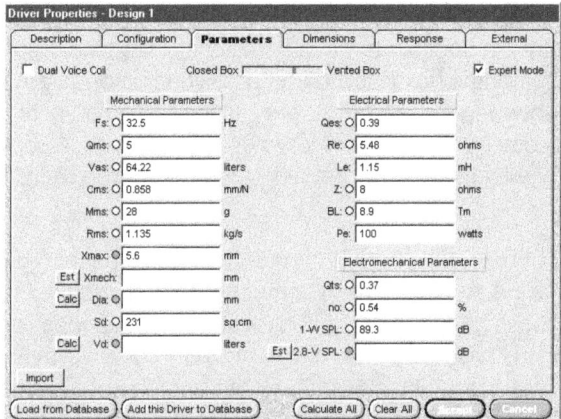

- Select the "External" tab. The speaker will be located about 6 feet (1.83 m) from the amplifier and so the speaker cables are of medium length. We shorted one end of the cable and measured its resistance to be 0.5 ohms. We entered this for Rx. Our amplifier has a damping factor of 100 for low frequencies so we entered 100 for DF. The program then calculated Rg to be 0.0808 ohms. Turn on the "Include Ext. Resistance" checkbox.

- Select the "Description" tab and enter "DRV 8" for the driver model name and "Acme" for the company name. The driver type should be set to "Standard one-way driver".

- Select the "Dimensions" tab. Select a "Round" outer shape and a "Cone" piston type. Enter the dimensions listed below:

    Dimensions (inches)
    Outside diameter (A):   8.125
    Mounting hole diameter (B):   7.8
    Cutout diameter (C):   7
    Magnet diameter (D):   3.62
    Front depth (F):   0.25
    Mounting depth (G):   3.65
    M-Vd (liters):   0.867

All of the driver information should now be entered. Click on the "Accept" button to close the Driver Properties window.

 *Sample 4: Passive Radiator Box Example*

Box Properties

We want our speaker to have a maximally flat bass response and BassBox Pro can help us by suggesting a box volume and passive radiator parameters that will hopefully produce just that. However, BassBox Pro's suggestion is only an estimation. We may need to adjust the design ourselves to achieve the desired response. Also, there is no guarantee that a passive radiator with the recommended parameters even exists. But it gives us a place to start. Let's begin:

- Click on the "Box" button of the design panel ([Ctrl]+[B]) to open the Box Properties window. Select the "Damping" tab and choose "Typical" for the amount of acoustic absorption. The "Use Classical box calculations" option will need to be turned off for you to do this. The reason we want to add a modest layer of fill to each wall inside the box is because we want to absorb possible standing waves and because we want to add a bit of damping.

    Note: It is usually best to set the damping setting before designing a box since the damping setting will affect the box calculations.

- We plan to flush-mount the woofer so select the "Interior" tab and set the driver mounting to "'Flush" in the "Box Volume Displaced by the Driver" section.

- Select the "Box Design" tab and set the box type to "Passive Radiator Box". This will enable the fourth tab and configure it as a "Passive Rad" (passive radiator) tab instead of a "Vents" tab. Finally, click on the "Suggest" button and let's see what BassBox Pro recommends.

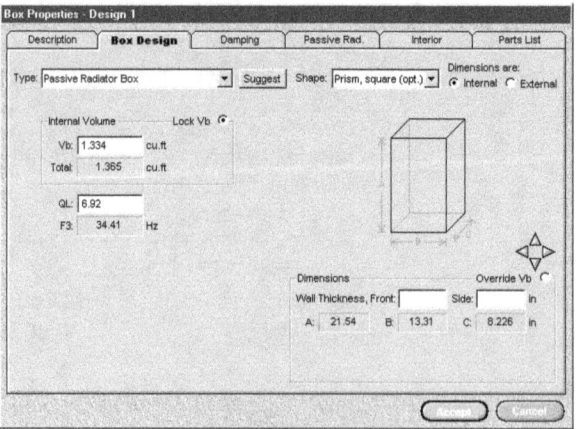

As you can see from the previous example, the program recommends a net internal box volume (Vb) of 1.334 cubic feet (37.78 liters). Next, select the "Passive Rad" tab and let's see what the suggested passive radiator parameters are. The "Passive

Sample 4: Passive Radiator Box Example **BassBox 6 Pro**

Rad" tab is shown below:

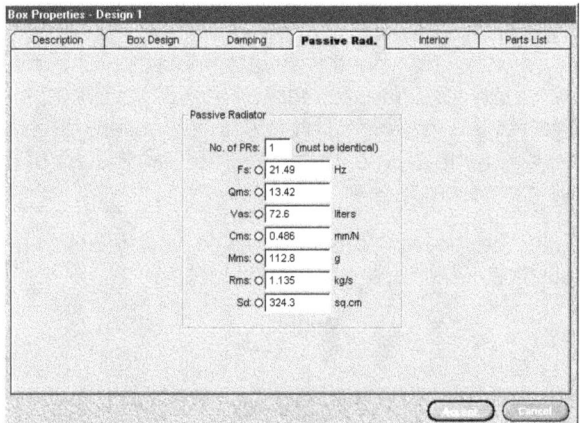

Notice that the "Passive Rad" tab includes status indicators just like the "expert" mode of the "Parameters" tab of the Driver Properties window. They show when one or more of the passive radiator parameters seems to be out of tolerance. They use the same tolerance settings as the "Parameters" tab. You can adjust them with the "Driver" tab of the Preferences window.

When a status indicator is yellow or red, a "Calc" button will usually appear so that you can calculate another value.

BassBox Pro arrived at the preceding box and passive radiator parameters by internally estimating the size and tuning of a maximally flat vented box and then calculating a passive radiator with a mass that is equivalent to the mass of air in the vent. The suggested diaphragm area (Sd) of 324.3 square inches indicates a 10 inch (254 mm) outside diameter. The big question now is, "Can we find a 10-inch passive radiator with these parameters?" If we cannot, what guidelines should we use to find a suitable substitute? The following list should be helpful:

- Large displacement volume. The passive radiator should be able to displace 2½ times as much air as the woofer. This can be achieved with a large excursion limit (Xmech) or a large diaphragm area (Sd) or a combination of both.
- Large compliance. The compliance of the passive radiator's suspension should be sufficiently high (usually several times higher than the woofer) or the passive radiator's output will be diminished.
- Low resonance. Typically the resonance is set an octave or more below the box cutoff frequency. This is usually achieved by increasing the mass of the diaphragm. Most passive radiators should also allow the designer to adjust the resonance by adding additional mass when necessary.

Sample 4: Passive Radiator Box Example

We were unable to locate a suitable 10-inch passive radiator but we did find an interesting 12-inch (304.8 mm) passive radiator that has a free-air resonance (Fs) that is approximately one octave below that of the driver. It is the Acme DRN 12. It also has a much larger compliance (Vas, Cms) but this should not be a problem because a large compliance is a desirable quality for a passive radiator. It also has a 1.22 inch (31 mm) excursion limit (Xmech). This is a very high limit and, together with the large diameter, should provide a large displacement volume. Let's enter the parameters of the DRN 12 and see what kind of response it produces.

- Enter the following passive radiator parameters (remember to set the units of each parameter before entering its value):

Model:	DRN 12
Fs (Hz):	16.4
Qms:	8.8
Vas (liters):	158
Cms (mm/N):	0.43
Mms (g):	220
Rms (kg/s):	2.56
Sd (sq.cm):	508

- For the next few steps, we will be switching back and forth between the graph window and the Box Properties window. If your Windows desktop is large enough, you should position the windows so that you can view them both. Otherwise, you will need to close and reopen the Box Properties window as needed.

We will use the combination graph window so select the "Display Mode > Single Window" command from the Graph menu. Then select the "Show Graph > Amplitude—Normalized" command from the Graph menu (or press [Ctrl]+[F1]) to open the graph window. Switch to the main window and click the "Plot" button ([Ctrl]+[1]) and let's examine the **normalized amplitude response** with the DRN 12.

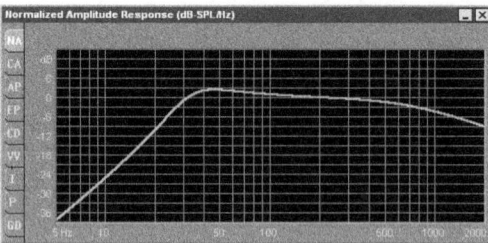

The graph is encouraging because the overall shape of the response isn't bad and there is only a modest 2 dB peak at 45 Hz. This indicates that the box resonance is too high and it can be easily lowered by adding mass to the passive radiator. Let's add 30 grams to the moving mass (Mms) so that it equals 250 grams as shown below.

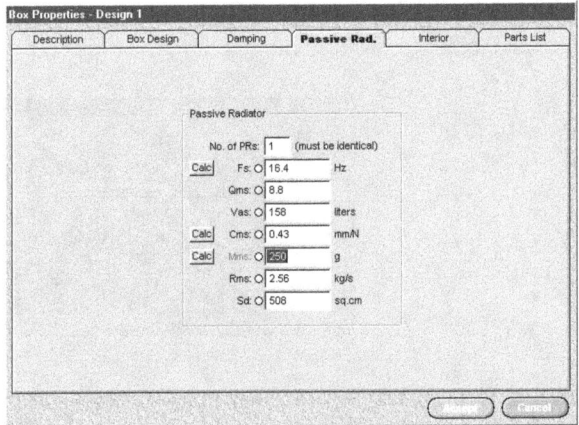

Notice that the Fs, Cms and Mms status indicators turn yellow when Mms is changed to 250 grams. This shows us that one or more parameters now need to be adjusted. *Note: The indicators would have turned red if we had made a larger change to Mms.* We know that the compliance (Cms) hasn't changed because we only changed the mass. But we do expect the resonance (Fs) to change so click on the "Calc" button next to Fs.

Fs now equals 15.35 Hz and the status indicator next to Cms has turned to green. However, the status indicators next to the mechanical Q (Qms) and the mechanical resistance (Rms) have now turned yellow. Changing Mms and Fs should not effect Rms but it will change Qms so click on the "Calc" button next to Qms.

Qms now equals 9.419 and all of the status indicators are green. When we change a parameter, these steps illustrate how the status indicators can help us recalculate all related parameters.

Let's plot the response again and see how it has changed. Change the plot color and replot the graph.

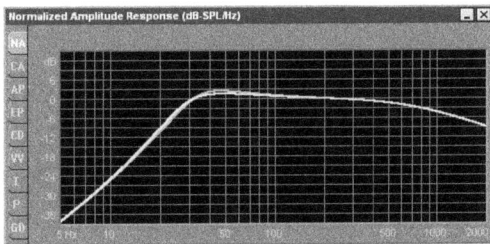

The previous graph shows that we're moving in the right direction because the response is smoother. Repeat the preceding steps to increase Mms another 30 grams to 280 grams and again change the plot color and replot the graph. The graph should look like the one below:

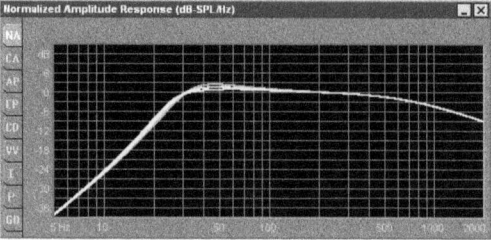

The response now looks very good. It has a maximally flat shape and a −3 dB frequency (F3) of 25.49 Hz. All we have to do to get this response is add a total of 60 grams to the diaphragm of the passive radiator. How we add this mass will depend on the passive radiator. Some allow you to add metal washers to a bolt on the back of the diaphragm where the voice coil former of a typical driver would be. Others allow you to epoxy lead shot to the back of the diaphragm. Still others allow you to add clay to the back of the diaphragm. In our case, we epoxied some lead shot. After doing this the passive radiator should now have the parameters shown below:

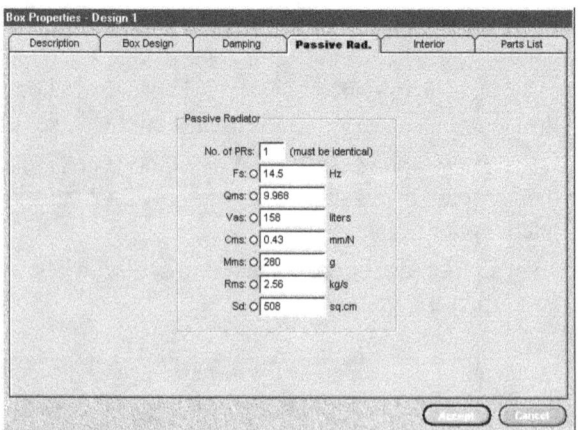

Sample 4: Passive Radiator Box Example

BassBox 6 Pro

Model:	DRN 12
Fs (Hz):	14.5
Qms:	9.968
Vas (liters):	158
Cms (mm/N):	0.43
Mms (g):	280
Rms (kg/s):	2.56
Sd (sq.cm):	508

Let's return to the box and calculate its dimensions. The front panel of our box must be large enough to contain the tweeter, woofer and passive radiator. We plan to mount the drivers as close together as possible to minimize phase cancellations. To do this we need the center-to-center spacing of the tweeter and woofer to be less than 9 inches (230 mm) because this is the wavelength of 1500 Hz, the crossover frequency of our speaker. We also plan to mount the drivers off-center so that each driver will be a different distance from each edge of the front panel. This will minimize diffraction problems. The illustration below shows what we have in mind:

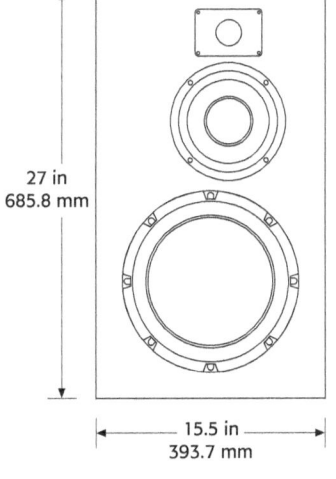
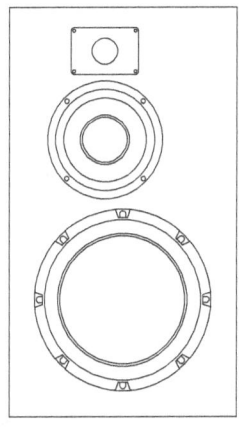

27 in
685.8 mm

15.5 in
393.7 mm

The driver mounting layout shown above requires that the front panel have a 27 inch (685.8 mm) height and a 15.5 inch (393.7 mm) width. These are external dimensions and assume a side wall thickness of ¾ inch (19.05 mm). Let's enter them:

- Select the "Box Design" tab. The default box shape for new single-chamber boxes is the optimum square prism. It forces the dimensions to fit a certain ratio in order to minimize the problem of "standing waves" in the box. We need to control all box dimensions because they may not fit the "optimum" ratio so select the "prism, square" box shape as shown below.

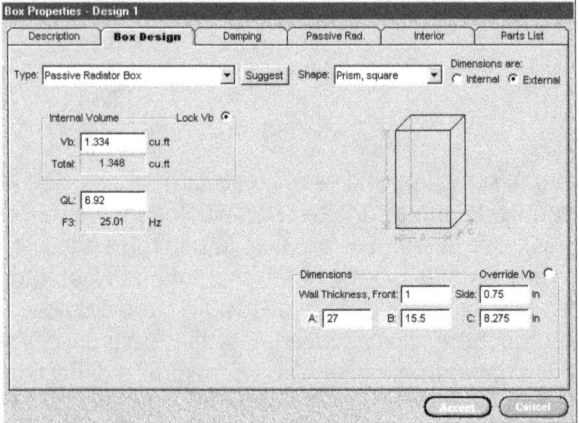

We will use external dimensions so set the dimensions to "External". We will use 1 inch MDF (medium density fiberboard) for the front panel so enter 1 inch (25.4 mm) for the front wall thickness. We will use ¾ inch MDF for the other walls of the box so enter 0.75 inches (19.05 mm) for the side wall thickness. Enter a height (A) of 27 inches (685.8 mm) and a width (B) of 15.5 inches (393.7 mm). Since the internal volume (Vb) is locked, BassBox Pro will calculate the final dimension. It calculates the depth (C) to be 8.275 inches (210.2 mm). The three-dimensional box drawing should look like the one in the illustration above.

Notice that the background of the "A" and "B" labels are highlighted with a lighter shade to show that these dimensions were entered and are therefore protected (locked) and can only be changed by us. In contrast, dimension "C" can be changed by the program whenever it sees that the box volume has changed.

We are finished with the Box Properties window for the time being. Close it by clicking on its "Accept" button.

Evaluating the Performance

So far everything looks good. The DRN 12 passive radiator worked nicely by adding 60 grams to its diaphragm and our design produces the response we want. But how does it perform in other areas? How loud will the speaker go? What is the driver's excursion? The performance graphs will answer these questions. We'll skip the Normalized Amplitude Response graph since it was discussed previously.

Since the graphs contain three plot lines, let's clear them and get a fresh plot. This is done with the "Clear > All Graphs" command from the Graph menu ([Ctrl]+[Y]). Finally, click on

Sample 4: Passive Radiator Box Example **BassBox 6 Pro**

the "Plot" button of the design panel (Ctrl+1).

The **Custom Amplitude Response** graph is selected with the "CA" graph tab (Ctrl+F2). It is very similar to the Normalized Amplitude Response graph except that it shows the sound pressure level that the speaker will produce on axis at 1 meter (3.3 feet) with the specified input power. We used 100 watts because this is the maximum power level of the DRV 8 and it is the maximum power that our amplifier will produce per channel into an 8 ohm load.

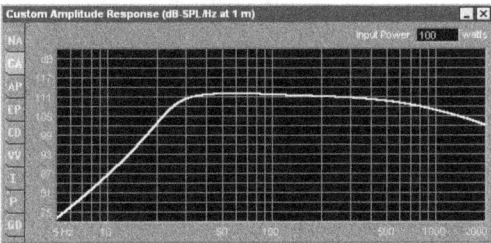

Notice the result: the speaker should produce up to 112 dB. This is good for an 8-inch woofer—especially since the power rating of both the woofer and the amplifier is a continuous RMS level and both can handle higher peak levels.

Also notice that there is a high-frequency roll-off caused by the inductive reactance of the woofer's voice coil. It is counteracted by the on-axis piston band response of the woofer. However, the high-frequency roll-off shown above does hint at the kind of off-axis response we can expect from the woofer. This was considered when we chose a tweeter and crossover frequency.

If desired, we can estimate the on-axis response by turning on the "Include > On-Axis Piston Band Response" option in the graph popup menu (Ctrl+T). The result is shown below:

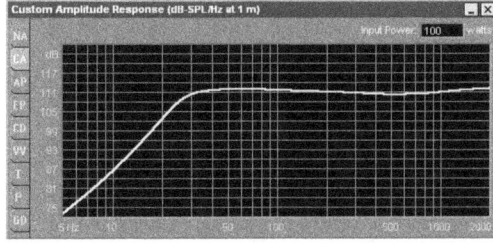

Notice how the high-frequency response has increased now that this option is included. However, it is important to remember that this increase in high-frequency response only occurs on-axis. The high-frequency response will be much lower off-axis as shown in the first graph.

The **Maximum Acoustic Power** graph is selected with the "AP" graph tab (`Ctrl`+`F3`). It shows how much acoustic power the speaker can produce before reaching its steady-state displacement limit or its thermal limit.

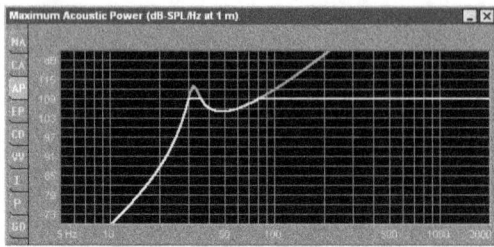

In our case, the speaker is thermal-limited down to 80 Hz. Except for a small peak at 32 Hz, the speaker is displacement-limited by Xmax below 80 Hz. However, the maximum acoustic power doesn't drop off rapidly until 30 Hz, just below the box resonance. This graph shows us that our 8-inch woofer will probably have excursion problems if we push it too hard below 30 Hz in spite of the assistance of the much larger passive radiator.

The **Maximum Electric Input Power** graph is selected with the "EP" graph tab (`Ctrl`+`F4`). It is derived from the Maximum Acoustic Power graph and it shows how much amplifier power the speaker can handle before reaching the first of either its displacement or thermal limits.

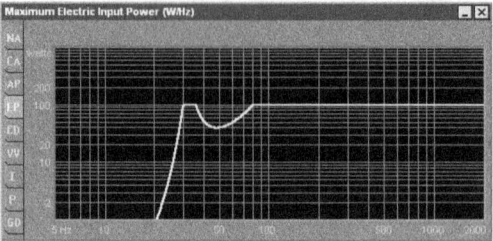

This graph shows us that the speaker should be able to handle a full 100 watts of power from the amplifier down to 80 Hz provided of course that the amplifier's output signal is not heavily distorted or "clipped". It would be wise to protect the driver below 30 Hz with a high-pass filter because of the rapid cutoff rate below this point. This would also protect the woofer from becoming "unloaded" by the passive radiator.

The **Cone Displacement** graph is selected with the "CD" graph tab (`Ctrl`+`F5`). It shows how far the diaphragm and voice coil of the woofer must travel with the specified input power. It also shows how far the diaphragm of the passive radiator must travel in response to the driver's motion. We used 100 watts for the input power because this is the maximum

Sample 4: Passive Radiator Box Example

power level of the DRV 8.

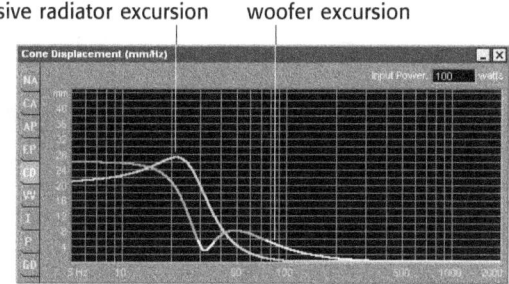

When we first plotted this graph, the plot lines were off-scale below 26 Hz. We wanted to see the shape of the entire plot line so we selected the "Scale > Vertical: Expanded" command from the graph popup menu. This doubled the vertical scale and produced the graph shown above.

Notice that there are two plot lines, one for the woofer and one for the passive radiator as labeled above. The woofer plot line changes intensity when it rises above the Xmax level of 5.6 mm to show when the woofer will exceed its peak linear excursion limit. Although this first happens at about 75 Hz it doesn't exceed Xmax too much until after the box resonance at around 29 Hz. The high excursion below 29 Hz underscores the need to add a high-pass filter to protect the woofer as mentioned previously. The passive radiator plot line looks great because it never exceeds its mechanical excursion limit of 31 mm.

Ideally, the motion of the woofer's diaphragm should not exceed Xmax (one way from its resting position). If it does, nonlinear distortion will increase. It could be argued that this is the weakest aspect of this design. However, several factors can reduce the seriousness of this problem. First, we do not plan to drive the speaker continuously at full power. And yet, at 50 watts, the 1 meter SPL will still be 109 dB and the excursion of the woofer will not exceed Xmax until 27 Hz. Second, the mechanical excursion limit (the distance the diaphragm can travel before reaching the limits of the suspension or before the voice coil former strikes the back plate) should be larger than Xmax. This allows the diaphragm to move farther than Xmax without risk of damage. Third the nonlinear distortion that increases when the excursion exceeds Xmax is subtle and is not audible to most listeners until it becomes severe.

If the excursion problem was reversed where the passive radiator exceeded its excursion limit rather than the woofer, the result would still be an increase in nonlinear distortion but the increase would be dramatic and sudden and it would sound different. This is because a well-made passive radiator has no Xmax—its excursion limit is Xmech, the mechanical excursion limit. When Xmech is reached, the passive radiator's diaphragm suddenly cannot move any farther and so the passive radiator cannot become any louder.

BassBox 6 Pro *Sample 4: Passive Radiator Box Example*

The **System Impedance** graph is selected with the "I" graph tab (Ctrl+F7). It shows the impedance of the speaker.

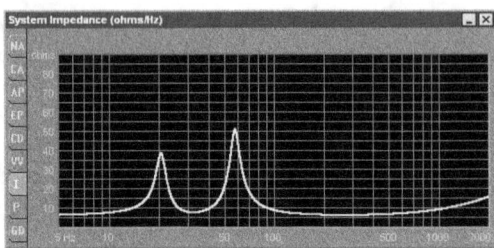

The first peak at 20.8 Hz is the resonance of the passive radiator in the box. The minima at 31.6 Hz is the system resonance of the box. The second peak at 57.7 Hz is the resonance of the driver in the box. *Note: The resonance of a driver or passive radiator is always at a higher frequency in a box than in free air (Fs).* It can sometimes be difficult to determine the precise location of a peak or minima. **Tip:** The phase of the impedance should pass through zero degrees at each peak and minima (unless the peaks are spaced closely together).

The System Impedance graph is useful for a couple of reasons. If the speaker will be connected directly to an amplifier, it shows the load that the amplifier must "drive". In this case the amplifier must be capable of driving an 8 ohm load. If the speaker will include a passive crossover network, the graph shows the impedance that the crossover network will "see". This is very important when designing the crossover network because it usually wants to see a flat impedance.

The **Phase Response** graph is selected with the "P" graph tab (Ctrl+F8). It shows how much the sound waves which emanate from the speaker will lag behind the input signal that feeds the speaker. This delay is expressed as a phase angle in degrees and it is literally the difference between the phase of the input signal and the phase of the output signal.

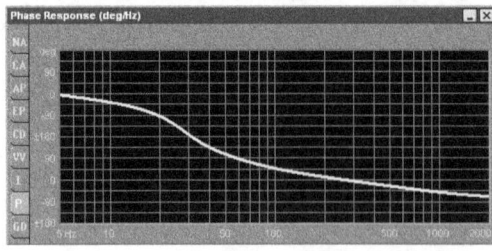

Notice that the low frequency sound waves never lag by more than 360°. This means that the output signal will not lag behind the input by more than one wavelength. Remember

Sample 4: Passive Radiator Box Example   BassBox 6 Pro

that a full wavelength requires 360° of phase rotation. *Note: The reason the phase response drops below 0° above 260 Hz is because of the inductive reactance of the woofer's voice coil.*

Ideally the phase response should be a perfectly flat line with 0° (zero degrees) of phase shift. Fortunately, a gradual change in phase angle will not significantly harm the fidelity. However, sharp and sudden changes in phase angle are a significant problem and should be avoided when possible because they result in poor transient response.

The **Group Delay** graph is selected with the "GD" graph tab ([Ctrl]+[F9]). It is very similar to the Phase Response graph because it also shows how much the sound waves which emanate from the speaker will lag behind the input signal that feeds the speaker. However, this graph expresses this information as a delay in milliseconds and it is derived from the slope of the phase response.

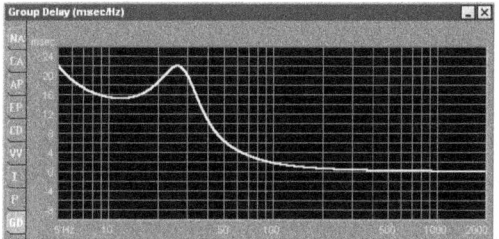

Ideally, there should be no delay. In practice, a gradual change will not significantly harm fidelity. However, sharp and sudden changes in group delay are a significant problem and should be avoided when possible. Our design has significant group delay which can diminish the transient response. This is another weak aspect of this design. Fortunately, the group delay does not begin to rise rapidly until it reaches the low-frequency portion of the response and at low frequencies, it is doubtful if most listeners will discern a problem.

*Note: The **Vent Air Velocity** graph (graph tab "V") was omitted because this is a passive radiator box design and therefore it has no vents.*

External Network
We plan to use a 1500 Hz passive crossover network which we designed with Harris Tech's passive crossover network design software, X•over Pro. If desired, we can include the low-pass portion of that crossover network in this box design. To do so, we return to the "External" tab of the Driver Properties window. The low-pass portion of the network includes a single parallel capacitor and series inductor. The capacitor was entered as "Cp1" and the inductor was entered as "Ls1" as shown on the next page. The remaining components were left blank since they were not used.

Sample 4: Passive Radiator Box Example

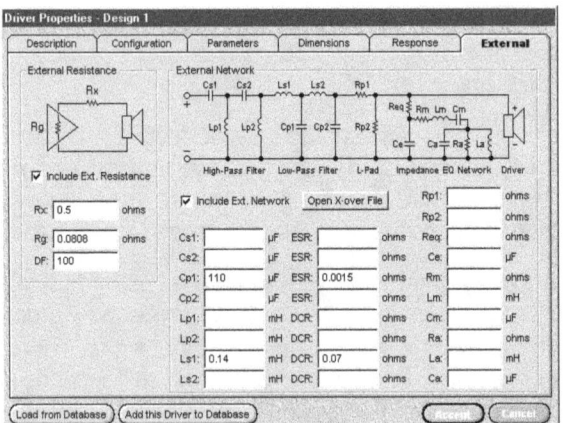

The parallel capacitor (Cp1) has a value of 110 µF and an ESR of 0.0015 ohms. The series inductor (Ls1) has a value of 0.14 mH and a DCR of 0.07 ohms.

Next, let's returned to the Normalized Amplitude Response graph. In order to see the high frequency response better, we expanded the horizontal scale by selecting the "Scale > Horizontal: 5Hz - 20 kHz" command from the graph popup menu. Then we cleared the graph and replotted it. The result is shown below.

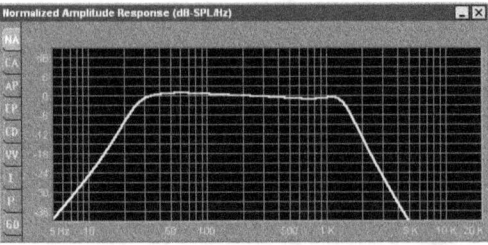

The crossover network was designed with a slightly off-axis response in mind so we did not include the on-axis piston band response in this graph. If the on-axis response had been deemed the most important, then we would have selected a different capacitor and inductor value for the low-pass filter (such as Cp1 = 250 µF and Ls1 = 0.05 mH).

The System Impedance graph on the next page depicts the impedance of the speaker, including the low-pass filter. Notice that the low-pass filter of the crossover network reduces

Sample 4: Passive Radiator Box Example BassBox 6 Pro

the second impedance peak. It now reflects the impedance that the amplifier will see with one exception. It is not able to show the effects of the tweeter and corresponding high-pass filter. *Note: If have you a licensed copy of X•over Pro, you can import this box design into it and view the response of the entire speaker, including the tweeter and full crossover network.*

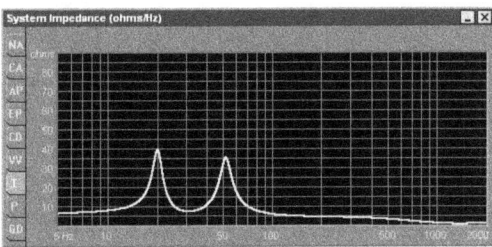

We are now finished with the Driver Properties window so it can be closed by clicking on its "Accept" button.

Conclusion

The last step in our design is to review the parts list and, if desired, create a printout. Open the Box Properties window by clicking on the "Box" button on the design panel (Ctrl + B). Select the "Parts List" tab. The drawing of each part and a list of dimensions should appear as below:

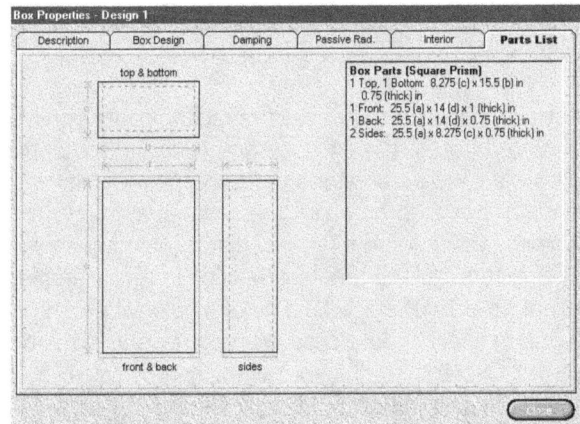

The illustration below shows the appearance of the finished speakers. They were laminated with a hardwood veneer. The inner walls were carefully braced to stiffen them and covered with a 1 to 1½ inch (25.4 to 38.1 mm) layer of Acousta Stuf to absorb standing waves.

Generally speaking, the passive radiators could have been mounted anywhere—even the back side—because the low frequencies produced by them are omnidirectional. And since the wavelengths are so long, we do not need to be concerned about phase problems if the passive radiators are mounted on a different side than the woofers.

In summary, we have seen that passive radiator boxes are more complicated to design than closed or vented boxes. The most common design problems center on obtaining or fabricating a suitable passive radiator and the proper adjustment of its moving mass. Usually the passive radiator needs to be able to displace at least 2½ times the volume of air as the woofer. This means that the passive radiator must have a larger diaphragm and/or have a much larger excursion limit than the woofer. In this example, the passive radiator, with its huge 31 mm Xmech, displaced plenty of air as evidenced by the Cone Displacement graph. However, this is often not the case. Many times passive radiators do not displace as much air as needed. Overcoming these challenging obstacles can be very rewarding when a successful speaker is completed.

This concludes the passive radiator box example. We hope that it has been helpful and that it has illustrated the types of decisions that must often be made by designers. Please review the other examples for additional information about box design. Additional tips on box construction are also available in Chapter 4 of the *Box Designer's Guide* earlier in this manual.

BassBox Pro Reference

This is the place to go when you need an explanation of one or more of the features of BassBox Pro. The previous sections of this manual focused on how speakers work and how to design a box. This Reference section focuses on BassBox Pro, itself.

Chapters

1 Menus 159
2 Design Wizard 169
3 Driver Properties 171
4 Box Properties 213
5 Room / Car Acoustic Properties 251
6 Evaluating Performance 255
7 Saving / Opening a Design 287
8 Printing a Design 291
9 Clearing / Closing a Design 305
10 Editing the Driver Database 307
11 Testing Drivers & Passive Radiators 317
12 Tools 327
13 Preferences 329

Reference: 1 Menus **BassBox 6 Pro**

1 Menus

Like many programs designed for Microsoft Windows, BassBox Pro contains a menu bar across the top of the main window. Most commands can be found in the menus and sub-menus which drop down from this menu bar. In addition, the graphs have a popup menu which is displayed when you right-click (🖱) on a graph.

File Menu

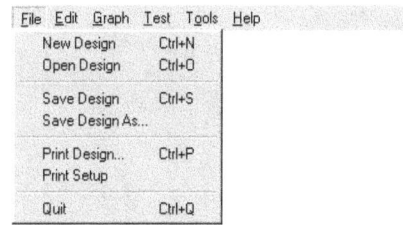

New Design
Begins a new speaker design, creating a new and empty design panel in the main window. (Keyboard shortcut: [Ctrl]+[N].)

Open Design
Opens a speaker design file which was previously saved to disk. You will be prompted for a file name and path. *Note: BassBox Pro can also open speaker design files from earlier versions of BassBox.* (Keyboard shortcut: [Ctrl]+[O].)

Save Design
Saves design changes to an existing speaker design file. You will <u>not</u> be prompted for a file name and path <u>unless</u> the design is new and has not yet been saved. (Keyboard shortcut: [Ctrl]+[S].)

Save Design As...
Saves a new speaker design or a copy of an existing design to a new file name and path. The file name should end with the extension ".bb6". If it does not, BassBox Pro will automatically append the extension.

Print Design...
Prints a speaker design or graphs with overlaid plot lines. A "Print a BassBox Design" window will open so you can configure the printout. (Keyboard shortcut: [Ctrl]+[P].)

Print Setup
Opens the Print Setup window so you can choose another printer and/or change the settings of the printer driver. *Important: Make sure the printer is set to print on a letter-size or A4-size page in portrait mode.* The Print Setup window can also be accessed from the "Print a BassBox Design" window.

Quit
Closes the program. You will be given a chance to save any unsaved changes before the program closes. (Keyboard shortcut: Ctrl+Q.)

Edit Menu

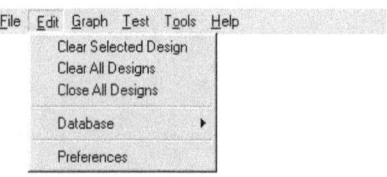

Clear Selected Design
Clears the selected speaker design only. After the design is cleared, its design panel will be empty as if it were a "new" design. The program will ask for confirmation before executing this command to prevent accidental data loss.

Clear All Designs
Clears all open speaker designs. After the designs are cleared, their design panels will be empty as if they were all "new" designs. The program will ask for confirmation before executing this command to prevent accidental data loss.

Close All Designs
Closes all open speaker designs. The program will ask for confirmation before executing this command to prevent accidental data loss.

Database > Edit Driver Data
Select this command to make changes to the drivers in the driver database. Drivers can be edited, added and deleted. (Keyboard shortcut: Ctrl+W.) *Caution: To prevent unwanted data loss it is a good idea to backup the database before editing it.* To do this make a copy of file "htaudio.mdb" in the BassBox Pro folder ("c:\Program Files\HT Audio\").

Database > Edit Company Data
Select this command to make changes to the company information in the driver database. Companies can be edited, added and deleted. *Caution: To prevent unwanted data loss it is*

Reference: 1 Menus **BassBox 6 Pro**

a good idea to backup the database before editing it. To do this make a copy of file "htaudio.mdb" in the BassBox Pro folder ("c:\Program Files\HT Audio\").

Database > Compact Database
Deleting drivers from the database does not decrease its size because it would require the entire database to be resaved, greatly slowing its operation. To compact the database after one or more drivers have been deleted, select the "Database > Compact Database" command. *Note: A backup copy of the uncompacted database will be made before the database is compacted. It is named "htaudio.bak". To restore it, delete "htaudio.mdb" and rename the backup copy by changing the "bak" extension to "mdb".*

Database > Repair Database
If the computer crashes while the driver database is being edited, it may become corrupted. If you ever receive an error message saying that the database is unrecognizable or is not a *Microsoft Access* database when you attempt to open it, it will probably need to be repaired. Select the "Database > Repair Database" command to repair it. *Caution: Some data may be lost when the database is repaired because corrupted portions that cannot be reconstructed will have to be discarded. Also note: A backup copy of the unrepaired database will be made before the database is repaired. It is named "htaudio.bak". To restore it, delete "htaudio.mdb" and rename the backup copy by changing the "bak" extension to "mdb".*

Preferences
Select the Preferences command to open the Preferences window and edit the default settings of BassBox Pro. They are saved in the BassBox Pro folder in file "bbxpref.ini".

Graph Menu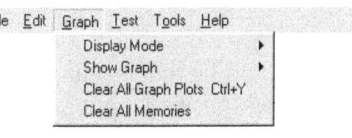

Display Mode
Two display mode commands are provided: "Single Window" and "Individual Windows". The "single window" mode combines all graphs into a single window where they can be individually accessed with a tab button. This mode is most useful for lower screen resolutions such as VGA and SVGA because it economizes space. The "single window" mode also uses less memory and system resources. In the "individual windows" mode each graph opens in its own window so multiple graphs can be viewed at the same time. The "individual windows" mode also provides a size option to allow each graph window to be enlarged to a fixed "large" size.

Show Graph > Amplitude—Normalized
Displays the Normalized Amplitude Response graph. This is often referred to as the "frequency response". It is normalized to zero dB so the plot lines of multiple designs can be overlaid and their shapes easily compared. (Keyboard shortcut: [Ctrl]+[F1].)

Show Graph > Amplitude—Custom
Displays the Custom Amplitude Response graph. This version of the amplitude response requires an input power or voltage. It displays the on-axis sound pressure level at 1 meter (3.28 feet) for the specified input power/voltage. It is useful in showing the sensitivity differences of different designs. (Keyboard shortcut: [Ctrl]+[F2].)

Show Graph > Max Acoustic Power
Displays the Maximum Acoustic Power graph. It shows the maximum acoustic power as a sound pressure level at 1 meter (3.28 feet). This is the total acoustic power radiated by the speaker in all directions before the linear excursion and thermal limits are reached. (Keyboard shortcut: [Ctrl]+[F3].)

Show Graph > Max Electric Power
Displays the Maximum Electric Input Power graph. This is the maximum amplifier power in watts that the speaker can handle before the linear excursion and thermal limits are reached. (Keyboard shortcut: [Ctrl]+[F4].)

Show Graph > Cone Displacement
Displays the Cone Displacement graph and it requires an input power or voltage. It displays the piston excursion of both drivers and passive radiators for the specified input power/voltage. The plot line of the driver excursion will be dimmed when it exceeds the maximum linear excursion (Xmax). (Keyboard shortcut: [Ctrl]+[F5].)

Show Graph > Vent Air Velocity
Displays the Vent Air Velocity graph and it requires an input power or voltage. It displays the air velocity inside the vent(s) for the specified input power/voltage. The plot line will be dimmed when it exceeds 10% of the velocity of sound in air to warn that turbulence and whistling may begin. (Keyboard shortcut: [Ctrl]+[F6].)

Show Graph > System Impedance
Displays the System Impedance graph. It shows the net impedance of the speaker including, if present, the effects of an external passive network (crossover network, impedance equalization network and/or L-pad). (Keyboard shortcut: [Ctrl]+[F7].)

Show Graph > Phase
Displays the Phase Response graph. It shows how far the output signal from the speaker

will lag behind the input signal. The signal delay is displayed as a phase shift. (Keyboard shortcut: [Ctrl]+[F8].)

Show Graph > Group Delay
Displays the Group Delay graph. It shows how far the output signal from the speaker will lag behind the input signal. The signal delay is displayed in milliseconds. (Keyboard shortcut: [Ctrl]+[F9].)

Clear All Graph Plots
Clears the plot lines from all graphs. (Keyboard shortcut: [Ctrl]+[Y].)

Clear All Memories
Clears all seven of the graph memories.

Test Menu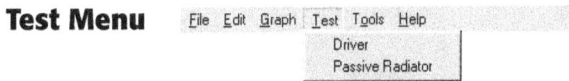

Driver
Runs a built-in procedure that can help with the testing of a driver. The illustrated delta-volume procedure shows how to test the small-signal parameters and performs all necessary calculations. Some basic test equipment is required (a list of required equipment is displayed at the beginning of the procedure along with a block diagram).

Passive Radiator
Runs a built-in procedure that can help with the testing of a passive radiator. The illustrated delta-volume procedure shows how to test the small-signal parameters and performs all necessary calculations. Some basic test equipment is required (a list of required equipment is displayed at the beginning of the procedure along with a block diagram).

Tools Menu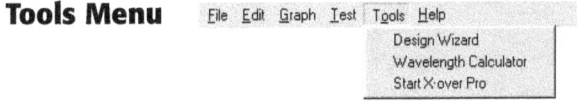

Design Wizard
Launches the Design Wizard. The BassBox Pro Design Wizard will help you design a box.

Wavelength Calculator
Opens the Wavelength Calculator window. Use it to calculate a wavelength from a frequency and visa versa. It can be very helpful when calculating the "standing wave" frequencies of a box (see page 327).

 **BassBox 6 Pro** *Reference: 1 Menus*

Start X•over Pro
Launches the X•over Pro program if you have purchased an X•over Pro license and the program is installed on your computer. X•over Pro is a passive crossover network design program (see page 328).

Help Menu
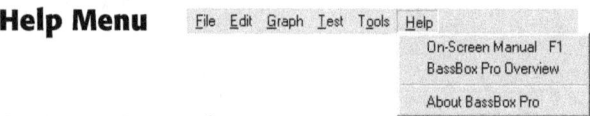

On-Screen Manual
Opens the on-screen manual to its Contents page. From it you can select from several topics. (Keyboard shortcut: F1 .)

BassBox Pro Overview
Opens the on-screen manual to its "An Overview of BassBox Pro" topic which introduces first-time users to the program.

About BassBox Pro
Opens the BassBox Pro title window and displays the program's serial number, user registration information and copyright notice.

Graph Popup Menu
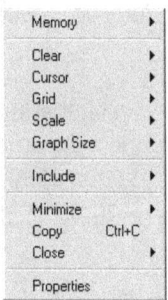

Memory > Store
Stores a "snapshot" of the most recently plotted design into one of seven graph memories so it can be replotted again later. The design parameters are stored into memory rather than the plot line data points. This enables all open graphs to be replotted when a memory is recalled—even if the graphs were closed when the design was originally stored into the graph memory. (Keyboard shortcuts: Shift + F1 to F7 .)

Memory > Recall
Replots one or all designs stored in the graph memories. (Keyboard shortcuts: Shift + Ctrl + F1 to F8 .)

BassBox 6 Pro

Memory > Clear
Clears one or all seven of the graph memories

Clear
Clears all plot lines from the selected graph or all open graphs. (Keyboard shortcuts: Ctrl+X to clear the selected graph only and Ctrl+Y to clear all graphs.)

Cursor > Show in Selected Graph
Activates the cursor in the selected graph. (Keyboard shortcut: Ctrl+U.)

Cursor > Show in All Graphs
Activates the cursors in all graphs.

Cursor > Link All Cursors
Links the cursors in all graphs ("individual windows" mode only) so they will move as one.

Cursor > Unlink All Cursors
Unlinks the cursors in all graphs ("individual windows" mode only) so they can be controlled separately.

Cursor > Hide in Selected Graph
Deactivates the cursor in the selected graph. (Keyboard shortcut: Ctrl+H.)

Cursor > Hide in All Graphs
Deactivates the cursor in all graphs.

Grid
Displays or removes the grid in the selected graph or all graphs.

Scale > Vertical: Normal
Displays the standard vertical scale in the selected graph.

Scale > Vertical: Expanded
Displays an expanded vertical scale in the selected graph.

Scale > Horizontal: 5 Hz - 2 kHz
Displays a 5 Hz to 2000 Hz horizontal scale in the selected graph.

Scale > Horizontal: 5 Hz - 20 kHz
Displays a 5 Hz to 20000 Hz horizontal scale in the selected graph.

Graph Size
Two sizes are provided ("individual windows" mode only): normal and large. The large size does not increase the range of either the vertical or horizontal scales but it does add more data points, improving the detail of the plot lines. This is because the plotting functions in BassBox Pro calculate a graph data point for each pixel column in the graph.

Include > Driver Acoustic Response
Includes the driver acoustic response—if it has been entered. The graphs affected by this option are: the Normalized Amplitude Response, Custom Amplitude Response and Maximum Acoustic Power graphs. (Keyboard shortcut: Ctrl+I.)

Include > Vent Resonance Peaks
Includes "pipe" resonance peaks. These peaks are a natural result of the vent(s) in vented and bandpass boxes. All nine graphs are affected by this option. (Keyboard shortcut: Ctrl+V.)

Include > Room/Car Acoustic Response
Includes the architectural/automotive acoustic response—if it has been entered. The graphs affected by this option are: the Normalized Amplitude Response, Custom Amplitude Response and Maximum Acoustic Power graphs. (Keyboard shortcut: Ctrl+R.)

Include > External Network Response
Includes the response of an external network—if it has been entered. This option affects all graphs **except** the Maximum Acoustic Power and Maximum Electric Input Power graphs. (Keyboard shortcut: Ctrl+E.)

Include > On-Axis Piston Band Response
Includes the estimated on-axis piston band response—if the diaphragm diameter (Dia) or diaphragm area (Sd) of the driver has been entered. The graphs affected by this option are: the Normalized Amplitude Response and Custom Amplitude Response graph. The "piston band" is the frequency band where the driver maintains a constant load versus frequency. This begins at the frequency whose wavelength is equal to the circumference of the piston and it extends upward in frequency. In the piston band, the on-axis response of most drivers will begin to increase with frequency at a rate of 6 dB/octave because the beamwidth or coverage angle of the driver will gradually narrow as the frequency increases. (Keyboard shortcut: Ctrl+T.)

Include > Diffraction Response Shelf
Includes the generalized effects of diffraction for a limited selection of box shapes (cube, square prism and optimum square prism). Diffraction is the bending of sound waves as

they pass near an edge or corner of a solid object. (Keyboard shortcut: Ctrl+F.)

Minimize
Hides either the selected graph or all graphs and places a graph icon on the Windows desktop. To restore a graph window, click on its icon.

Copy
Creates a copy of the selected graph in the Windows clipboard so that it can be pasted into another application (like a word processor or page layout program). This command actually "prints" to the clipboard, creating a printable version of the graph with a white background. The color of the plot lines is preserved, except for white which is changed to black. Be aware, however, that some colors like yellow may be very light. (Keyboard shortcut: Ctrl+C.)

Close
Closes either the selected graph or all open graphs.

Properties
Opens the Graph Properties window, displaying the graph settings.

2 Design Wizard

The easiest way to begin a design is to use the BassBox Pro Design Wizard. It acts like a smart assistant to help you use BassBox Pro, taking you to the relevant property windows of the program so information about a design can be entered in an organized progression.

The Design Wizard can be launched from the BassBox Pro title window by clicking the "Run Design Wizard" button or it can be launched from the main window by selecting "Design Wizard" from the Tools menu.

The first step of the Design Wizard is to decide where to begin: with the driver or the box. If you already have a driver in mind or you want to select the driver first, select the "Start with driver" option. If the box already exists or you have a box design in mind, select the "Start with box" option.

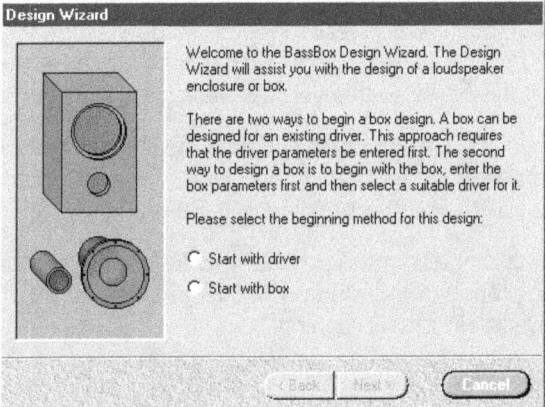

BassBox Pro Design Wizard

Notice that a "Cancel" button is available to stop the Design Wizard. Clicking it will close the Design Wizard causing any data that was entered into it to be lost. *Note: If the Design Wizard was run from the title window, you will also be given the choice of closing BassBox Pro.*

If in the previous step you decided to begin with the driver, the Design Wizard will need to know whether you want to manually enter information about the driver or whether you want to load it from the program's driver database.

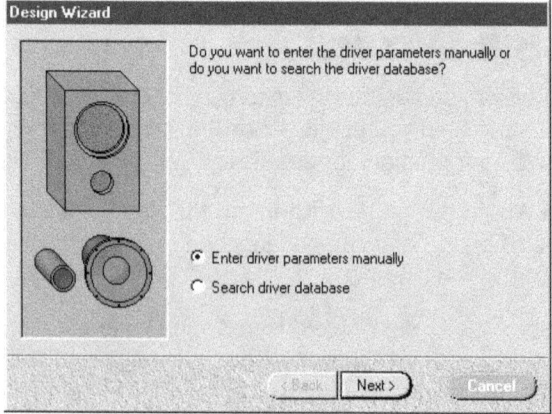

Selecting "Enter driver parameters manually" as shown above, will cause the Design Wizard to lead you through the Driver Properties window where you will be prompted for the driver parameters. Selecting "Search driver database" will cause the Design Wizard to open the Driver Locator window so that you can search for a driver.

From here on, the Design Wizard will use the standard property windows of BassBox Pro so that you can enter the required information when prompted. With its on-screen explanations, the Design Wizard's operation is self-explanatory.

The following chapters of the *BassBox Pro Reference* portion of this manual will describe each of the property windows in detail. Please refer to them if you have further questions while the Design Wizard is running.

3 Driver Properties

The Driver Properties window holds a lot of information about the driver and the way it is configured. To open it, click on the "Driver" button of the desired design panel in the main window ([Ctrl]+[D]). A tab sheet layout is used to help organize the information.

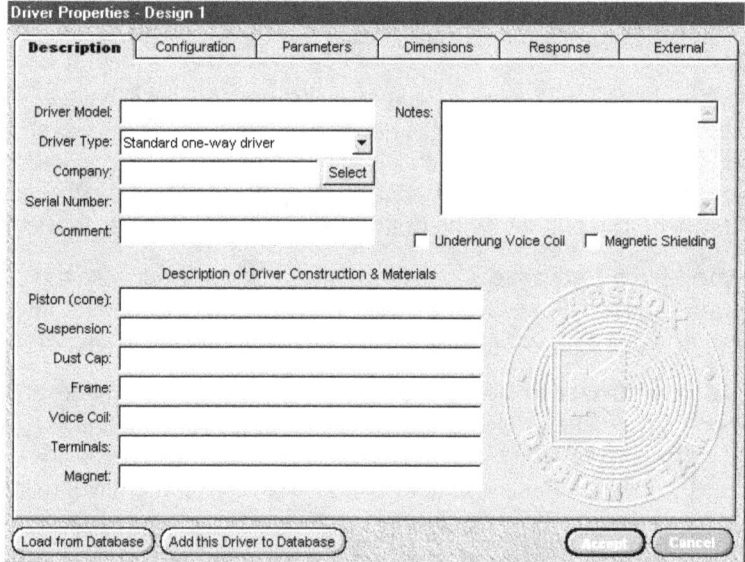

The driver information can come from one of three different sources:

1. **Driver Database** BassBox Pro includes a huge driver database with information for thousands of drivers. This is the first place to look for driver information. The driver database is opened with the "Load from Database" button at the bottom of the Driver Properties window. It can be easily searched by manufacturer, model, driver parameters or box parameters. Once a driver is located, its information is "loaded" into BassBox Pro by clicking on the "Load" button of the Driver Locator window.

2. **Manual Entry** Driver information can be entered manually by the user. This is a convenient way to enter data direct from a manufacturer's data sheet or website. It also allows the user to enter measured Thiele-Small parameters. Information can be entered manually into any of the tabs of the Driver Properties window. *Note: BassBox Pro also includes a built-in procedure for measuring Thiele-Small parameters (basic test equipment is required). When this procedure is used, BassBox Pro will automatically calculate the driver parameters and allow you to load them into a design.*

3. **Import Parameters** BassBox Pro also provides limited import capability for driver

Thiele-Small parameters. These parameters can be imported from a "txt" (T-S) file created with the CLIO measurement system, a "txt" (T-S) file created with either a DATS (Dayton Audio Test System) or WT3 (Dayton Audio Woofer Tester 3), an "ats" file created with the LAUD version 3.12 (or later) measurement system, or a "log" file created with the "Export to BassBox" command of a WT2 (Smith & Larson Audio Woofer Tester 2) measurement system. They can also be imported from a BassBox 6 Pro/Lite "bb6" design file. The import feature is accessed with the "Import" button of the "Parameters" tab of the Driver Properties window.

When the Driver Properties window is first opened, it will have a "Close" button in the bottom right corner. After changes have been made, the "Close" button will be replaced with an "Accept" button and a "Cancel" button (shown on the previous page). Most of the windows in BassBox Pro work this way to help you realize the consequences of closing a window. The two buttons in the bottom left corner are describe next:

Load from Database Opens the Driver Locator window so the driver database can be searched for a driver. Once a driver is found it can be "loaded" into a design. When this happens, all of its information will be copied to the Driver Properties window.

Add this Driver to Database This button is available whenever sufficient driver information has been entered into the Driver Properties window. It is used to add a driver to the database and it is provided as a convenience for those occasions when new drivers are entered. However, there is better way to add to or edit the driver database (see Chapter 10).

Next, let's examine the windows related to the driver properties. We'll begin with the Driver Locator window. Then we'll discuss each tab of the Driver Properties window, itself.

Driver Locator

The purpose of the Driver Locator window, as its name suggests, is to locate a driver in the driver database. However, the contents of the driver database cannot be edited with the Driver Locator window. (See Chapter 10 to learn how to edit the driver database.)

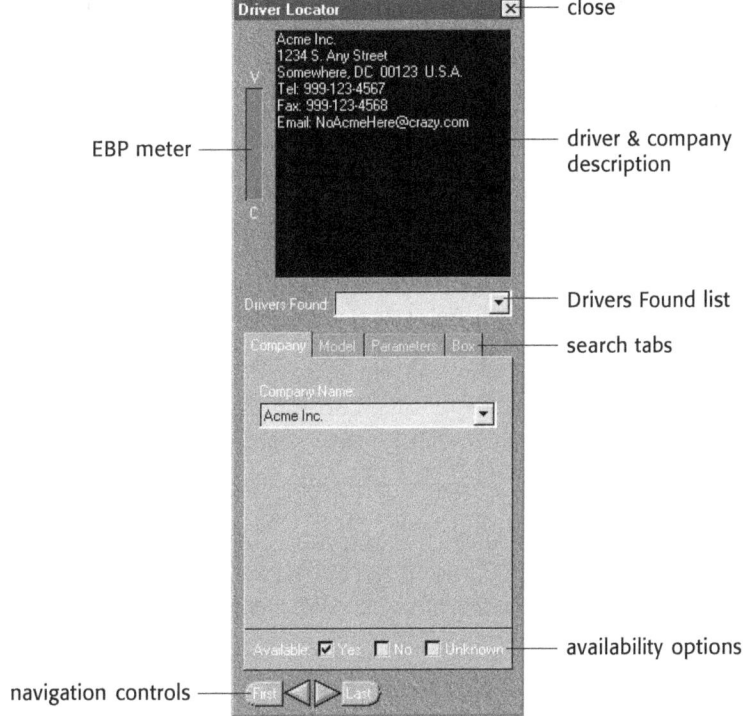

EBP meter The EBP meter shows the suitability of a driver for either a closed box or a vented box. If a driver is more suited for a vented box, the indicator will appear high on the meter near the "V" (vented) label at the top. Conversely, if a driver is more suited for a closed box, the indicator will appear low on the meter near the "C" (closed) label at the bottom.

EBP is an acronym for efficiency bandwidth product and it is calculated by dividing Qes (the driver's electrical Q) into Fs (the driver's free-air resonance). If Qes is unknown, BassBox Pro will estimate EBP with Fs / Qts. A driver is considered better suited for a closed box when its EBP is 50 or less and better suited for a vented box when its EBP is 100 or more. Pausing the mouse pointer over the EBP indicator will cause its value to appear as shown on the next page.

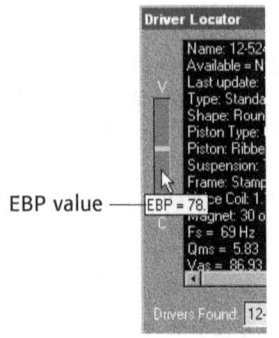

EBP value

The EBP meter is only a general guide—it is not a rule. There are many exceptions where, for example, a driver with a low EBP will still work well in a vented box and visa versa. For a more complete discussion of box and driver selection, see Chapter 3 in the *Box Designer's Guide* earlier in this manual.

Note: The "Parameters" tab of the Driver Properties window also has an EBP meter.

Navigation controls The navigation controls at the bottom of the Driver Locator are rarely used during a search because they step through the database one record at a time. The "First" button selects the first driver record in the database. The "<" button selects the previous driver record in the database. The ">" button selects the next driver record in the database. The "Last " button selects the last driver record in the database.

These controls are useful when the Edit Database Driver Data window is open and you need to step through the driver records. **Caution:** Using one of the navigation controls will cause the results of a search (which are located in the "Drivers Found" list) to be cleared.

The normal way to choose a driver after a search is to select it from the "Drivers Found" list.

Close The "Close" button serves two purposes. The first is obvious—it closes the Driver Locator window. The second purpose is not obvious—it requires the use of the ⇧Shift key and it toggles the size of the Driver Locator. We'll explain.

The Driver Locator window is available in two sizes as shown on the next page. You can switch between the small and large sizes by holding down the ⇧Shift key and clicking the "Close" button. Select the size that looks best with your computer display. *Note: Most of the illustrations in this manual use the large size.*

Driver & company description Information about the selected company and driver are listed in this text box. The driver information, including its parameters, are always listed first. If no driver has been selected, only the company will be listed.

Drivers Found list After a search has been run, the resulting drivers will be listed in the "Drivers Found" list. Use this list to select a driver to be viewed in the "Driver & company description" text box above it. After a driver has been selected, a "Load" button (shown below) will appear in the lower right corner of the Driver Locator window.

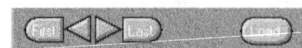

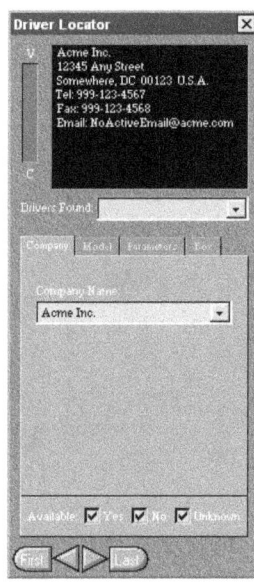
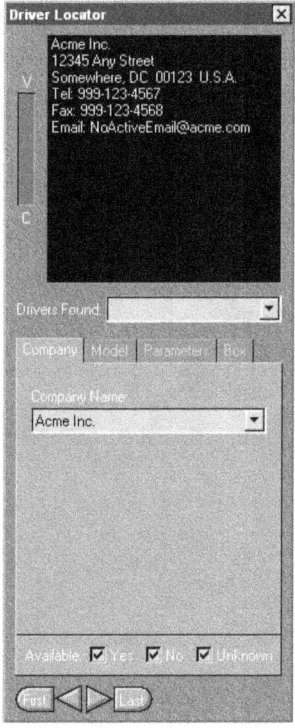

small size   large size

Clicking on the "Load" button will cause the selected driver to be loaded into the design and the Driver Locator window will close.

Search tabs There are four ways to search the driver database and each method is represented with a tab. You can search by company, model and driver or box parameters. Each of these search methods will be discussed in detail on the following pages.

Availability options The driver database includes both old and new drivers. Having the parameters of old drivers is useful in case you ever need to design a new box for an old driver. However, most of the time you will probably want to use a driver that is currently available. For this reason, each driver in the database has a setting which identifies its availability. You can use the "Availability" search options to control which ones will be included in a search.

For example, check only the "Yes" checkbox if you only want to include drivers that are currently available (available = yes). Check all three checkboxes if you want to include all drivers in the search. *Note: There is an "Unknown" setting because the availability of some drivers is unknown. Contact Harris Tech at support@ht–audio.com if you have more information about a driver whose availability is unknown.*

 **BassBox 6 Pro** *Reference: 3 Driver Properties*

Search by Company

This is one of the easiest search methods. With the "Company" search tab selected, choose a company name from the "Company Name" drop-down list. A message window will open stating how many drivers were found for the selected company as shown below.

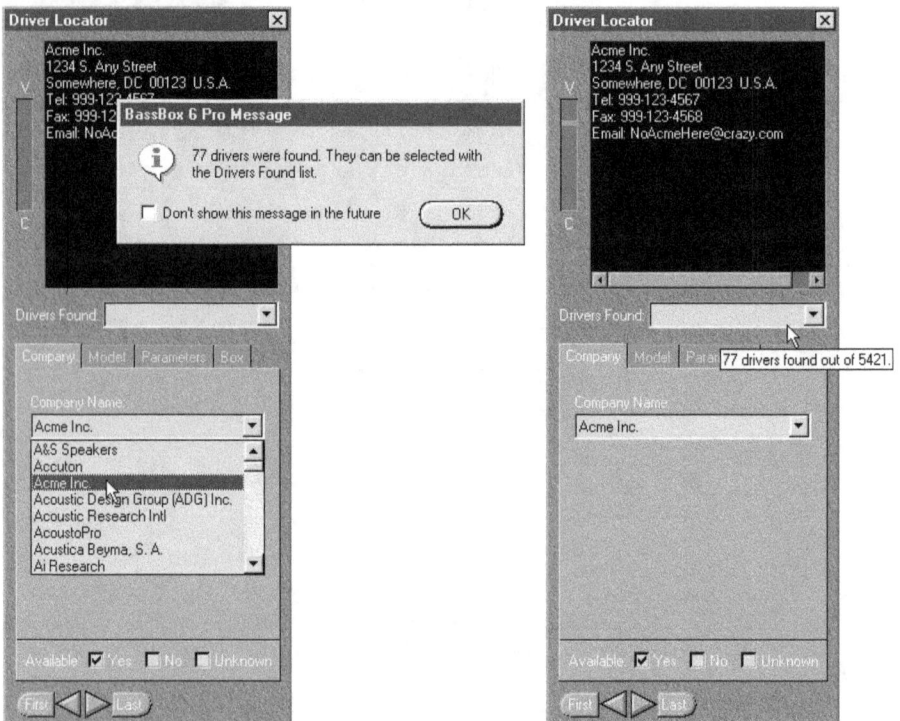

After a search, moving the pointer over the "Drivers Found" list will cause the number of drivers found and the total number of relevant drivers in the database to be listed. *Note: In BassBox Pro only the total number of woofers or "open back" drivers are reported. In X•over Pro this number is larger because it also includes the closed back midrange drivers and tweeters.*

Immediately after a search, no drivers have been selected yet so only the company information is listed in the description box at the top of the window. Use the "Drivers Found" drop-down list to select a driver. This is shown on the next page.

Caution: Do not use the manual navigation controls (First, <, >, Last) at the bottom of the window after a search. Doing so will clear the "Drivers Found" list and deselect the company. Use them when you want to step through the database one record at a time.

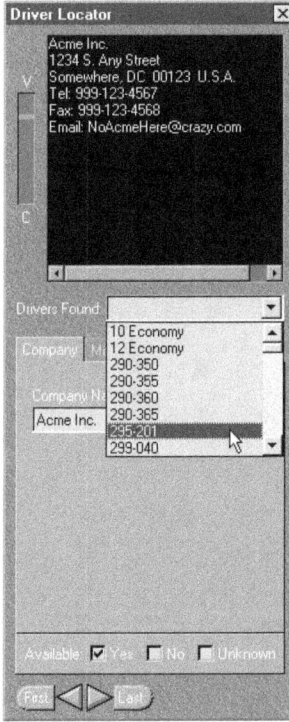
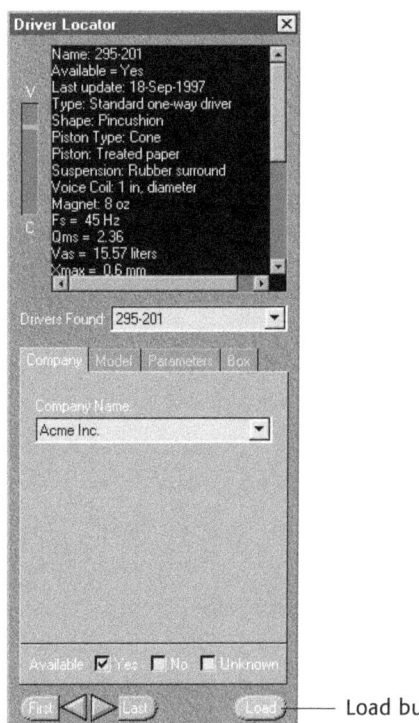

Load button

All of the drivers that were found during the search will be listed in the "Drivers Found" list. Selecting one of them causes its description, parameters and company information to be displayed in the description box. However, it does not load the driver into the speaker design or the Driver Properties window.

Selecting a driver from the "Drivers Found" list also causes the "Load" button to appear in the bottom right corner of the window. Clicking the "Load" button will load the driver into the Driver Properties window and close the Driver Locator window.

BassBox 6 Pro   *Reference: 3 Driver Properties*

Search by Model Name

This is another easy-to-use search method. With the "Model" search tab selected, enter a model name into the "Model Name" input box. Then click the "Search" button at the bottom of the window.

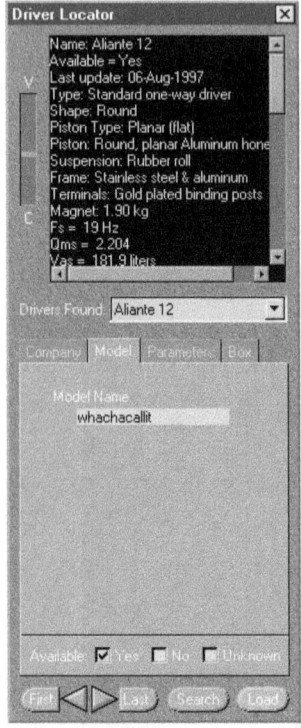

The model name is not case sensitive. The search engine will find all drivers whose model name begins with the letters and/or numbers you enter. For example, entering a single letter like "a" will produce a list of every driver beginning with "a" or "A" such as:

A11EC80-02F	A13WG-01-04 - LF
A13WH-01-04 - LF	A17WG-01-04 - LF
A25FU14-57F-Q	A25F020-53F

Wildcard searches are not supported. Most characters entered into the "Model Name" box like hyphens, asterisks, commas and periods are treated literally. However, spaces are ignored.

Search by Driver Parameters

This search method is very flexible and offers many ways to narrow the search. However, it requires more decision making on the user's part. With the "Parameters" search tab selected, begin by selecting the driver parameters you want to use in the search and then configure their units. Up to five driver parameters can be specified. Changing parameters or their units is accomplished by clicking on the parameter labels and their respective unit labels. Any of the twenty-one driver parameters used in BassBox Pro can be selected (Fs, Qms, Vas, Cms, Mms, Rms, Xmax, Xmech, Dia, Sd, Vd, Qes, Re, Le, Z, BL, Pe, Qts, ηo, 1-W SPL, 2.8-V SPL). In the illustration below, Fs, Dia, Cms, Pe and Qts are selected. For example, to change Cms to Vas, simply click the Cms label. Each time a parameter label is clicked, it will change to the next available parameter.

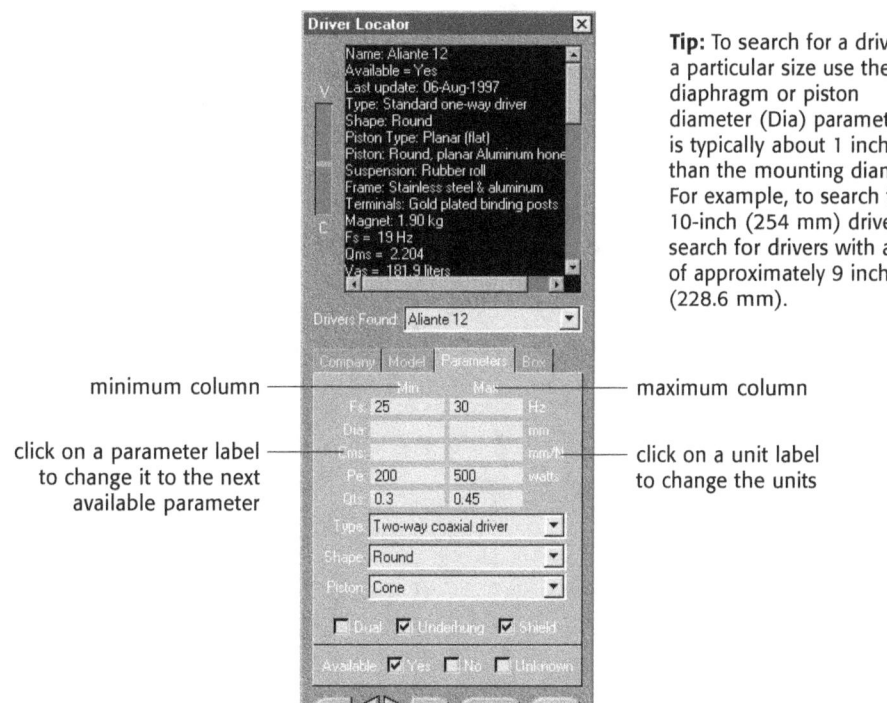

Tip: To search for a driver of a particular size use the diaphragm or piston diameter (Dia) parameter. It is typically about 1 inch less than the mounting diameter. For example, to search for a 10-inch (254 mm) driver, search for drivers with a Dia of approximately 9 inches (228.6 mm).

minimum column — maximum column

click on a parameter label to change it to the next available parameter — click on a unit label to change the units

Next, enter a minimum ("Min") and maximum ("Max") value for each parameter. Selected parameters will be ignored during a search if a minimum and maximum value have not been entered for them. In the example above, Dia and Cms will be ignored. Remember to set the units before entering the minimum and maximum values.

Select the desired driver type, shape and piston type from their respective drop-down lists. Finally, select "Dual" if you want to search for a dual-voice coil driver. Select "Underhung" if you want to search for a driver with a voice coil that is shorter than the gap of the magnet. (Underhung voice coils can behave with more linearity than a typical overhung voice coil but they are less common because they tend to be more expensive—especially if a long excursion is also required.) Select "Shield" if you want to search for a driver with magnetic shielding so it can be used near video monitors, computer monitors or televisions.

Click the "Search" button at the bottom of the window to begin the search.

Some recommendations for driver selection are listed next. They are intended only as generalizations.

Drivers for Closed Boxes
- Qts greater than 0.3.
- Low resonance (Fs). High moving mass (Mms) and compliance (Vas, Cms).
- Large Xmax (small woofers: 2-4 mm; large woofers: 5-8 mm).
- Moderately strong motor (BL).
- EBP less than or equal to 50.

Drivers for Vented Boxes
- Qts from 0.2 to 0.5.
- Moderate resonance (Fs) and moving mass (Mms).
- High compliance (Vas, Cms) is not necessary but can be beneficial.
- Moderate Xmax (small woofers: < 2 mm; large woofers: < 5 mm).
- Strong motor (BL).
- EBP greater than or equal to 100.
- Avoid woofers with excessive unknown losses.

Drivers for Single-Tuned Bandpass Boxes
- Qts greater than 0.3. Higher Qts values have less passband ripple but they also have a narrower passband.
- Low resonance (Fs). High moving mass (Mms) and compliance (Vas, Cms).
- Large Xmax (small woofers: 2-4 mm; large woofers: 5-8 mm).
- Moderately strong motor (BL).
- Compatible with a wide range of EBP values.

Drivers for Double-Tuned Bandpass Boxes
- Qts greater than 0.4. Higher Qts values have less passband ripple but they also have a narrower passband.
- Moderate resonance (Fs) and moving mass (Mms).
- High compliance (Vas, Cms) is not necessary but can be beneficial.
- Moderate Xmax (small woofers: < 2 mm; large woofers: < 5 mm).
- Strong motor (BL).

Drivers for Passive Radiator Boxes
- Qts from 0.2 to 0.35.
- Moderate resonance (Fs) and moving mass (Mms).
- High compliance (Vas, Cms) is not necessary but can be beneficial.
- Moderate Xmax (small woofers: < 2 mm; large woofers: < 5 mm).
- Strong motor (BL).
- EBP greater than or equal to 100.
- Avoid woofers with excessive unknown losses.

In general, smaller boxes require a lower Qts and smaller Vas than do larger boxes of the same type. For information on selecting a box, see Chapter 3 in the *Box Designer's Guide* earlier in this manual.

Search by Box Parameters

This search method is presently available only for closed and vented box types. With the "Box" search tab selected, select a box type from the Box list. If necessary, change the box volume (Vb) units by clicking on the units label ("cu.ft" in this example). Then enter the minimum ("Min") and maximum ("Max") box internal volume for parameter Vb. Next, enter a minimum and maximum −3 dB cutoff frequency for parameter F3. Finally, click on the "Search" button to begin the search.

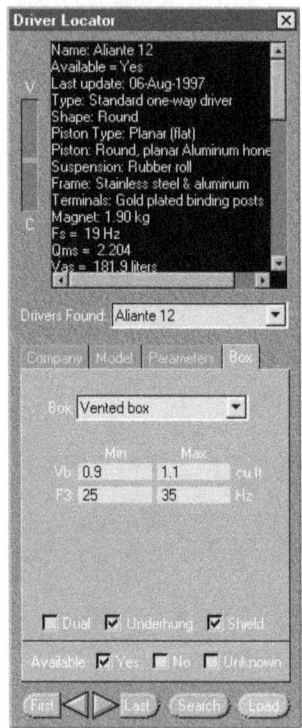

The search engine will attempt to find all drivers that produce a smooth amplitude response with the specified internal box volume and achieve a cutoff frequency in the specified range. If a closed box type was selected, drivers will be rejected if EBP is greater than 75 or Qts is less than 0.3. If a vented box type was selected, drivers will be rejected if EBP is less than 75 or Qts is greater than 0.5.

Reference: 3 Driver Properties BassBox 6 Pro

Description

The "Description" tab contains optional descriptive information about the driver.

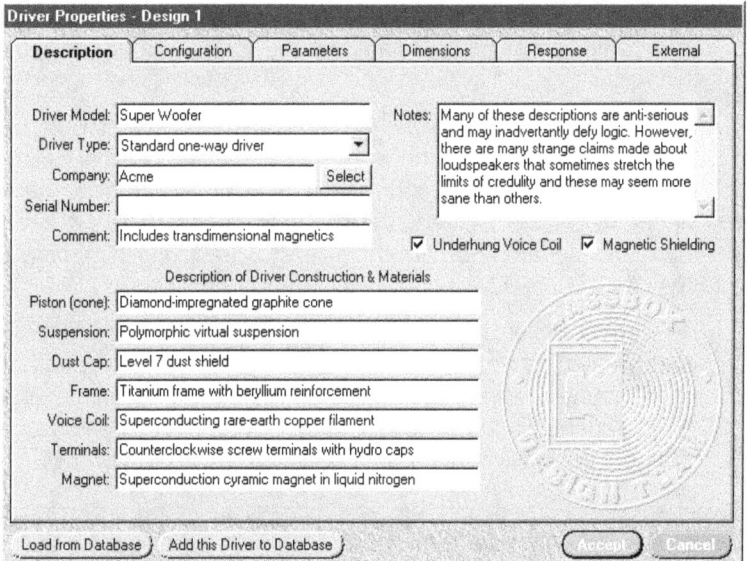

Driver Model The model name of the driver can be up to 30 characters long. Keep in mind that in the driver database a single manufacturer is not allowed to have more than one driver with the same model name.

Driver Type Four driver types are provided: standard one-way drivers, two-way coaxial drivers, two-way coincident drivers and three-way drivers. *Note: The driver shape is selected on the "Dimensions" tab. The round driver shape is shown below:*

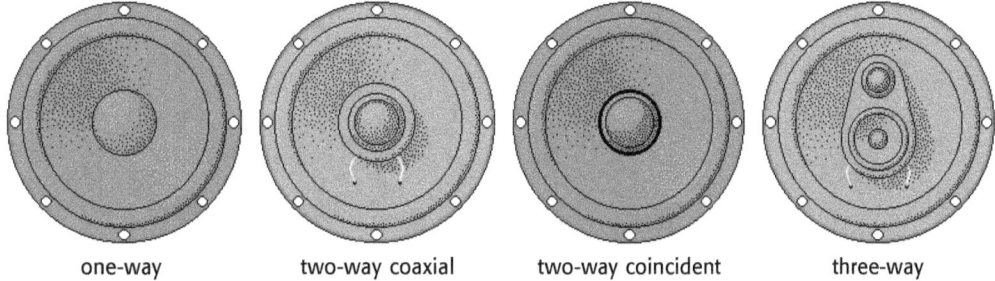

one-way   two-way coaxial   two-way coincident   three-way

One-way drivers are the most common type of driver and contain only a single piston. Two-way and three-way drivers actually consist of more than one driver combined together in various ways as described below:

Coaxial two-way drivers have an additional tweeter. Both drivers are mounted like two wheels on an axle with the tweeter mounted in front. The magnet assemblies of the drivers are completely separate and the sound from the tweeter arrives ahead of the woofer.

Coincident two-way drivers have an additional tweeter. But the tweeter is mounted in the acoustical center of the woofer. The magnet assemblies are combined and the sound from the tweeter and the woofer emanate at the same time and from the same apparent location. Because the sound from both drivers is "coincident" these drivers have superior time-domain characteristics, resulting in minimal phase cancellations in the crossover region.

Three-way drivers have an additional midrange driver and tweeter. The midrange driver and tweeter are usually mounted side by side in a common bracket. This assembly is then mounted in front of the woofer. Neither the midrange driver or tweeter are coaxial or coincident with the woofer and the sound of all three driver elements will arrive at different times with the woofer lagging behind the other two.

Two-way coaxial drivers and three-way drivers are commonly used in automotive applications where full-range sound from an extremely small space is desired. Two-way coincident drivers are not very common and are sometimes used in recording studio monitors of various types. They are also used in some higher-quality home hi-fi speaker systems.

Company The company name can be up to 40 characters in length and there are two ways to enter it. The preferred way is to use the "Select" button and choose a company name from the driver database. This insures that the spelling matches the spelling in the database in case the driver is added to the database later. The second way to enter the name is to type a name directly into the company input box.

Serial Number The serial number can be up to 30 characters long.

Comment The comment can be up to 50 characters long. It is a single-line note that will be displayed in the design properties list of each design in the main window.

Piston (cone), Suspension, Dust Cap, Frame, Voice Coil, Terminals, Magnet These can each be up to 50 characters long and provide specific descriptions about the driver, its construction and features.

Notes This is a multi-line note that is not limited in length and can contain more lengthy comments about the driver.

Underhung Voice Coil Check this checkbox when the driver has an underhung voice coil. Overhung voice coils are the most common type because they are typically less expensive to design and manufacture. With an overhung voice coil the height of the voice coil is greater than the height of the gap of the magnet. An underhung voice coil is just the opposite. The height of its voice coil is less than the height of the gap of the magnet so the entire voice always stays in the gap up to the maximum linear excursion (Xmax). This usually gives underhung voice coils an advantage in terms of linearity. But it can make them expensive when a large excursion is needed.

Magnetic Shielding Typical drivers cannot be placed near a television, computer monitor or video monitor because the magnetic field of the magnet assembly of the driver will bend the path of the electrons in the picture tube causing the picture to distort. To avoid this problem, many drivers are now available with magnetic shielding. Check this checkbox if the driver has magnetic shielding.

Note: The driver description does not have to be entered in order to complete a speaker design. However, it can make a design much more understandable to others.

Reference: 3 Driver Properties

Configuration

The "Configuration" tab can be ignored if only one low-frequency driver will be used in the box. If more than one is used, use this tab to set their mechanical and electrical configurations. This will also cause "net" parameter values to appear on the "Parameters" tab.

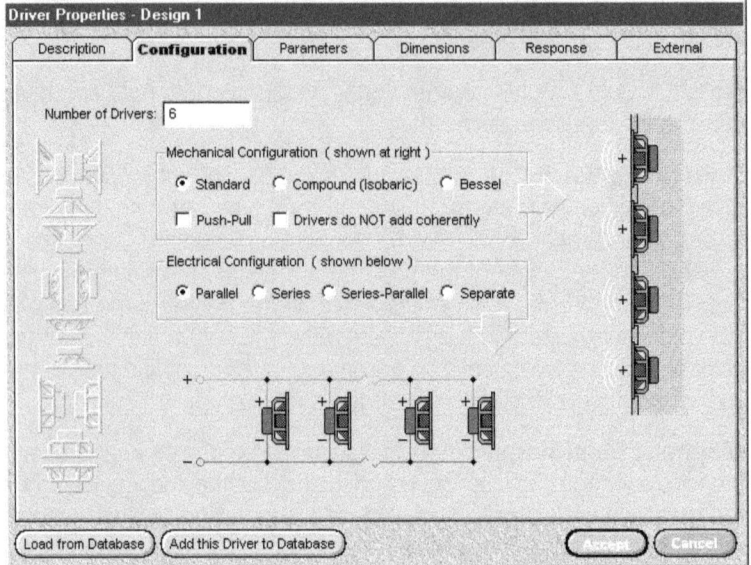

Number of Drivers BassBox Pro allows you to specify any number of drivers for any of the box types with one exception. The Bessel configuration requires either 5 or 6 drivers and it requires that a closed box be used. *Note: Although Bessel arrays with more than 6 elements are possible, BassBox Pro models only 5 and 6 element Bessel arrays.*

Mechanical Configuration Refers to the way the drivers are mounted in the box. Four choices are provided:

Standard

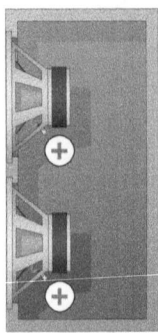

Standard This is the simplest way to mount multiple drivers. When the Push-Pull parameter is turned off, the drivers are all mounted facing the same direction and are wired with the same polarity. The box size must increase with each additional driver. Every doubling of drivers causes the 1-watt sensitivity (1-W SPL) to increase 3 dB if the parallel electrical configuration is chosen and decrease 3 dB if the series electrical configuration is chosen. The 2.83-volt sensitivity (2.8-V SPL) will increase 6 dB in parallel and remain unchanged in series.

Reference: 3 Driver Properties **BassBox 6 Pro**

Compound

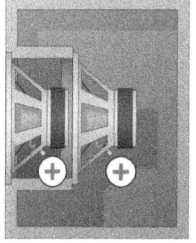

Compound (Isobaric) This configuration mounts pairs of drivers closely together with a small isobaric (constant pressure) sealed chamber between them.

As such, there must be an even number of drivers. The most unique feature and the greatest advantage of this configuration is that the required box volume for a pair of compound drivers is half that of a single driver. The wiring polarity depends on the way the drivers are mounted as well as the setting of the Push-Pull parameter. Compared to a single driver, a pair of compound drivers will have 3 dB less sensitivity at 1 watt (1-W SPL). This is the same for both parallel and series electrical configurations. The 2.83-volt sensitivity (2.8-V SPL) for two compound drivers will be the same as a single driver if the parallel electrical configuration is chosen and will decrease 6 dB if the series electrical configuration is chosen.

Compound

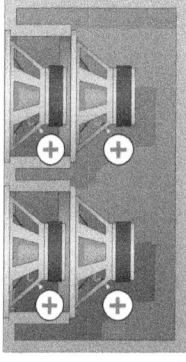

How does BassBox Pro handle more than two compound drivers? It assumes that only two drivers are connected to each isobaric chamber. For example, four drivers would be divided into two compound pairs as shown at right. Six drivers would be divided into three compound pairs and so on.

Push-Pull The standard and compound mechanical configurations each have a push-pull option. It is only available when there are an even number of drivers (2, 4, 6, 8, etc.). When the push-pull option is used, every other driver is mounted in a reverse direction and is wired with opposite polarity. Examples are shown below:

Standard Push-Pull

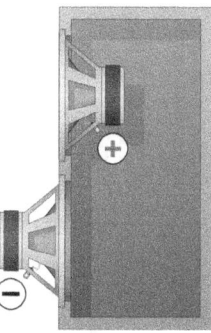

Compound Push-Pull

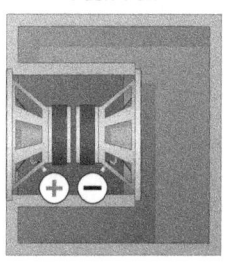

Compound Push-Pull

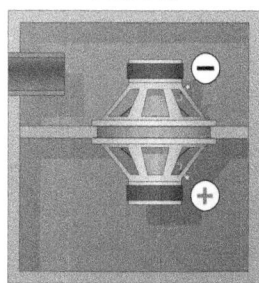

There are two different ways to mount a pair of compound push-pull drivers. One way (the middle illustration on the previous page) mounts them in an isobaric chamber with each driver pointing away from the other. The second way (the right illustration on the previous page) mounts them facing each other on a common wall. Of course, there is a hole in the wall so that the space between the two cones forms the isobaric chamber. The second method is popular with compound bandpass box designs.

Why use the push-pull option? Because it is a very effective way to reduce some kinds of distortion. Since the pistons of half of the drivers in a push-pull configuration move "in" while the other half move "out", many nonlinearities are cancelled. The result is a dramatic reduction in even-order distortion.

Bessel This is a specialized mounting and wiring method that is available only when there are 5 or 6 drivers and only for closed box designs. The drivers are mounted in a line array and should be mounted as close together as possible with one important exception—a Bessel array with 6 drivers must leave a single blank space in the middle that is the size of one driver.

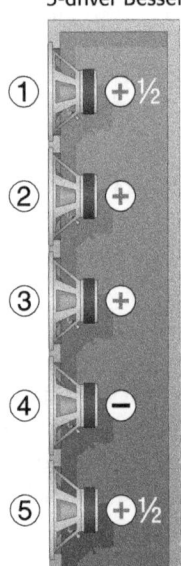
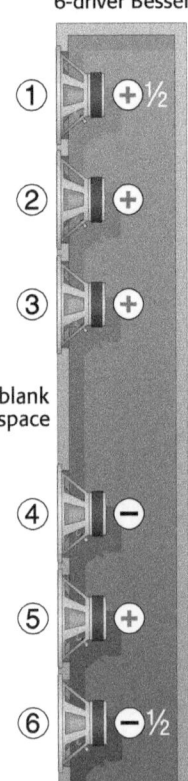

When a speaker has multiple drivers it is normal for there to be phase cancellations which cause the coverage pattern and sound quality to suffer. The unique feature of a Bessel array is that the phase cancellations are mostly eliminated, resulting in a combined coverage pattern that is nearly identical to a single driver. However, the pattern needs some distance to combine. One "rule of thumb" is to use a working distance that is 20 times the length of the array. This assumes the length of the array is measured from the center of the two outside drivers.

The real secret of a Bessel array is in its unique combination of mounting and wiring. A unique series-parallel method of wiring is required which controls the polarity and level of each driver so that their signals combine properly.

Drivers do NOT add coherently When multiple drivers are used, the sound waves emanating from them will combine to create a composite sound wave that is louder. BassBox Pro assumes that the drivers are all the same kind and, in most cases, that they are driven with identical signals so that their sound waves will usually sum coherently. This means that the net sound level will increase 6 dB with every doubling of drivers.

However, there may be occasions when the sound waves do not add coherently and the "Drivers do NOT add coherently" option should be turned on. In these cases the net sound level will increase only 3 dB for every doubling of drivers. The following list describes situations when coherent additions will <u>not</u> happen. Turn on the "Drivers do NOT add coherently" option for these situations:

- The drivers will not sum coherently if they are wired separately and are driven with different signals. For example, two woofers are mounted in a common cabinet but one is driven from a left stereo signal and the other is driven from a right stereo channel.

- The drivers will not sum coherently if they are mounted too far apart. Their center-to-center spacing should be no greater than one quarter (¼) wavelength for the frequencies in their passband. This is usually not a problem for subwoofers because they are driven only with low-frequencies having long wavelengths. For example, many subwoofers use a crossover frequency of 100 Hz or less—the wavelength of 100 Hz is 136 inches (345 cm). As long as the drivers reproduce frequencies that are not higher than 100 Hz then the drivers in the subwoofer can be mounted as far as 34 inches (86 cm) apart because this is one quarter of 136 inches.

- Coherent summing is not relevant to compound configurations because only one of each isobaric pair of drivers will radiate sound outside the box. The "Drivers do NOT add coherently" feature will be disabled when a "Compound" mechanical configuration is chosen.

The drivers are always assumed to sum coherently when a Bessel array is chosen.

Electrical Configuration   Describes the driver wiring. Four choices are provided.

Parallel   This is usually the best electrical configuration because it offers the best electrical isolation between drivers. However, it has one serious drawback: if too many drivers are used, the net impedance seen by the amplifier can become extremely low and, if the amplifier does not have sufficient current headroom, it may blow a fuse, activate its protective circuitry or fail if it has no protective circuitry. For example, four 8 ohm drivers wired in parallel have a net impedance of 2 ohms. Four 4 ohm drivers wired in parallel have a net impedance of only 1 ohm! Four examples of parallel wiring are shown below:

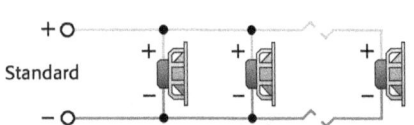

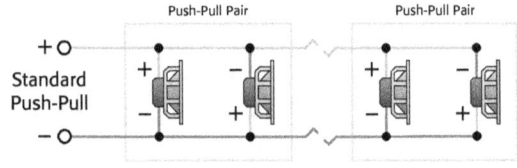

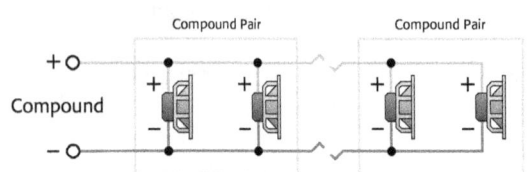

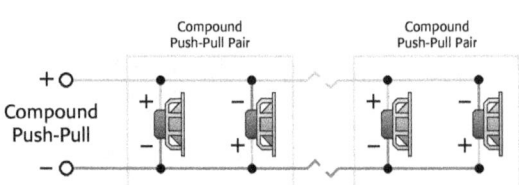

Notice that the polarity of every other driver is reversed in the push-pull example.

Series   This is usually the least desirable electrical configuration because the drivers have the greatest interaction. An "open" circuit anywhere in the wiring (such as a blown driver) will prevent sound from reaching all of the remaining drivers. However, it has one advantage: the net impedance seen by the amplifier increases with each additional driver. For example, two 8 ohm drivers wired in series have a net impedance of 16 ohms. Four 8 ohm drivers wired in series have a net impedance of 32 ohms. Be aware that as the impedance rises, the drivers draw less power from the amplifier. Four examples of series wiring are shown at the top of the next page:

Reference: 3 Driver Properties BassBox 6 Pro

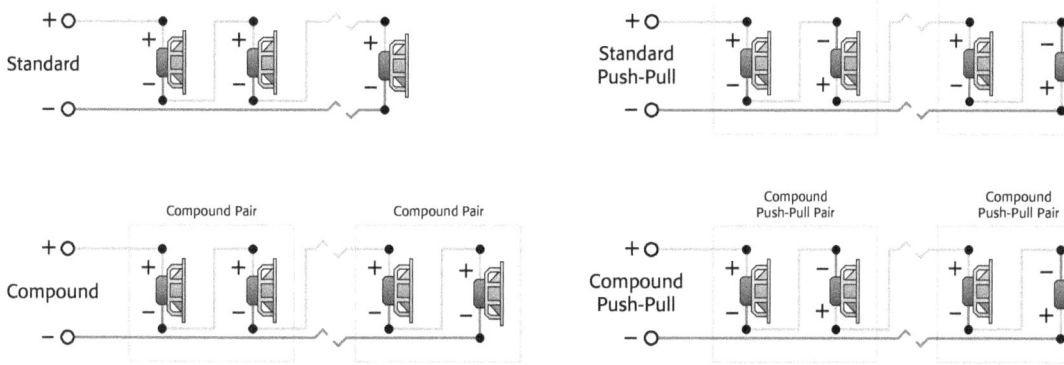

Series-Parallel This is a combination of both series and parallel wiring which is sometimes used to prevent the impedance from going too low or too high. BassBox Pro assumes that the drivers are paired in series and that each pair is in turn wired in parallel with all the other pairs (except for Bessel configurations). As a result, there must be a minimum of four drivers and the number of drivers must be even to use this configuration. Four examples of series-parallel wiring are shown below:

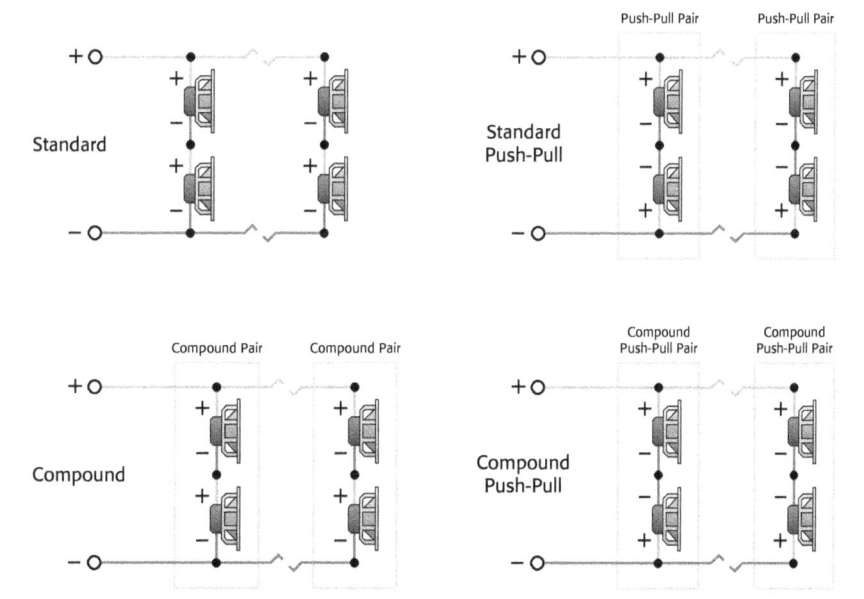

Bessel designs require special series-parallel wiring in order to control the level and polarity of each driver. The wiring for 5-driver and 6-driver Bessel configurations are shown below:

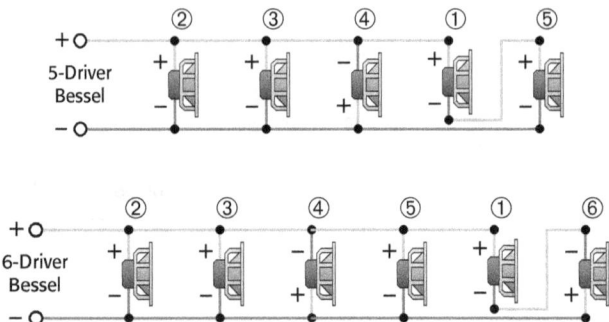

Separate Select this configuration if the drivers will not be connected together but will instead be each driven by a separate amplifier channel. When the "Separate" option is first selected, the "Drivers do NOT add coherently" option will also be turned on since the program has no way of knowing if the drivers will be driven with the same signal. If they are and their sound waves will sum coherently, then the "Drivers do NOT add coherently" option should be turned off. *Note: When the "Separate" configuration is selected, the "Drivers do NOT add coherently" option will determine whether some of the graphs show the net response or the response of just one driver. See the descriptions of each graph in Chapter 6 for more details.*

Reference: 3 Driver Properties BassBox 6 Pro

Parameters

The "Parameters" tab contains the Thiele-Small and electromechanical parameters for the driver. "Thiele-Small", often abbreviated "T-S", refers to Neville Thiele and Richard Small who first popularized modern methods of speaker modeling.

EBP Notice in the illustration below that the "Parameters" tab includes an EBP indicator. EBP is the efficiency bandwidth product and it is calculated by dividing the driver's free-air resonance (Fs) by its electrical Q (Qes). If Qes is unknown, EBP will be estimated with the total Q (Qts).

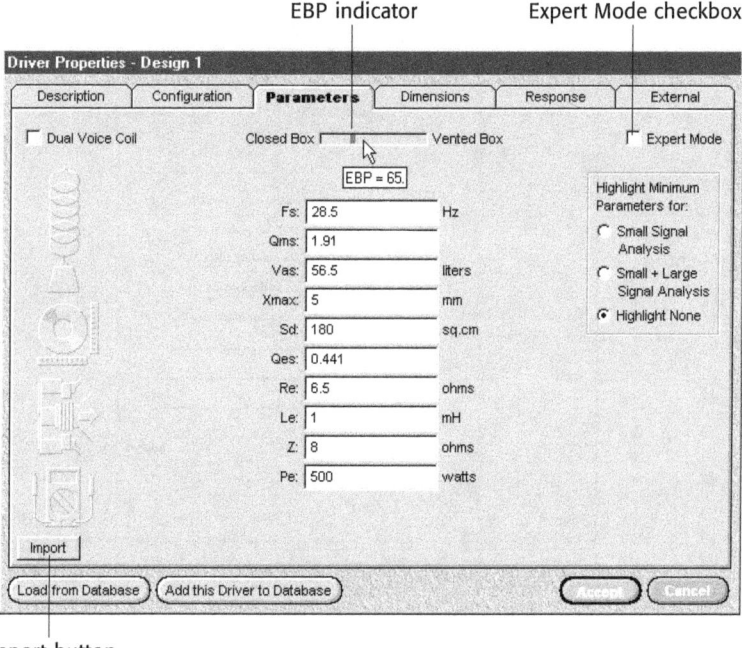

Import button

The significance of EBP is its ability to indicate how suitable a driver is for a closed box versus a vented box. Drivers with EBP values less than or equal to 50 will usually work better in a closed box. Drivers with EBP values greater than or equal to 100 will usually work better with a vented box. Pausing the mouse pointer over the EBP indicator will cause the EBP value to be displayed as shown above. Use the EBP indicator as a general guide only—there are always exceptions.

Mode BassBox Pro supports a total of 21 parameters but not all of them are required at the same time. For this reason, two modes are provided for the "Parameters" tab, "normal" mode and "expert" mode. They are selected by turning on and off the "Expert Mode" checkbox in the upper right corner of the "Parameters" tab as shown in the illustration above. The default mode can be set with the "Driver" tab of the Preferences window. Let's examine each mode in detail:

Normal Mode (The "Expert Mode" checkbox is turned OFF.) The "normal mode" presents a less complex version of the "Parameters" tab with only 10 frequently used driver parameters. The minimum parameters for a complete analysis (both small and large signal) are included and a "Highlight Minimum Parameters for:" option is provided to identify subgroups of these parameters for a small-signal and/or large-signal analysis. In the example below, the "Small Signal Analysis" setting is selected:

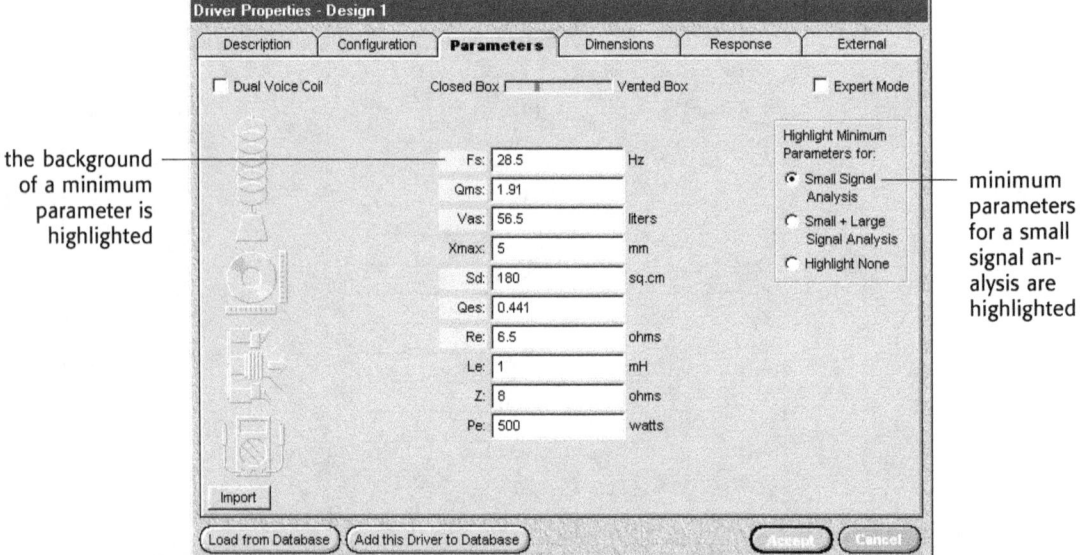

the background of a minimum parameter is highlighted

minimum parameters for a small signal analysis are highlighted

Notice that the background of Fs, Qms, Vas, Sd, Qes and Re is displayed with a lighter shade of grey to mark them as the minimum parameters for a full small-signal analysis. *Note: The absolute minimum set of parameters is Fs, Vas and Qts. With them, BassBox Pro can display a Normalized Amplitude Response graph. However, many of the other performance graphs included in a small signal analysis will not be available unless more information is entered.*

Small-Signal & Large-Signal Analysis
If you are new to speaker design you might not be familiar with the phrases "small-signal" and "large-signal" analysis. Small-signal analysis examines the speaker at low power levels to determine the overall character of its sound such as its amplitude (frequency) response. Large-signal analysis examines the speaker at high power levels to determine how loud it will go with such things as cone excursion and vent air

velocity. Although each analysis shares the need for some parameters, they also each require unique parameters as highlighted with the "Highlight Minimum Parameters for:" option above.

In the "normal mode" a few of the parameters can be switched with others that are not visible. Qms and Qes can be switched with Qts and BL. And Sd can be switched with Dia. To switch one of these parameters, simply click on its label.

The default parameters for the normal mode (Qms, Qes, Sd or Qts, BL, Dia) can be set with the "Driver" tab of the Preferences window.

Expert Mode (The "Expert Mode" checkbox is turned ON.) The "expert mode" presents a complete set of 21 Thiele-Small and electromechanical parameters. They are grouped according to their type (Mechanical, Electrical and Electromechanical).

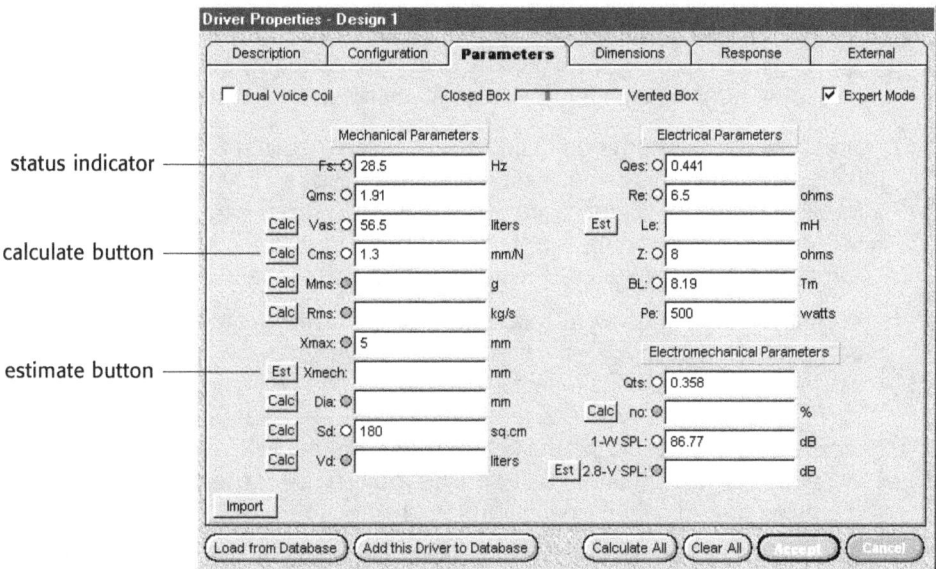

A key feature of the "expert mode" is its automatic tolerance testing of many parameters. A status indicator shows whether a parameter passed or failed. The tolerance settings for the tests can be set with the "Driver" tab of the Preferences window. This self-analyzing feature can quickly reveal "typos" and other mistakes immediately when they are made. It also provides a convenient way to see if a manufacturer's data contains mistakes or has been "fudged."

Note: Three parameters, Xmech, Le and Pe, do not have status indicators because they cannot be verified.

The color code for the parameter status indicators is:

Grey – there is not enough information available to test the parameter.
Green – the parameter is probably correct.
Yellow – the parameter appears to be slightly out of tolerance.
Red – the parameter is probably incorrect.

When the status indicators of multiple parameters turn red it does not mean that they are all necessarily incorrect. When two or more related parameters fail a test, BassBox Pro can't know which one is incorrect and so it must flag them all. The default tolerance values are ±3% for yellow and ±7% for red.

What are some of the reasons that cause a parameter value to be out of tolerance?

- The wrong units were selected before the parameter was entered.
- The decimal place is incorrect.
- The value has been rounded too much.
- An error was made when the driver was tested by the manufacturer.
- An error was made when the parameter was calculated by the manufacturer.

Unfortunately, it is common to find one or more "bogus" parameters in driver specifications. Most often this is the result of "honest" mistakes. Even some of the drivers in the BassBox Pro driver database are out of tolerance. This is why critical speaker designers use driver databases only as a starting point for design and, after obtaining some sample drivers, will measure their parameters, themselves.

Even if you are not doing a "critical" design, there will be times when you simply can't get good data because it isn't available. The drivers may no longer be in production or, as strange as it sounds, a manufacturer may not be willing to release the information. In such cases you, too, will need to measure the driver parameters, yourself. There are inexpensive devices on the market to do this, like the DATS from Dayton Audio and the Woofer Tester 2 from Smith & Larson. Or you can use the test procedure in this program and measure them manually (see Chapter 11).

Of course, you can always choose a driver from a manufacturer who is able and willing to provide accurate and complete data about their products.

Calc, Est Whenever a parameter is blank or out of tolerance, the "expert mode" will check to see if there is enough information available to calculate or estimate its value. If there is, a "Calc" or "Est" button will appear to the left of the parameter label. Note that special circumstances exist for the Qms, Qes and Le parameters.

A "Calc" button will appear beside the Qms and/or Qes parameters whenever their values can be calculated from the other parameters. However, if either Qms or Qes is blank and there are not enough known parameters to calculate them, then an "Est" button will appear. Clicking it will cause the Driver Q Estimator window shown on the next page to open.

This window enables you to estimate Qms and/or Qes from the shape of the driver's free air resonance peak from the impedance graph supplied by the driver's manufacturer. Follow the on-screen instructions to estimate one or both of the Q parameters. *Note: The total Q (Qts) that results from Qms and Qes will also be calculated and displayed.*

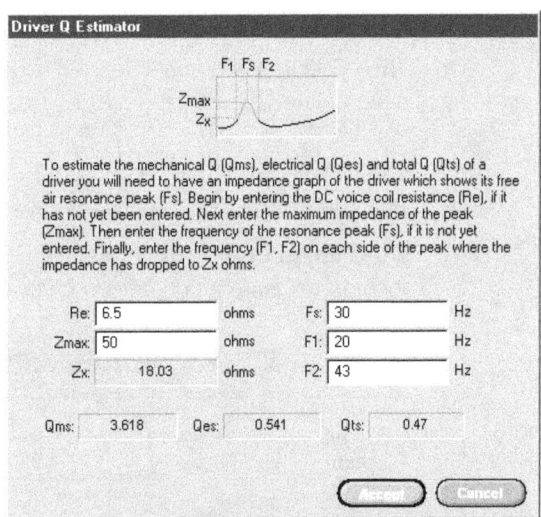

The Le parameter cannot be calculated from the other parameters. So, when Le is blank, an "Est" button will appear. Clicking on it opens the Driver Inductance Estimator window shown at right.

This window enables you to estimate Le from the shape of the high-frequency impedance rise from the impedance graph supplied by the driver's manufacturer. Follow the on-screen instructions to estimate Le.

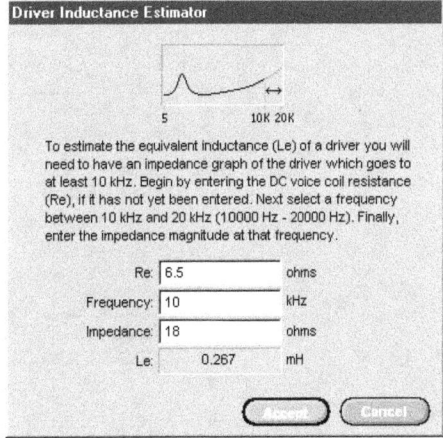

Import Use the "Import" button to import driver Thiele-Small parameters from a "txt" (T-S) file created with the CLIO measurement system, a "txt" (T-S) file created with either a DATS (Dayton Audio Test System) or WT3 (Dayton Audio Woofer Tester 3), an "ats" file created with a LAUD version 3.12 (or later) measurement system, or a "log" file created with the "Export to BassBox" command of a WT2 (Smith & Larson Audio Woofer Tester 2) measurement system. You can also use this feature to import driver properties from an existing BassBox 6 Pro/Lite "bb6" design file (see Chapter 7).

Calculate All, Clear All A "Calculate All" and "Clear All" button (shown below) are available below the "Parameters" tab when the "expert mode" is selected.

 *Reference: 3 Driver Properties*

The "Calculate All" and "Clear All" buttons do just what their names imply, calculate all possible unknown parameter values (except Le and Pe) or clear all parameter values.

The remainder of this section on the "Parameters" tab will describe the inputs for the "expert mode" since it covers all possible inputs.

Dual Voice Coil Used to identify drivers with dual voice coils. It expands the parameters to include separate inputs for each voice coil wiring method. An example is shown below:

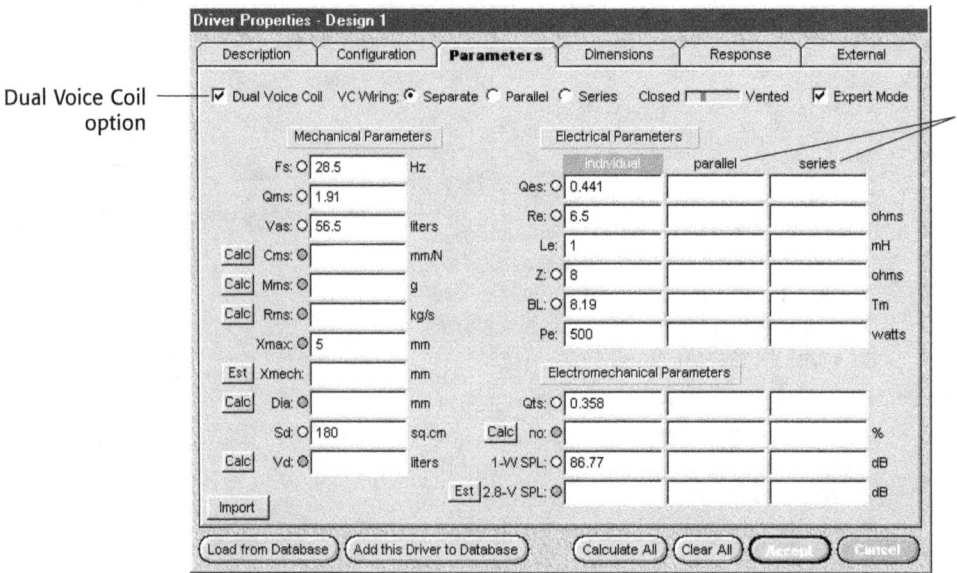

Use the "VC Wiring" option to select the desired wiring method. Use the "Separate" setting if each voice coil will be connected to a separate amplifier channel. Use "Parallel" if the voice coils will be wired in parallel and connected to the same amplifier channel. Use "Series" if the voice coils will be wired in series and connected to the same amplifier channel.

The electrical and electromechanical parameters will have a different value depending on the wiring method used for the dual voice coil. The mechanical parameters are not affected by the dual voice coil's wiring method. Some of the electrical and electromechanical parameters can be estimated if a value for at least one wiring method has been entered for the parameter. To estimate a value, double-click (🖰🖰) in the input box of the empty parameter. For example, double-clicking in the Re input box in the "parallel" column of the above illustration will result in an estimated value for Re of 3.25 ohms because the "parallel" value of Re is usually one half that of each separate (individual) voice coil.

The parameters for the selected wiring method will be used for all internal calculations. *Note: When the dual voice coil option is enabled, the status indicators of the electrical and electromechanical parameters will display only the status of the selected wiring method. And the "Calc" and "Est" buttons of the electrical and electromechanical parameters will only calculate the values of the selected wiring method.*

Mechanical Parameters

Fs The free air resonance frequency of the driver. Units: Hz.

Qms The mechanical resonance magnification of the driver at Fs.

Vas The volume of air equal to the mechanical compliance of the driver's suspension. Units: liters, cu.cm, cu.m, cu.ft or cu.in.

Cms The mechanical compliance of the driver's suspension. Units: μm/N, mm/N, cm/N, m/N or in/lb.

Mms The mass of the diaphragm or piston including the air load. Units: g, kg or oz.

Rms The mechanical resistance of the driver resulting from its suspension losses. Units: kg/s, lb/s or mohms.

Xmax The maximum linear excursion of the driver. It is measured in one direction from rest. Units: mm, cm, m or in. (P-P or peak-to-peak values must be halved.)

Xmech The maximum mechanical excursion of the driver. It is measured in one direction from rest: Units: mm, cm, m or in. (P-P or peak-to-peak values must be halved.)

Dia The diaphragm or piston diameter of the driver. It is usually measured from the middle of the surround as shown below. Units: mm, cm, m or in.

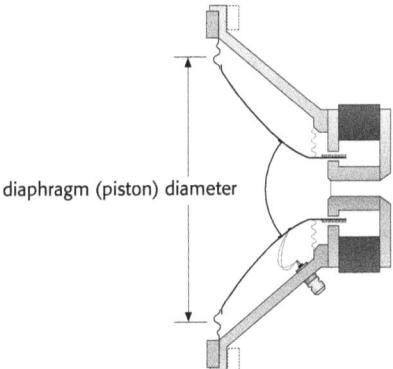

Sd The diaphragm or piston area of the driver. Units: sq.mm, sq.cm, sq.m or sq.in.

Vd The displacement volume of the diaphragm or piston at Xmax. Units: liters, cu.cm, cu.m, cu.ft or cu.in.

Reference: 3 Driver Properties

Electrical Parameters

Qes The electrical resonance magnification of the driver at Fs.

Re The DC voice coil resistance of the driver. Units: ohms.

Le An inductance value that is approximately equivalent to the upper frequency inductive reactance of the driver's voice coil. Units: mH.

Z The nominal impedance of the voice coil of the driver. Units: ohms.

BL The motor strength of the driver. Units: Tm, N/A, Tft or lb/A.

Pe The maximum electrical input power that the driver can handle. Units: watts.

Electromechanical Parameters

Qts The total resonance magnification of the driver at Fs.

ηo (eta zero) The half-space reference efficiency of the driver. Units: percent.

1-W SPL The sensitivity of the driver at 1 meter (3.3 feet) with 1 watt input. Units: dB.

2.8-V SPL The sensitivity of the driver at 1 meter (3.3 feet) with 2.83 volt input. Units: dB.

The units of many of the parameters can be easily changed. Simply click on the unit label as shown below. The units will advance each time their label is clicked.

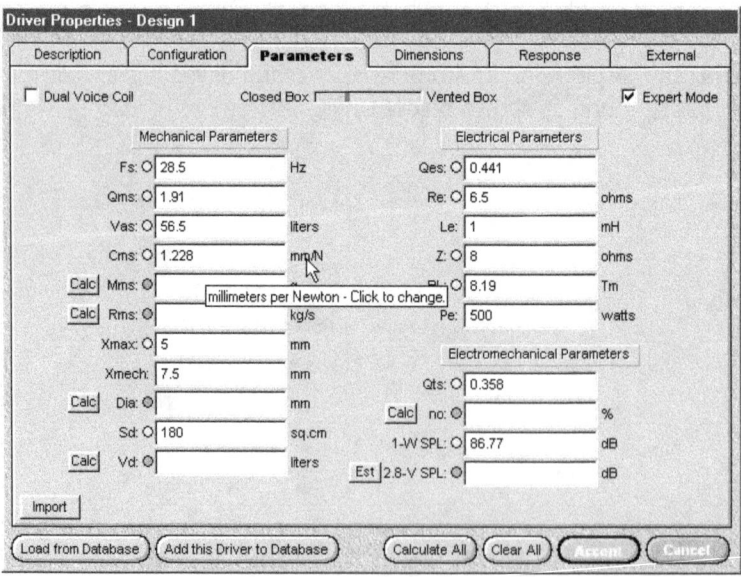

Reference: 3 Driver Properties **BassBox 6 Pro**

Dimensions

The "Dimensions" tab contains the shape and mounting dimensions of the driver.

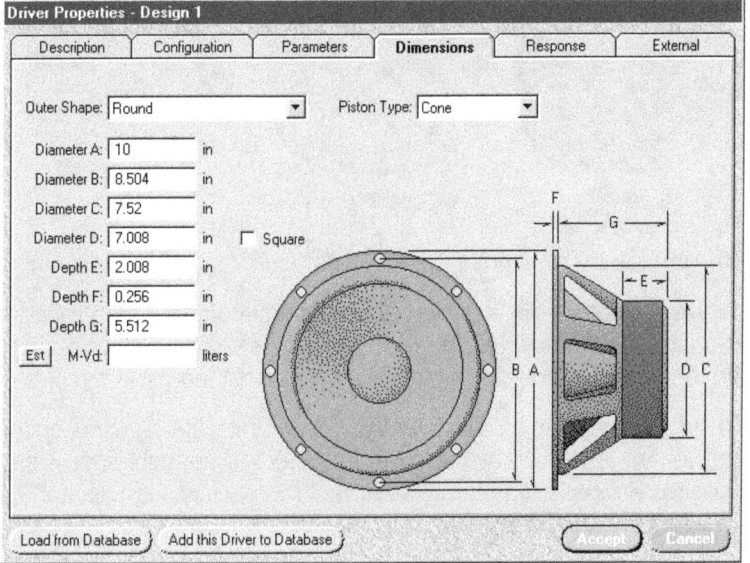

Outer Shape Several driver shapes are available as shown below:

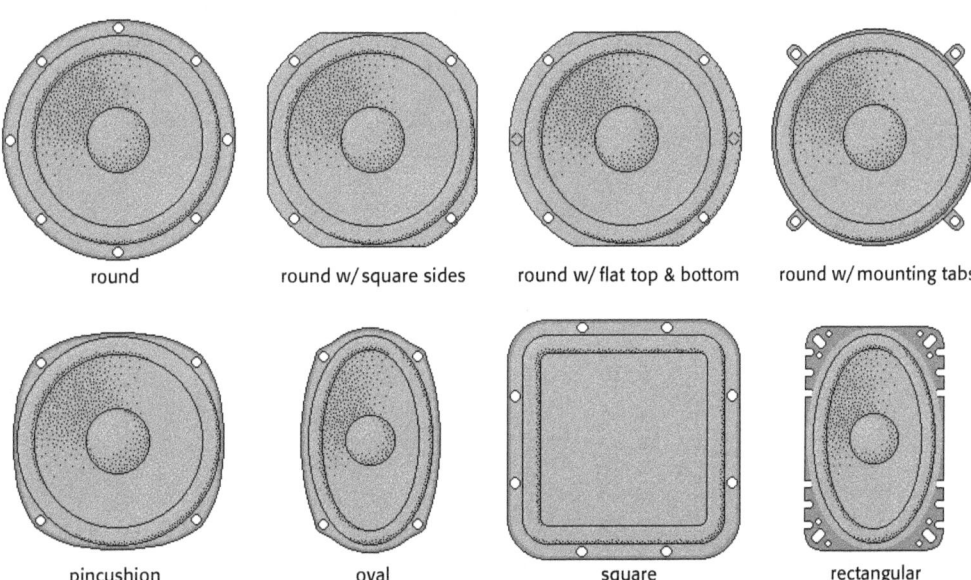

Piston Type Several piston types are available: cone, planar, concave and convex.

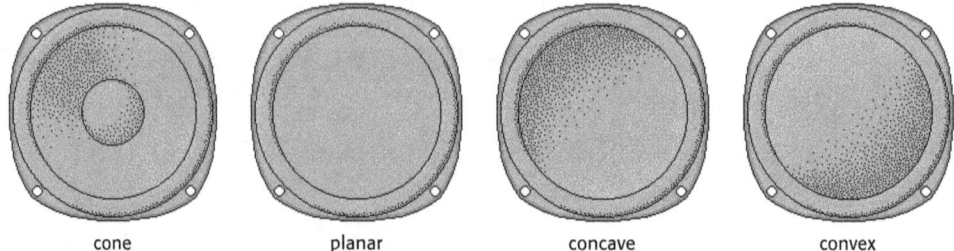

Dimensions The various mounting dimensions of the driver. Units: mm, cm, m or in.

Square Occasionally a driver will have a square rather than round magnet. Check the "Square" checkbox when this is so. When it is checked, dimension "D" will be assumed to be the length of one side of the magnet rather than the diameter.

M-Vd The mounting volume of the driver. Units: liters, cu.cm, cu.m, cu.ft or cu.in. This is the volume displaced by the driver in the box. If enough dimensions have been entered, BassBox Pro can estimate this volume. When this is possible, an "Est" button will appear beside the "M-Vd" label.

Note: The mounting dimensions of a driver do not have to be entered in order to complete a speaker design. However, it is best to enter them because BassBox Pro can check them against the box dimensions to make sure that the driver will fit in the box. If the driver dimensions are available, and the driver is too big for the box, a warning message will appear and relevant parameters will turn to a red color.

Response

The "Response" tab contains the normalized acoustic response of the driver (it must be normalized to the predicted response of the driver). It is very important that the acoustic data entered for the driver not include the response of a test box or baffle. The acoustic data should include only the +/– variations which occur when the measured response differs from the predicted Thiele-Small response.

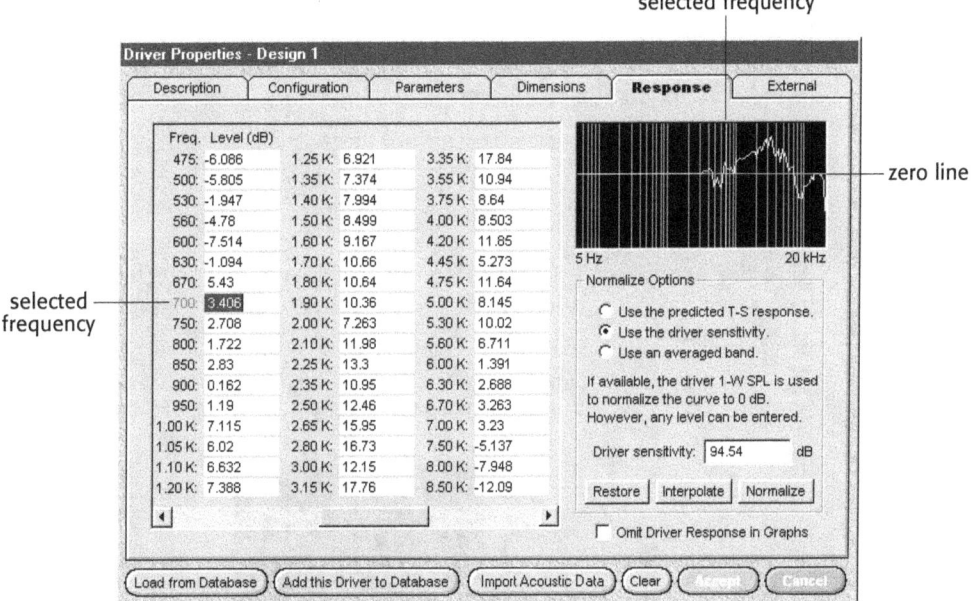

Why enter the driver's acoustic response? Because it can improve the accuracy of several performance graphs by showing how acoustic anomalies will affect the overall system.

There are two ways to enter the acoustic response. It can be manually entered using the individual "Level" input boxes for each data point or it can be imported from one of several measurement systems.

The preview graph in the upper right corner shows the shape of the response while you edit or import data. Notice also that the preview graph includes a "zero line" to help you adjust the overall level of the data so that it is normalized. A red vertical line is also provided to show what part of the response will be changed if the input box of the selected frequency is changed.

Level The normalized amplitude response in dB for each acoustic data point. The response of the driver can be manually entered or edited with the "Level" input boxes. There are 134 data points from 5 Hz to 20 kHz. Use the horizontal scroll bar to access the "Level" input boxes for all of the data points. *Note: You do not have to enter a value for every acoustic data point. You can leave some blank and then use the "Interpolate" button later to estimate the missing values.*

Normalize Options The acoustical data must be normalized to the predicted response of the driver. (Zero dB represents the predicted response of the driver as defined by its Thiele-Small parameters and, if present, its box parameters.) This usually involves two-steps:

1. Adjust the overall level of the acoustic data so that the flat region of the response curve is level with the zero line in the preview graph. This can be accomplished with the second, third or fourth normalization options listed on the following pages.
2. Subtract the predicted response from the acoustic data. This can be accomplished with the first normalization option listed below.

Each normalization option is described next:

Use the predicted T-S response When this option is selected, the predicted response of the driver will be added to the preview graph with an orange plot line as shown below.

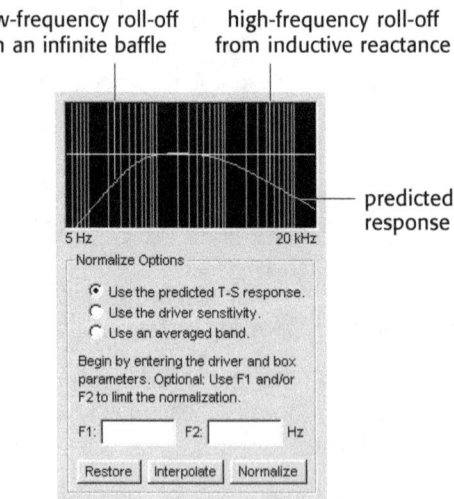

The predicted response should approximately match the shape of the acoustic response curve as shown in the left example in the illustration on the next page. If it does not, then one of two problems exits: 1) either there is an error in the driver or box parameters, or 2) there is a problem with the acoustic data.

The predicted response includes the box response if box parameters are entered. Otherwise it represents the driver in an infinite baffle. The predicted response can be changed by changing the driver parameters on the "Parameters" tab or by changing the box parameters on the "Box Design" tab of the Box Properties window. **Important:** The box parameters should duplicate the test box used when the driver was measured. *Note: An infinite baffle can be simulated by selecting a closed box and using a huge box volume (Vb) and huge QL value.* If necessary, remember to restore the driver and box parameters after normalizing the acoustic response.

When the "Normalize" button is clicked, the difference between the predicted response and the measured acoustic response will be calculated. This "difference" will be entered as the "normalized" acoustic response. The illustration below shows the same acoustic data before and after normalization:

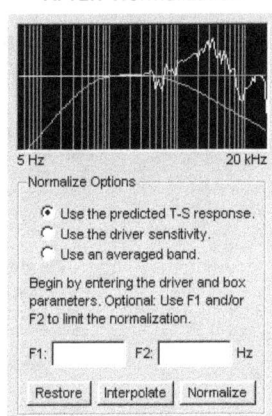

If desired, this method of normalization can be limited with F1 and F2. Use F1 to prevent low-frequency data from being normalized. Use F2 to prevent high-frequency data from being normalized. For example, enter "500" into F1 if you do NOT want acoustic data below 500 Hz to be normalized. Enter "7000" into F2 if you do NOT want acoustic data above 7 kHz to be normalized. Either or both F1 and F2 can be left blank if you do not want to limit normalization.

Tip: Do not leave the "Use the predicted T-S response" option selected after you have finished normalizing the acoustic data because it can slow the operation of the program on some computers. This is because BassBox Pro recalculates the predicted response every time a change is made to a driver or box parameter. Instead, select either the second or third normalization option before you leave the "Response" tab after you have finished preparing the acoustic data.

Use the driver sensitivity This normalization option does not change the shape of the acoustic response like the first option. Instead, it adjusts the overall level of the acoustic data. For example, an acoustic measurement may represent a 1-watt, 1-meter sound pressure level. In cases like this, the acoustic response may not be visible in the preview graph because the acoustic level is too high. For example, the flat portion of the response may be between 85 and 95 dB. Use this normalization option to adjust the level of the acoustic data so that it is level with the zero line in the preview graph. This is best done before normalizing the data to the predicted response.

Depending on the nature of the acoustic data, it can sometimes be normalized to the zero line with the driver sensitivity rating. However, any level can be used—even a negative level if the acoustic data is below the zero line. Let's examine an arbitrary example:

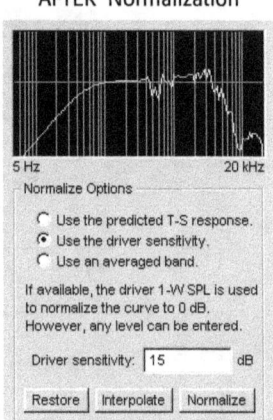

In the above example, the flat region of the acoustic data is about 15 dB above the zero line as shown on the left. To correct this, replace the "Driver sensitivity" value with "15" and click the "Normalize" button to subtract 15 from each of the acoustic data points. The result is shown above in the example on the right.

Use an averaged band This normalization option does not change the shape of the acoustic response like the first option. Instead, it adjusts the overall level of the acoustic data like the second option. It does this by subtracting the average level of a selected frequency band from the acoustic data. The start and end of the frequency band is entered into F1 and F2 as shown below:

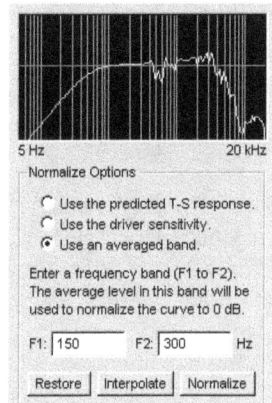

In the above example, F1 = 150 Hz and F2 = 300 Hz so the acoustic data from 150 to 300 Hz will be averaged and this level will be subtracted from each acoustic data point when the "Normalize" button is clicked.

Use a single level (not shown) The final option for normalizing the acoustic data is not listed in the "Normalize Options" box. Like the second and third options, it adjusts the overall level of the acoustic data without changing the shape of the response. It is executed by double-clicking ( ) on any of the "Level" input boxes. When you do this, a message box will appear to confirm that you want to normalize the acoustic data by subtracting the selected level from each acoustic data point.

Restore The last three acoustic response curves are remembered so that you can "undo" changes when you attempt to normalize the acoustic data. Click the "Restore" button to return to a previous response curve. Clicking the "Restore" button again after the oldest curve has been restored will return you to the most recently stored curve. An acoustic response curve is stored into this "undo" buffer when either the "Interpolate" or "Normalize" button is clicked. The "undo" buffer is not affected by changes to individual acoustic data points with the "Level" input boxes.

Interpolate Clicking the "Interpolate" button will cause BassBox Pro to search the acoustic data for all zero values. It will then replace the zero values with estimated values by calculating the slope of surrounding non-zero data points.

Normalize Click the "Normalize" button to execute the selected normalization option. These options were described on the previous pages and usually cause the level of the acoustic data to be adjusted to either a predicted response or to the zero line. *Note: The "Normalize" button has one special feature when the "Use the predicted T-S response" option is selected. This feature is activated by holding down the ⇧Shift key when the "Normalize" button is clicked to cause the predicted response to be added to the acoustic data rather than subtracted from it as is normal.*

Omit Driver Response in Graphs Check this checkbox to cause the graphs to ignore the driver acoustic data. This can also be controlled with a checkbox in the Graph Properties window and directly from the graph window by right-clicking (🖱) on the graph and turning off "Include > Driver Acoustic Response" in the popup menu.

Import Acoustic Data Frequency-domain data from several popular measurement systems (including Brüel & Kjaer, CLIO, IMP, LMS, MLSSA, OmniMic, Sample Champion, Smaart, TEF®-20 and TrueRTA) can be imported with the "Import Acoustic Data" button. Clicking it will open the Import Acoustic Response File window so you can select the data file. Several of the file types are identified by the file name extension. Use the "Files of type" drop-down list to select the file type as shown below:

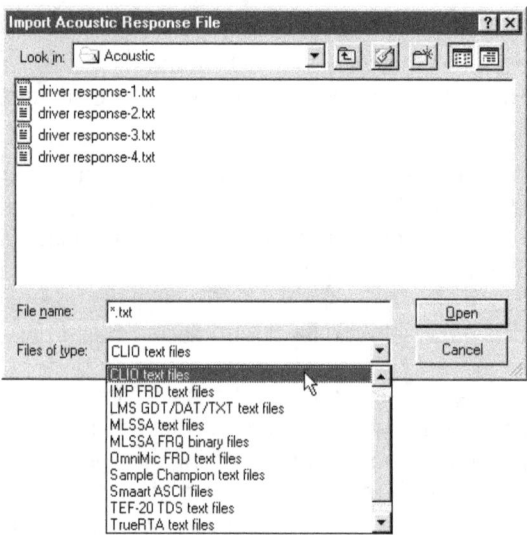

Note: The above Import Acoustic Response File window will vary in appearance based on the version of Windows. For example, the "Files of type" list may not be labelled and may be located to the right of the "File name" input box in Windows 8, 7 and Vista. It is important that the file type be selected with this list so it's type will be correctly identified.

The program will open a Select Data File Type window to prompt the user to select the file type if a file was chosen which the user forgot to identify by the "Files of type" list and BassBox Pro cannot identify it by its file name extension.

The acoustic data import filter works in the following manner:

- The import filter will interpolate between data points if some are missing.
- The import filter will average data points if there are extra ones.
- The import filter will give you the option of extrapolating the beginning or end of the data if it begins at a frequency above 5 Hz or ends below 20 kHz. The slope between the first or last five data points will be used if you choose to extrapolate.
- The import filter for some file types will automatically adjust the overall level of the data so that it will be visible in the preview graph.

Note: In most cases, the data will need to be normalized after it has been imported. Please read the preceding instructions about normalization.

Caution: Most acoustic measurements do not include accurate data at very low frequencies. For example, many measurements are grossly inaccurate below 100 Hz. The lowest usable frequency of the measurement will depend on the measurement's frequency resolution, "time window" and the location of nearby reflecting surfaces. These problems can sometimes be overcome with special measurement techniques like the "near field" and "ground plane" techniques. If the low-frequency portion of your acoustic data is not accurate, you should enter zeros for that portion of the acoustic data after the rest of the data has been normalized.

Clear Use this button to set all "Level" settings to zero dB.

Note: The acoustic response of a driver does not have to be entered in order to complete a speaker design. Most manufactures do not provide it, leaving it up to you, the designer, to measure it yourself.

External

The "External" tab contains information about the resistance of external wiring or a passive network.

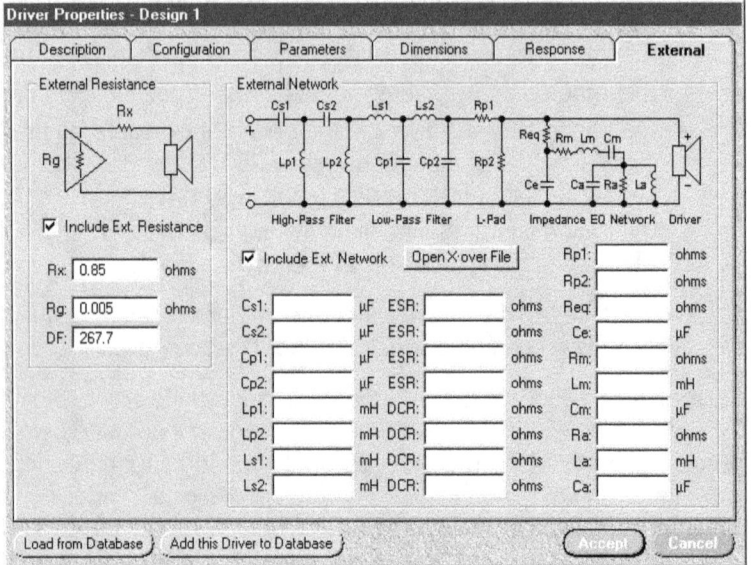

Include Ext. Resistance External resistance can have a dramatic effect on the response of a speaker design. This option allows you to include these effects in the design. When they are included, a "net" value for Qes, BL and Qts will be displayed in the "Parameters" tab. They show how the external resistance affects these parameters. *Note: The "Include Ext. Resistance" checkbox will be disabled when Rx and Rg are zero.*

Rx The series resistance of the speaker cables and terminals. It does not include the resistance of the crossover network.

Rg The source resistance of the amplifier. This is usually very low with most modern amplifiers. If the driver's impedance (Z) is known, DF is calculated when Rg is entered.

DF The damping factor of the amplifier. If the driver's impedance (Z) is known, Rg is calculated when DF is entered.

Include Ext. Network The effects of an external passive network can be included in the design. This enables the graphs to show the total system response with a passive crossover network, impedance equalization circuit and/or L-pad. The inclusion of the external network response in the graphs can also be controlled with a checkbox in the Graph Properties window and directly from the graph window by right-clicking (🖱) on the graph and using the popup menu.

The circuit diagram above shows the position of each component in the circuit. The crossover network section can accommodate a 1st-4th order high-pass, band-pass or low-pass filter.

Open X•over File BassBox Pro is not a circuit designer—values for the network components cannot be calculated for you. However, our companion program X•over Pro is a filter designer and the "Open X•over File" button allows you to import a filter or network design from it. Clicking this button will open a dialog box to prompt you to select an X•over Pro design file. These files end in the file name extension "xo3". *Note: If it is installed, X•over Pro can be launched from the Tools menu of the BassBox Pro main window (see page 328).*

Cs1, Cs2, Cp1, Cp2 These are the series and parallel capacitors in a passive crossover network or filter. Units: mF, µF and pF. (Click on the units label to change units.)

ESR The equivalent series resistance is required for all capacitors in the crossover network or filter. The ESR is not required for the impedance equalization circuit. Units: ohms.

Lp1, Lp2, Ls1, Ls2 These are the parallel and series inductors or coils in a passive crossover network or filter. Units: mH.

DCR The DC resistance is required for all inductors in the crossover network or filter. The DCR is not required for the impedance equalization circuit. Units: ohms.

Rp1, Rp2 The series (Rp1) and parallel (Rp2) resistors of an L-pad. Units: ohms.

Req, Ce, Rm, Lm, Cm, Ra, La, Ca The components of an impedance equalization circuit. Such circuits are used to make the impedance of the driver appear "flat" to the crossover network so it can function properly. The impedance equalization network included with BassBox Pro can accommodate a wide variety of designs and our X•over Pro program can design them. Req and Ce are used to help equalize the inductive reactance of the voice coil. Rm, Lm and Cm are used to help equalize the mechanical resonance of the system. Ra, La and Ca are used to help equalize the acoustic resonance of the system.

4 Box Properties

The Box Properties window holds the information about the box. To open it, click on the "Box" button of the desired design panel in the main window ([Ctrl]+[B]). A tab sheet layout is used to help organize the information.

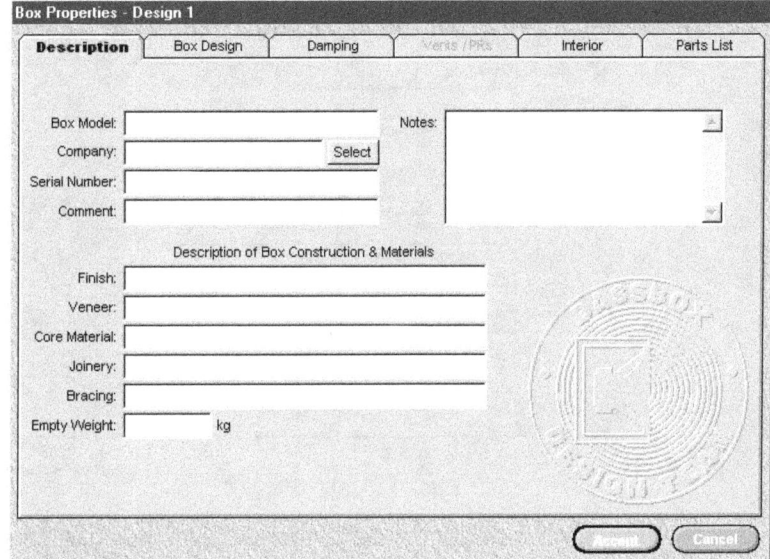

The box information can come from one of two different sources:

1. **Manual Entry** Box information can be entered manually by the user. This is a convenient way to enter the dimensions or volume of an existing box. The dimensions of an existing vent or port can also be entered and, in many cases, the box tuning can be calculated from them.

2. **Calculated Parameters** BassBox Pro can "suggest" a box volume for many drivers. If a vented, bandpass or passive radiator box is selected, BassBox Pro can also suggest vent tuning or passive radiator parameters. A "suggested" box often provides a good starting point for speaker design.

When the Box Properties window is first opened, it will have a "Close" button in the bottom right corner. After changes have been made, the "Close" button will be replaced with an "Accept" button and a "Cancel" button (shown above). Most of the windows in BassBox Pro work this way to help you realize the consequences of closing a window.

Next, let's examine each of the tabs of the Box Properties window.

Description

The "Description" tab contains optional descriptive information about the box.

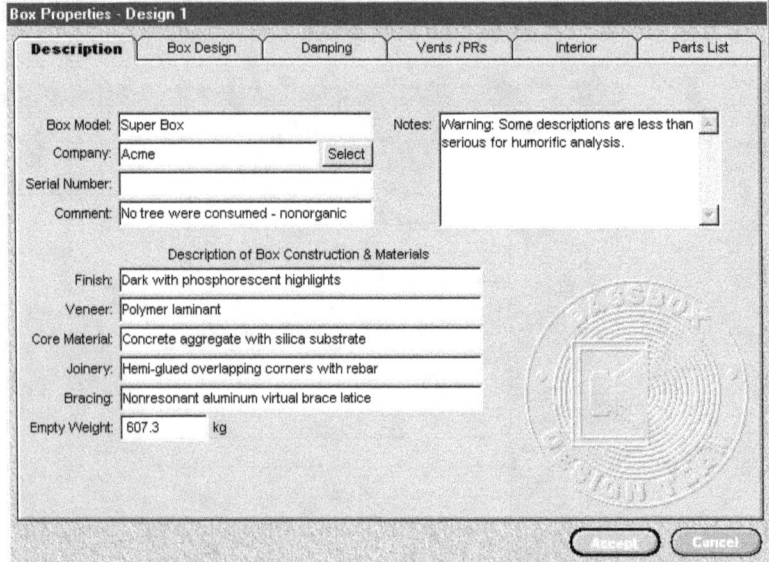

Box Model The model name of the box can be up to 30 characters long.

Company The company name can be up to 40 characters in length and there are two ways to enter it. The preferred way is to use the "Select" button and choose a company name from the driver database. The second way to enter the name is to type a name directly into the company input box.

Serial Number The serial number can be up to 30 characters long.

Comment The comment can be up to 50 characters long. It is a single-line note that will be displayed in the design properties list of each speaker design in the main window.

Finish, Veneer, Core Material, Joinery, Bracing These can each be up to 50 characters long and provide specific descriptions about the box, its construction and finish.

Empty Weight The weight of the box without the driver(s). Units: g, kg, oz and lb.

Notes This is a multi-line note that is not limited in length and can contain more lengthy comments about the box.

Note: The box description does not have to be entered in order to complete a speaker design. However, it can make a design much more understandable to others.

Reference: 4 Box Properties   BassBox 6 Pro

Box Design

The heart of the Box Properties window is the "Box Design" tab. It contains the main parameters for the box such as the type of box, its volume, shape and dimensions.

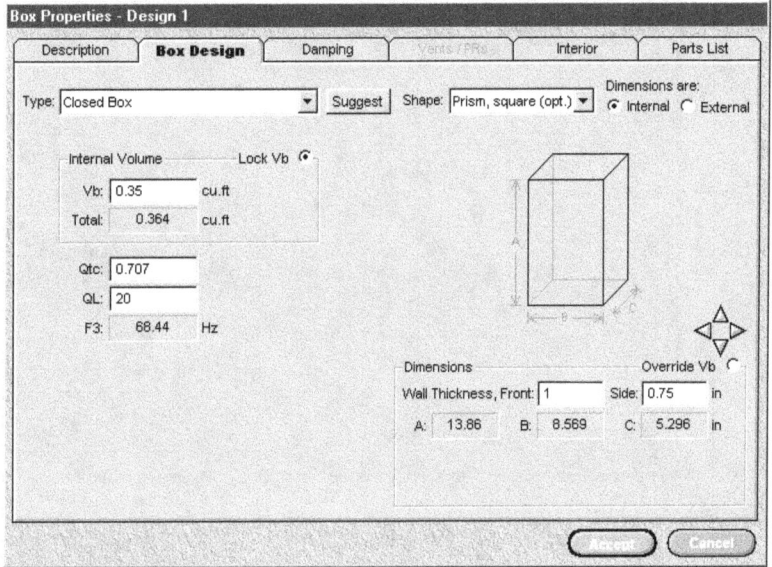

Seven box types are available from the "Type" drop-down list:

- closed box
- vented box
- vented box with an active high-pass equalization filter
- single-tuned bandpass box
- parallel double-tuned bandpass box
- series double-tuned bandpass box
- passive radiator box

The characteristics of each box type were discussed in Chapter 3 ("How to Choose a Box") of the *Box Designer's Guide* earlier in this manual. Please refer to it for a detailed description of each box type along with their strengths and weaknesses.

The following series of illustrations show how the box parameters in the left side of the "Box Design" tab are configured differently for each box type.

closed box

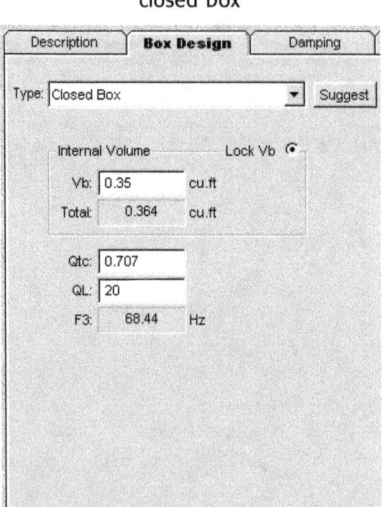

vented box

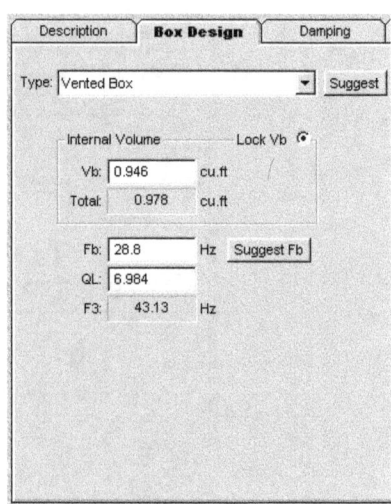

vented box w/ active high-pass filter

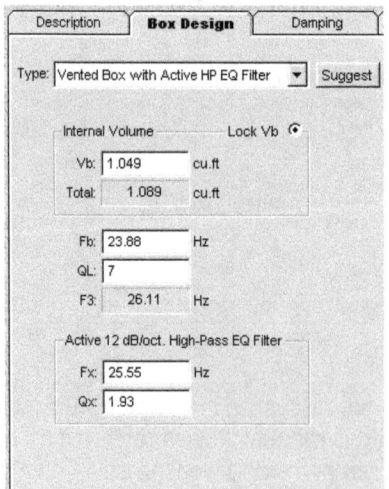

single-tuned bandpass box

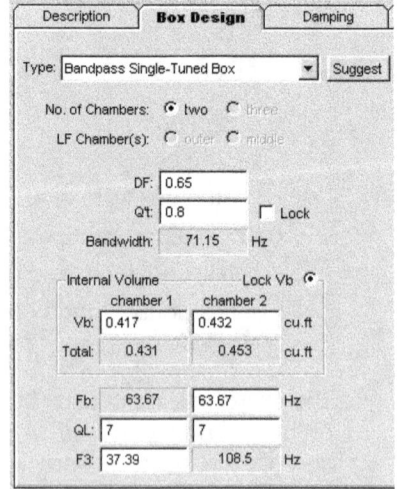

Reference: 4 Box Properties BassBox 6 Pro

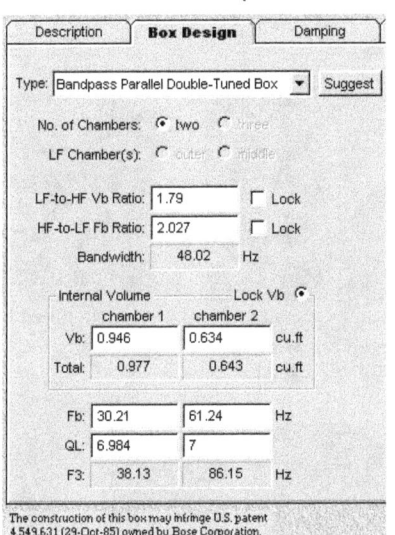

double-tuned bandpass box

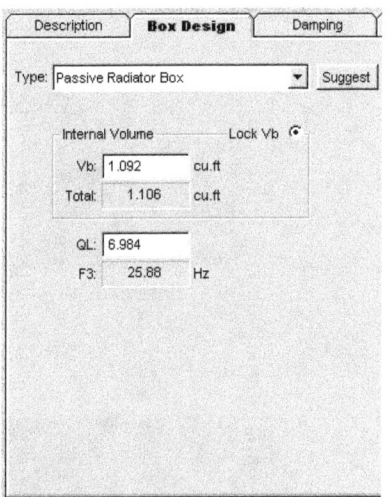

passive radiator box

Notice that some parameters have white input boxes and others have grey ones. A grey input box indicates a value that was calculated rather than entered. Also notice that the closed, vented and passive radiator box types have only a single column of parameters while many of the parameters of the bandpass box types have two sets of parameters, one set for each chamber in the box. We'll see later that bandpass boxes with three chambers have three sets of parameters.

After the box type has been selected, the internal box volume (Vb) or chamber volumes must be calculated or entered. There are three ways to do this:

- The "optimum" internal volume can be calculated with the "Suggest" button.
- Any desired internal volume can be entered manually with the Vb input boxes.
- The box dimensions can be entered with the Dimensions "Override Vb" option turned on and the Internal Volume "Lock Vb" option turned off. This will cause Vb to be calculated from the box dimensions.

All possible options and inputs in the left side of the "Box Design" tab are described next:

Type A drop-down list which selects the box type. The choices are: closed box, vented box, vented box with active high-pass equalization filter, single-tuned bandpass box, parallel double-tuned bandpass box, series double-tuned bandpass box and passive radiator box.

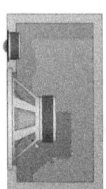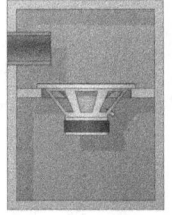

closed box   vented box and vented box with an active HP EQ filter   single-tuned bandpass box

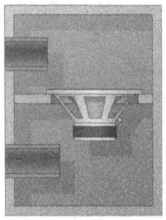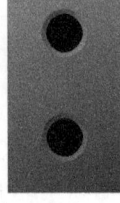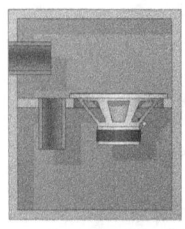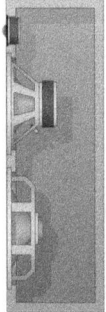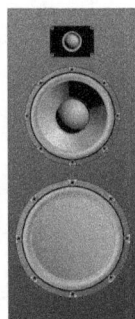

parallel double-tuned bandpass box   series double-tuned bandpass box   passive radiator box

Reference: 4 Box Properties   BassBox 6 Pro

Suggest Use the "Suggest" button when you want BassBox Pro to suggest a box volume and, if appropriate, box tuning setting(s). Its function varies slightly depending on the type of box that is selected as explained below.

What makes a speaker "optimal" can vary widely depending on its intended use. One box size rarely "fits" all applications. For critical listening, a speaker usually needs to have a "maximally flat" response and a larger box size can be tolerated. For live music creation, a speaker usually needs to have a high output and a small box size is usually desired for portability.

Closed boxes and standard Vented boxes

If a closed box or standard vented box is selected, the "Suggest" button will cause a Design Priority window to open. It enables you to control the way that a "suggested" box is calculated.

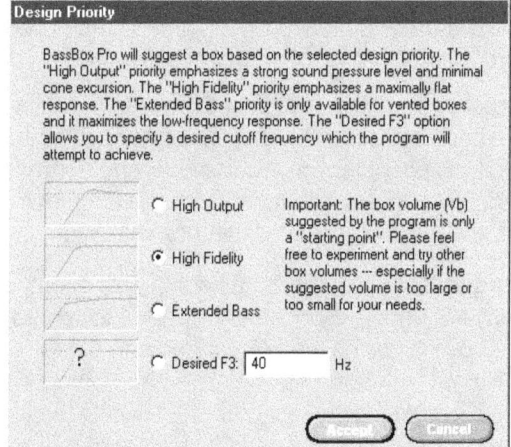

The Design Priority window shown above is configured for a vented box. (The "Extended Bass" option is disabled for closed boxes.)

Four design "priority" options are provided:

High Output This priority is usually desired for live music creation and live sound reinforcement applications. It often produces smaller box sizes than the "High Fidelity" option. Closed boxes will typically have a total system Q (Qtc) of approximately 1.5 and a 4.5 dB peak at the box resonance. Vented boxes will typically have a 3 dB peak at the box resonance. But these are generalizations only—the actual response will depend on the driver parameters.

Since the "High Output" option results in a smaller box size with a higher total system Q, it will also produce less diaphragm or cone excursion. This will enable the driver to handle more power before reaching its excursion limit (Xmax). However, this reduced excursion comes at a price—more rapid changes in phase response and group delay—so the speaker will suffer from decreased transient response. Fortunately, this problem is subtle and not usually a problem for live sound applications where the listening environment can overshadow these subtleties.

High Fidelity This priority is usually best for critical listening applications such as home hi-fi, recording studios and reference monitors. It attempts to produce a speaker with a "maximally flat" response which often results in box sizes that are larger than the "High Output" option. Closed boxes will typically have a total system Q (Qtc) of 0.707. However, you can set the default Qtc value for this option with the "Box" tab of the Preferences window. Vented boxes will have a B4 (4th-order Butterworth) alignment if the driver's Qts is close to 0.4. Drivers with lower Qts values will tend toward a QB3 (quasi 3rd-order Butterworth) alignment and drivers with higher Qts values will tend toward a C4 (4th-order Chebychev) alignment.

The goal of the "High Fidelity" option is to produce a speaker that creates sound waves with a uniform sensitivity versus frequency so that audio signals are reproduced as faithfully as possible. The required box size is strongly influenced by the total Q of the driver (Qts). Drivers with low Qts values will require small box volumes and drivers with high Qts values will require large box volumes.

Extended Bass (vented boxes only) This priority is useful when you want to "push" the low-frequency response of a speaker as low as possible. The bass response is extended by increasing the box volume and tuning it to a slightly lower frequency than the "High Fidelity" option. The increase in box size will depend on the total Q of the driver (Qts) with lower Qts values producing the largest increases. For example, a driver with a Qts of 0.3 will see a twofold increase in box volume compared to the "High Fidelity" option. While a driver with a Qts of 0.5 will see an increase in box volume of only 1.4 times.

The goal of the "Extended Bass" option is to extend the bass response with a minimal increase in ripple. "Ripples" are the small dips and peaks that sometimes occur in the amplitude response when a vented box is tuned below its "maximally flat" cutoff point. The amount of ripple and the amount of bass extension will depend on the driver parameters. Generally, the ripple will increase and the bass extension will decrease as the driver Qts increases. **Caution:** A disadvantage of an "Extended Bass" design is increased diaphragm or cone excursion. Be sure that the driver can handled the increased excursion.

Desired F3 This priority allows the user to specify the −3 dB cutoff frequency. For example, enter 40 Hz if you want the program to suggest a box whose response drops to

–3 dB at 40 Hz. (Many live sound reinforcement speakers are designed to cutoff at 40 to 50 Hz in order to prevent low-frequency feedback problems.) When this option is used, BassBox Pro will begin with the smallest possible box size and will slowly increase it until the desired F3 is achieved. This may take a few minutes. The F3 value will change while the program tries each increment in box size so that you can watch its progress. A warning message will be displayed if the program is unable to design a box with the desired F3, either because the F3 value is too high or too low.

Note: BassBox Pro calculates F3 by tracking the system response in a descending frequency direction until the first –3 dB level is found. This may or may not be below the "knee" of the response curve. This is a little different than other methods which calculate F3 as the –3 dB point relative to the "knee" of the response. BassBox Pro uses the method it does because it shows the absolute –3 dB frequency rather than a relative one. To find the –3 dB frequency below the "knee" of the response, use the cursor in the Normalized Amplitude Response graph.

Important: There are occasions when the compliance (Cms and Vas) and/or total Q (Qts) of a driver are relatively low and the "suggested" box would be too small for the physical size of the driver. When this happens, BassBox Pro will automatically increase the box size so that the driver will fit in it.

Vented boxes with active high-pass equalization filters, Bandpass boxes and Passive Radiator boxes

If a vented box with an active high-pass equalization filter, a bandpass box or a passive radiator box is selected, the "Suggest" button will cause the program to immediately attempt to design a box with a "maximally flat" amplitude response for high fidelity.

The program calculates the maximally flat box from the driver parameters and whether or not the response is truly flat will depend on these parameters. However, there are occasions when the compliance (Cms and Vas) and/or total Q (Qts) of a driver are relatively low and a maximally flat box would be too small for the physical size of the driver. When this happens, BassBox Pro will automatically increase the size of vented and passive radiator boxes so that the driver will fit in them. However, passive radiators, themselves, will not be checked because their dimensions are not entered into the program. You will need to manually check the dimensions of all bandpass boxes to make sure that the suggested chamber volumes are large enough for their drivers.

Is a maximally flat response the best? The answer depends on the intended purpose of the speaker. It is often considered to be the best for high fidelity playback because the speaker should reproduce the original audio signal as accurately as possible. However, for a musical instrument speaker, the maximally flat response is probably not the best. Often musical instrument speakers have non-flat response curves whose shape depends on the sound quality and power handling that is desired by the musician. A maximally flat response may also produce a narrow bandwidth for a bandpass box. Often a little experimentation with the box or chamber volumes (Vb) and tuning frequencies (Fb) will allow you to achieve a wider bandwidth. However, as the bandpass bandwidth increases, so does the ripple in the passband region of the response and the efficiency decreases.

No. of Chambers (bandpass boxes only) The total number of chambers in the box. A bandpass box with a single driver can have only two chambers. It is possible to have three chambers (shown below) when more than one driver is specified in the "Configuration" tab of the Driver Properties window. When three chambers are possible, this option will be activated. *Note: Chamber 3 of a triple chamber box is always identical to Chamber 1.*

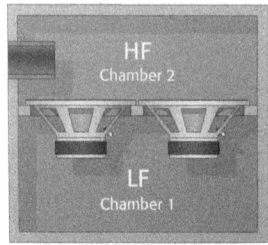

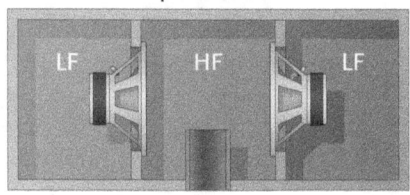

LF Chamber(s) (bandpass boxes only) Identifies which chamber(s) will control the lower frequency (LF) limit of the design. This option is only available when the "No. of Chambers" option above is set to three. With a triple-chamber bandpass box, the outer two chambers are normally used to control the low frequency limit of the design. The middle chamber is used to control the upper frequency limit. This is easily visible in the illustration below because it is a single-tuned bandpass box and the low frequency chambers do not have vents.

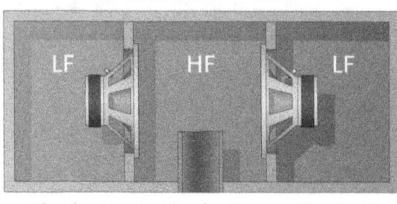

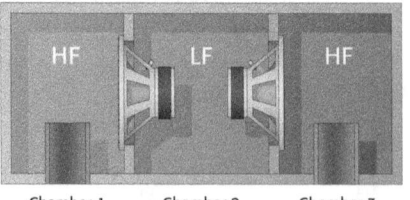

The "LF Chambers" option allows you to reverse this, causing the outer chambers to control the upper frequency limit and the middle chamber to control the lower frequency limit.

Why would you ever want to do this? Probably not very often for a single-tuned bandpass box. However, with a double-tuned bandpass box, all chambers have a vent. If the low-frequency vents are too long and don't fit in the smaller outer chambers, it sometimes helps to swap the chambers.

Note: With two-chamber bandpass boxes, Chamber 1 is always considered the low-frequency chamber and Chamber 2 is always considered the upper-frequency chamber.

DF (single-tuned bandpass boxes only) The system damping factor. Typical values range from 0.7 to 0.4. Small DF values produce a wider bandwidth, lower low-frequency cutoff, poorer transient response and more passband ripple. Large DF values produce a narrower bandwidth, higher low-frequency cutoff, better transient response and less passband ripple.

Q't (Q prime T, single-tuned bandpass boxes only) The total Q of the driver suspension with the load of the rear chamber. The value of Q't (pronounced "Q-prime-T") must be larger than the value of the Qts of the driver. Typical values range from 1.0 to 0.4. Small Q't values produce a wider bandwidth, lower low-frequency cutoff and less efficiency. Large Q't values produce a narrower bandwidth, higher low-frequency cutoff and greater efficiency. The value of Q't can be locked.

LF-to-HF Vb Ratio (double-tuned bandpass boxes only) It is sometimes helpful to specify a volume ratio between the low-frequency to upper-frequency chambers when experimenting with a bandpass design. When the LF-to-HF Vb Ratio value is locked by checking the "Lock" checkbox next to it, the program will automatically calculate the Vb values of other chambers whenever one of them is changed.

HF-to-LF Fb Ratio (double-tuned bandpass boxes only) It is sometimes helpful to specify a tuning ratio between the upper-frequency to low-frequency chambers when experimenting with a bandpass design. When the HF-to-LF Fb Ratio value is locked by checking the "Lock" checkbox next to it, the program will automatically calculate the Fb values of other chambers whenever one of them is changed.

Lock (Q't or Ratio Lock control—bandpass boxes only) When checked, the "Lock" checkbox prevents the associated parameter next to it from changing when other related parameters are changed. This has the effect of forcing certain other parameters to be recalculated.

Bandwidth (bandpass boxes only) The width of the passband. It is calculated by subtracting the –3 dB low-frequency cutoff point from the –3 dB upper-frequency cutoff point.

Vb The net internal volume of the box or chamber. This is the volume that remains after subtracting the volume inside the box that is displaced by internal objects like the driver, vent and bracing. Units: liters, cu.cm, cu.m, cu.ft or cu.in.

Total The total internal volume of the empty box or chamber. It is calculating by adding to Vb all the volume that was displaced by all internal objects. The total internal volume is the volume used for all box dimension calculations. Units: liters, cu.cm, cu.m, cu.ft or cu.in.

Lock Vb This radio-style button controls whether or not the net internal box/chamber volume (Vb) will be allowed to grow or shrink as the box dimensions change. Vb should be "locked" when you manually enter its value or whenever you calculate its value with the "Suggest" button. **Locking Vb automatically unlocks the "Override Vb" control and allows unlocked box dimensions to change whenever Vb changes.** The unlocked box dimensions will be automatically recalculated whenever Vb changes.

What happens when you manually change Vb?
Because Vb is "locked" when it is manually changed, one or more "unlocked" box dimensions will be recalculated and the vent(s), if present, will be adjusted. For example, suppose that you have a square prism closed box whose internal height x width x depth is 24 x 12 x 18 inches with a Vb of 3 cubic feet. If you manually change Vb to 2 cubic feet and the box depth is an unlocked dimension, it will be recalculated to equal 12 inches. This will result in a box whose internal dimensions are 24 x 12 x 12 inches. *Note: Individual box dimensions can be locked and unlocked by clicking on the label to the left of each box dimension.*

Attempting to manually edit Vb when it is not "locked" will result in its being immediately "locked" so that the box dimensions will not affect the value you enter. (The "Override Vb" control in the Dimensions frame will also be "unlocked", allowing the unlocked box dimensions to change according to the value of Vb.)

Qtc (closed boxes only) The system total Q. Usually Qtc should not be less than the total Q (Qts) of the driver. Many designers consider the optimum value for Qtc to be 0.707. The value for Qtc that is used for the maximally flat closed box calculations can be set with the "Box" tab of the Preferences window. Some commonly used Qtc values are listed below along with a description:

   Qtc = 0.5   produces a "Critically Damped" closed-box response.
   Qtc = 0.577 produces a "Bessel" or D2 closed-box response.
   Qtc = 0.707 produces a "Butterworth" or B2 closed-box response.

The Qtc value is inversely proportional to the system damping. In other words, as the Qtc value is reduced, the system damping increases. Closed boxes with a lower Qtc value produce a "tighter" sound because they are damped more. You might conclude from this that a low Qtc (such as the critically damped 0.5 value) would often be preferred. However, this is not the case. To some listeners a low Qtc sounds a bit "thin" and they prefer a higher value—even as high as 1.0 or higher. Although Qtc values above 0.707 will produce a response bump at the system resonance frequency, the increased level produces a "warmer" sound which some prefer.

Fb (vented and bandpass boxes only) The box resonance or tuning frequency. It is controlled by both the internal box volume (Vb) and, if a vent is present, the vent dimensions. If the vent dimensions are locked, increasing Vb or increasing the length of the vent (Lv) will lower Fb and visa versa. In most cases, the driver parameters do not affect Fb. Units: Hz.

Note: The Fb of a sealed low-frequency chamber in a single-tuned bandpass box is the system resonance of the driver and chamber since a vent is not present. In this case the value of Fb is affected by the driver parameters.

Suggest Fb (standard vented boxes only) BassBox Pro can recommend a value for Fb if enough information has been entered into the design. It attempts to suggest a value that will result in a smooth amplitude response.

QL The Q of the box leakage losses. The lower the value, the more the losses and visa versa. A high number like 1000 indicates very little leakage loss. When QL has not been entered, it is estimated in the background based on the value of the box volume (Vb). This is done by interpolating between the default small and large box settings in the "Box" tab of the Preferences window. *Note: There are separate default values for closed boxes and vented boxes.* QL can be as low as 5 or less for a very large "lossy" box or as high as 20 for a small well-constructed "tight" box. A QL of 7 is considered to be typical for a vented box of modest size and average quality. A value of 20 is considered typical for a closed box of modest size that uses a driver with a solid dust cap. Also, using the "Suggest" button to calculate a box volume will cause the program to estimate and enter QL for the calculated box or chamber volume(s).

F3 The –3 dB half-power frequency for the design. F3 is sometimes referred to as the "corner" or "cutoff" frequency. However, this is not always true for BassBox Pro as explained in the note below. It is the point along the response curve where the level has dropped 3 dB from the average level in the passband.

Note: BassBox Pro calculates F3 by tracking the system response in a descending frequency direction until the first –3 dB level is found. This may or may not be below the "knee" of the response curve. This is a little different than other methods which calculate F3 as the –3 dB point relative to the "knee" of the response. BassBox Pro uses the method it does because it shows the absolute –3 dB frequency rather than a relative one. To find the –3 dB frequency below the "knee" of the response, use the cursor in the Normalized Amplitude Response graph.

Fx (vented boxes with an active high-pass equalization filter only) The resonance or tuning frequency of the active 2nd order (12 dB/octave) high-pass equalization filter.

Qx (vented boxes with an active high-pass equalization filter only) The Q of the active 2nd order (12 dB/octave) high-pass equalization filter. A value of about 2 produces a 6 dB peak at Fx and is common with B_6 and similar designs.

This concludes the description of the box parameters on the left side of the "Box Design" tab. Let's examine the right side of the "Box Design" tab next.

 *Reference: 4 Box Properties*

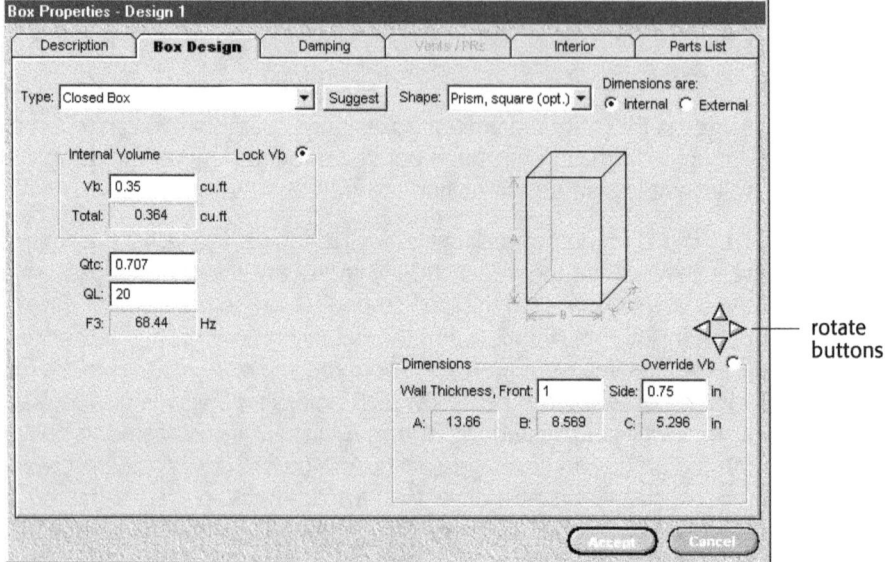

rotate buttons

The box dimensions are the central focus of the right side of the "Box Design" tab. When a closed box, vented box or passive radiator box is first selected, BassBox Pro will preselect the "optimum square prism" for its box shape (shown above). The "bandpass prism" is the default selection for a bandpass box. At any point in the design you can select another shape. Be aware that selecting a different shape will reset the dimensions. For most box shapes, this means that they will be cleared.

Notice how the dimension information is organized. At the top are the "Shape" drop-down list and the "Internal" and "External" options. In the middle is the three-dimensional scaled drawing of the box and beneath the drawing are the box dimensions. Each option and input is described next.

Shape The "Shape" drop-down list is used to select a box shape. The available shapes will vary depending on the box type that is selected. This is shown in the list below:

Single-chamber box shapes for closed, vented and passive radiator boxes:

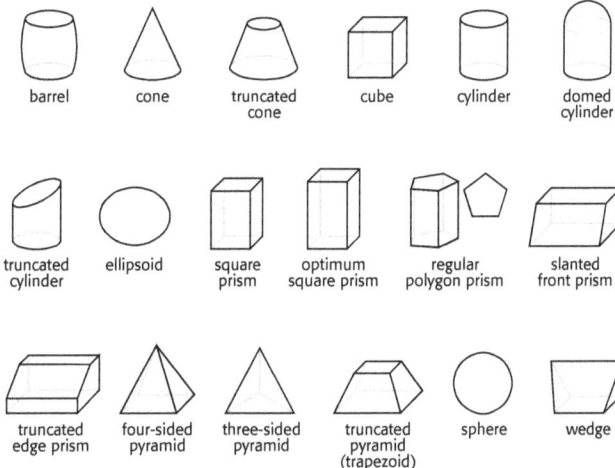

The optimum square prism has a couple of unique features. First, its three dimensions are calculated to have a 1.62 : 1 : 0.62 ratio. This ratio spreads the "standing wave" frequencies which can develop inside the box so that they will not add together and become louder. Standing waves are the result of sound reflecting back and forth between parallel walls. (The box shape with the worst standing wave problem is the cube.) Second, a cluster of rotate buttons is available so that you can choose how this ratio is applied to the height, width and depth of the box (see the illustration on the previous page).

Multi-chamber box shapes for bandpass boxes:

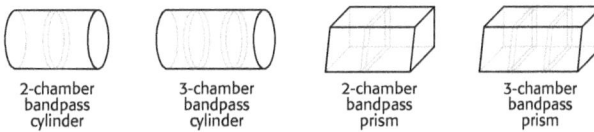

The bandpass prism is similar to the slanted front prism shape of the single-chamber boxes. Being able to slant the front wall makes the shape more versatile.

Internal/External The box dimensions will be interpreted as either internal dimensions or external dimensions, depending on the setting of this option. Once the dimensions have been entered, you can alternate between the two modes freely. The three-dimensional scaled box drawing will also represent either the internal or external shape based on the setting of this option.

Dimensions With the dimension labels on the box drawing, it is easy to see what the dimension labels represent. In addition, you can pause over a dimension label in the "Dimensions" frame with the mouse pointer and the balloon help will describe it as shown below:

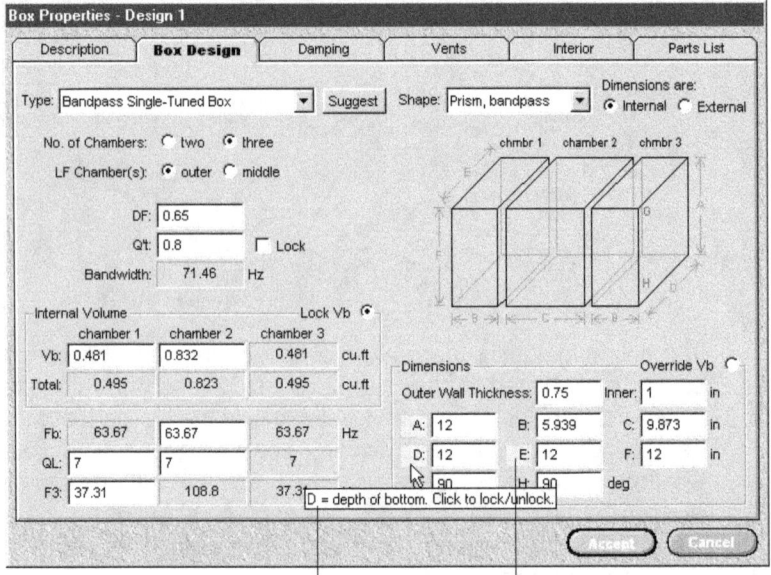

balloon help describes each dimension   a label with a light grey background indicates a locked dimension

In the above example, the pointer is paused over the label of dimension "D" and the help balloon says, "D = depth of bottom. Click to lock/unlock." Notice that the background of the "D" label is a lighter shade of grey as are the labels of dimensions A and E. This lighter background color signifies that the dimension is "locked" and cannot be changed by BassBox Pro. Of course, you can manually change its value whenever you want. However, "locked" dimensions are protected from recalculation whenever the net box/chamber volume (Vb) changes.

A box dimension can be locked or unlocked by clicking on its label. This is why the balloon help for dimension "D" said "…Click to lock/unlock." The selected box shape will determine how many and what combinations of dimensions can be locked at any given time. If the

Reference: 4 Box Properties   BassBox 6 Pro

net box/chamber volume (Vb) is locked, at least one box dimension will have to remain unlocked so that it can change if and when Vb changes. Why is this required? Because the volume produced by the box dimensions must also produce Vb. The two must agree. If the "Lock Vb" control is turned off, then its counterpart, the "Override Vb" control, will be turned on to show that the box dimensions will now determine the value of Vb.

Note: Often an unlocked box dimension will become locked as soon as you start to type a number into its input box. If the dimension is not available for locking, a message will appear to notify you that you will need to unlock another dimension before this one can be changed. This is because box dimensions that are not locked cannot be changed manually—they can only be changed by BassBox Pro. It might help to remember that "locking" a box dimension (or even Vb) is a way of taking control of its value so that BassBox Pro cannot change it. Instead, it gives you manual control over it.

In addition to box dimensions, the "Dimensions" frame also includes the wall thickness setting(s) for the box. Many of the box shapes have two wall thickness settings. For example, some of the single-chamber box shapes have one thickness setting for the front wall and another for all the remaining walls.

Override Vb This radio-style button controls whether or not the unlocked box dimensions will be allowed to grow or shrink as the net internal box/chamber volume (Vb) changes. The "Override Vb" control should be turned on when an already existing box is entered into the program. **Turning on the "Override Vb" control automatically unlocks Vb, allowing it to change in response to manual changes in the box dimensions.** When this is the case, Vb will be automatically calculated by subtracting the internal displacement volumes on the "Interior" tab from the total internal box/chamber volume as defined by the box dimensions.

What happens when you manually change a box dimension?
If the "Override Vb" control is turned on a new value for Vb will be calculated and the vent(s), if present, will be adjusted. Suppose, for example, that you have a square prism closed box whose internal height x width x depth is 24 x 12 x 18 inches with a Vb of 3 cubic feet. If you manually shorten the depth to be 12 inches then Vb will be recalculated to equal 2 cubic feet and the box height and width will not change.

If the "Override Vb" control is turned off (Vb is locked instead) the unlocked box dimensions will be adjusted so that Vb remains the same. Lets return to the preceding example and assume that the unlocked box dimension is the height. Now, when the depth is shortened from 18 inches to 12 inches, the height will be recalculated to equal 36 inches so that Vb remains 3 cubic feet. *Note: If you ever want to change which box dimensions are locked and unlocked, simply click on their labels (to the left of each input box). BassBox Pro will recognize when a dimension is unlocked and will recalculate it when sufficient information is present.*

Damping

The "Damping" tab contains the damping setting(s) for the box. The "damping" is the amount of acoustic absorption material, "fill" or "stuffing" that is placed inside the box. This setting is only available when the "Use Classical box calculations" checkbox at the top of the "Damping" tab is turned off (not checked).

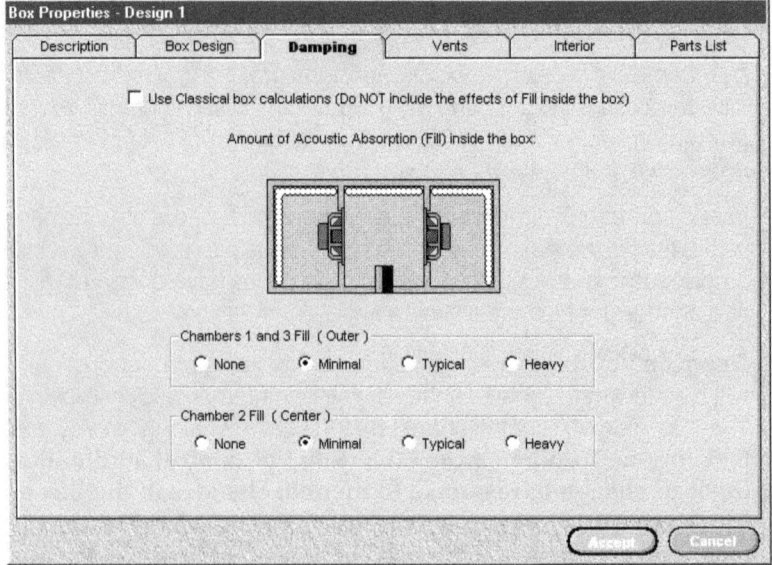

Why should a box be damped? If the speaker is a subwoofer and will only be used to produce low frequencies, then it should probably not be damped because most damping materials will have very little effect on frequencies below 100 Hz and virtually no effect below 50 Hz. However, the woofer in a two-way speaker must often operate to 1500 Hz or higher before crossing over to a tweeter. In this case, damping material or "fill" inside the box can have a significant effect. The rest of this discussion will assume that this is the case.

The presence of "fill" or "stuffing" inside a box will have two principal effects:

- The acoustical absorption inside the box will increase.
- The box damping will increase, making the box seem bigger than it really is.

So, why should a box be damped? First, a box that is damped with an acoustic absorbing material will have less of a problem with standing waves inside the box because the "fill" will absorb and attenuate them. How well it does this will depend on the frequency of the standing waves and the absorptive properties of the "fill" at those frequencies. Remember that one of the main purposes of a box is to prevent the sound waves emanating from the back side of the driver from interacting with sound waves emanating from the front of the driver. Second, a properly damped box will produce a smoother response and will respond better to transients. This contributes to a third benefit: A box with a proportionally large amount of "fill" will sound bigger than it really is. In other words, a smaller box with lots of "fill" will damp the driver in similar fashion as a larger box with no "fill". This trick has often been used with closed boxes to achieve a smooth response from a smaller box.

BassBox Pro will attempt to compensate for the apparent increase in box size whenever you click on the "Suggest" button of the "Box Design" tab to design a box. This means that BassBox Pro will recommend a smaller box size than other programs. This feature can be disabled with the "Use Classical box calculations" checkbox at the top of the "Damping" tab.

What makes a "good" damping material?
Most damping materials share the following desired characteristics in varying amounts:

- Absorb sound waves in the desired frequency range.
- Have a low density and not displace significant air volume inside the box.
- Have a high specific heat (that is, does <u>not</u> absorb heat easily).
- Resist environmental fluctuations (humidity, temperature, ozone, etc.).
- Does not impede the flow of air inside the box.

Typical damping materials include: fiberglass, long-fiber synthetics (Acousta Stuf, Dacron, polyester, etc.), acoustic foam (Sonex, etc.) and long-fiber wool. *Note: There are many different types of fiberglass batting. "Acoustic grade" fiberglass is the best type because it is designed specifically for these kinds of applications.*

Use Classical box calculations (Do NOT include the effects of Fill inside the box)

Checking this checkbox control will cause BassBox Pro to ignore the damping settings and the damping controls will disappear as shown below:

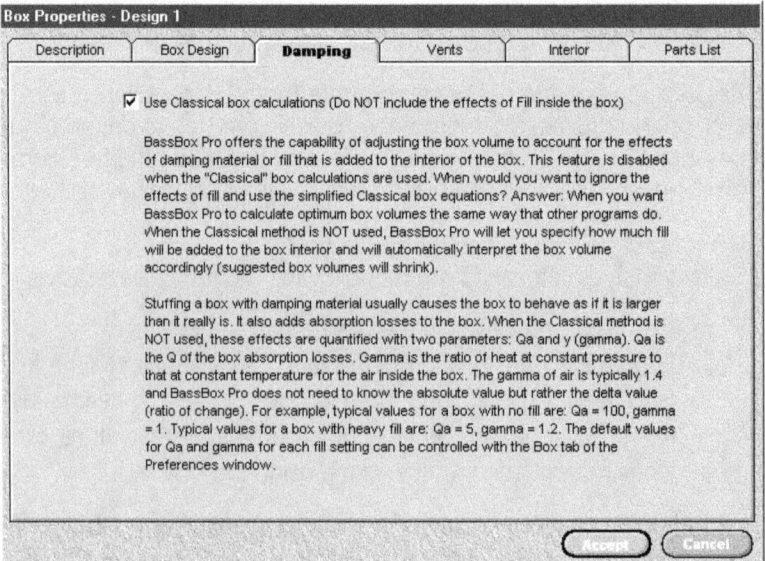

In place of the damping controls, a description of BassBox Pro's approach to damping compensation is displayed. *Note: The simplified "classical" box calculations ignore the effects of damping material inside the box by nullifying the Qa and γ parameters explained below.*

Now let's assume that the "Use Classical box calculations" checkbox is NOT checked and let's return to an explanation of how BassBox Pro compensates for the presence of "fill".

The two effects of "fill" (increased acoustic absorption in the box and increased damping resulting in larger apparent box size) are quantified with the Qa and γ (gamma) parameters. Qa is the Q of the absorption losses. Gamma is the ratio of heat at constant pressure to that at constant temperature for the air inside the box. BassBox Pro has a default setting for each of these parameters for each of the four damping settings (none, minimal, typical and heavy). The default values are listed next and can be controlled on the "Box" tab of the Preferences window (see pages 337-339).

The four damping settings are: none, minimal, typical and heavy.

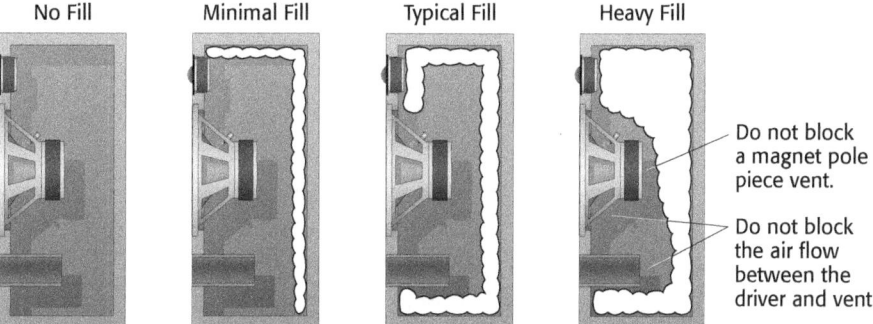

None No absorptive material is added to the interior of the box. (Qa = 100, γ = 1)

Minimal A modest layer of absorptive material is placed on just one of each pair of opposing walls. This option is used by designers who want to absorb some of the standing waves and reflections inside the box but who also want to minimize the effects of absorption on the apparent box size. (Qa = 50, γ = 1.04)

Typical All walls are covered with about 1 to 1½ inches (25 to 40 mm) of absorptive material. This option is used by designers who want to aggressively absorb standing waves and reflections inside the box and can tolerate a moderate increase in the apparent box size. (Qa = 10, γ = 1.08)

Heavy The box is stuffed full of absorptive material. Care must be taken not to restrict the operation of the driver or, if a vent or passive radiator is present, the air flow between the driver and the vent or passive radiator. This option is used by designers who want both an aggressive reduction in internal standing waves and reflections and want the apparent box size to increase as much as possible. (Qa = 5, γ = 1.2)
Note: Most box designers avoid using a "heavy" amount of fill for boxes with vents because large amounts of fill can decrease the output of the vent.

The chambers of a bandpass box have separate damping settings with one exception—the two outer chambers (Chambers 1 and 3) of a triple-chamber box share a common setting.

Closed boxes often use a heavy amount of fill while vented and bandpass boxes often use a minimal amount of fill. A default fill setting for three box types (closed, vented/bandpass, passive radiator) can be adjusted on the "Box" tab of the Preferences window.

Vents

The "Vents" tab is not available for closed boxes and its contents will differ, depending on the box type. For vented and bandpass boxes, it contains the vent parameters. For passive radiator boxes, it contains the passive radiator parameters. This section will focus on its function with the vent parameters. A separate section follows to cover its role with passive radiators. Let's begin with some typical design goals for a vent:

- For faithful sound reproduction, a vent should tune the box to a resonance frequency that both extends the low frequency response of the driver and maintains a smooth response curve free of excessive ripple. This goal does not apply to musical instrument speakers that are used in the creation of sound.

- The overall size of a vent should be modest when compared to the overall size of the box. There are two reasons for this: First, the vent should not affect the box's ability to trap the unwanted sound waves that emanate from the rear of the driver. Second, the vent should not affect the compliance of the air inside the box. The vented box model in BassBox Pro does not work with extreme situations where, for example, a vent is larger than the box.

- The total cross-sectional area of the vent should be large enough so that the velocity of air in and out of the vent does not exceed 10% of the velocity of sound in air. (Some designers prefer to keep it below 5%.) This prevents the turbulence and "whistling" that is sometimes heard from small vents. This requirement can sometimes result in huge vent sizes when the driver is large in size and has a large Xmax. However, drivers are usually not operated at the extremes of Xmax and so a more conservative size is sometimes selected. And there is a reason for not having a vent that is too large. This is explained in the next two points. The table below provides some general recommendations based on the diameter of the driver:

Recommended Vent Cross Section Size for Nominal Music Playback

Driver Diameter		Minimum Vent Diameter/Area				Better Vent Diameter/Area			
		diameter		area		diameter		area	
inches	mm	inches	mm	sq.in	sq.cm	inches	mm	sq.in	sq.cm
4	100	1	25	0.8	4.9	2	51	3.1	20
5	130	2	51	3.1	20				
6	150	2	51	3.1	20	3	76	7.1	46
8	200	3	76	7.1	46	4	102	13	81
10	250	4	102	13	81				
12	300	4	102	13	81	6	152	28	182
15	380	6	152	28	182				

The Vent Air Velocity graph can help you evaluate the vent size. It is described in Chapter 6 ("Evaluating Performance").

- The size of the vent should be small enough to avoid problems with "pipe" resonance. Typically, this is not a concern with a standard vented box because the pipe resonance is masked by the direct sound response of the driver. However, it can be a serious problem for a bandpass box because all of its sound waves emanate from the vent(s). A single-tuned bandpass box is depicted in the graph below:

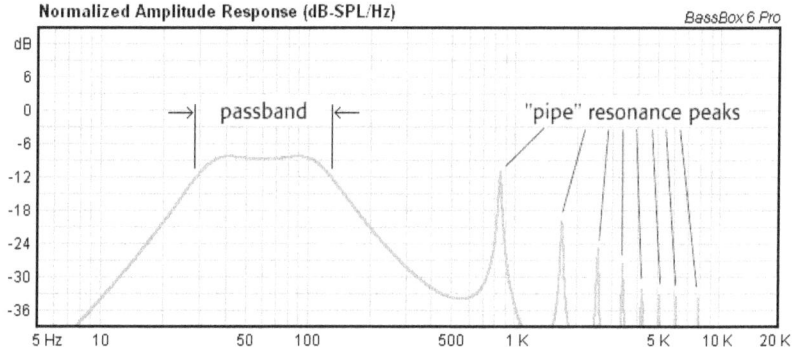

This pipe resonance phenomenon results from the natural tendency of a vent to resonate and produce additional harmonic waves. In extreme circumstances (very large and long vents) this can also have an adverse effect on a standard vented box.

By making the vent smaller, the pipe resonance and its harmonics are pushed to higher frequencies, reducing the likelihood of problems in the operating range of the driver. Large vents will usually have a greater problem with pipe resonance because their length is much greater than their diameter. The problem can often be overcome with prudent use of a crossover network. Bear in mind that the pipe resonance peaks often look worse than they sound. This is because the peaks are sometimes super-narrow and are therefore difficult to measure and difficult to hear. The larger, wider peaks are worse because they have more energy and can be very audible.

- The total interior surface area of the vent should be small enough to avoid excessive vent losses due to the viscosity friction of the air as it moves through the vent. There are three ways to minimize the interior surface area: First, choose a vent shape with the least surface area. Second, use one vent instead of multiple vents. Third, avoid vents that are very large.

The vent shape with the least interior surface area is a cylinder so a single round vent is usually the best shape. For example, rectangular vents have more interior surface area than round vents.

- Finally, the vent should fit in the box. This can be a design-breaking problem when trying to mix a small box size with a low tuning frequency (Fb). This is because the resonance frequency that results from a vented box is a product of both the internal volume of the box (Vb) and the size of the vent. If Fb and the diameter of the vent (Dv) are not allowed to change, then decreasing the box volume will require the vent to be longer and visa versa. When a vent is too long to fit inside a box, the only alternatives are to increase the box volume or increase Fb (or both). In general, the end of the vent tube inside the box should be at least one vent diameter away from the nearest wall.

The first example below shows the layout of the "Vents" tab for a vented box.

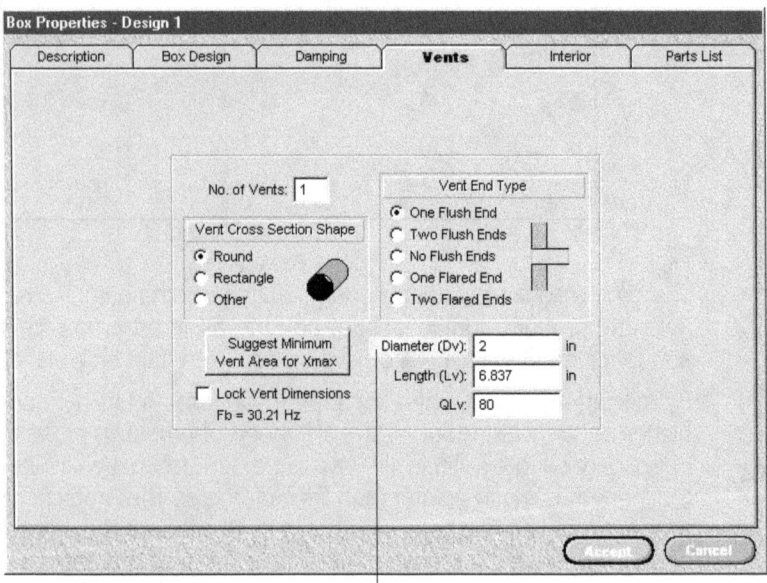

a label with a light grey background indicates a locked dimension

The two main dimensions of a typical vent are its diameter (Dv) and length (Lv). Notice in the example above that the "Diameter (Dv)" label has a lighter background than the "Length (Lv)" label. This signifies that the diameter is protected or "locked" and that BassBox Pro will not change it. The vent length label has a normal background and will be recalculated as required in order to produce the desired resonance (Fb). This can be easily reversed—the vent length (Lv) can be locked by clicking on its label. This will unlock the vent diameter, allowing it to be recalculated by the program as required.

Reference: 4 Box Properties

The second example below shows the layout for a double-tuned bandpass box with three chambers. Notice that only one set of vents (LF – low frequency or HF – high frequency) can be edited at a time and are selected with the "Select a Vent" buttons near the top of the window.

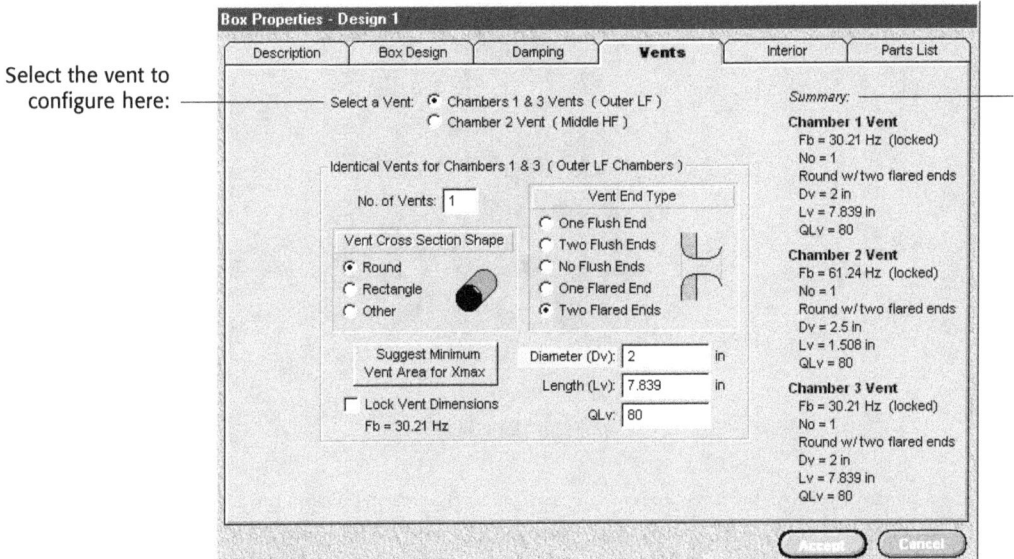

Select the vent to configure here:

A summary of all vents is displayed here.

Select a Vent This option is only available for double-tuned bandpass boxes. Use it to select the vent for editing. Notice that double-tuned bandpass boxes with three chambers use identical vents for the two outer chambers.

No. of Vents Enter the number of vents. All of the vents included in this number are expected to be identical. Multiple small-diameter vents are sometimes used in place of a single large-diameter vent. However, using multiple small vents will not decrease the required vent length (Lv) because the vent length is dependent on the total cross-sectional area of the vents. In fact the required vent length will increase slightly because the end-correction for the smaller-diameter vents will be less than that of a single larger-diameter vent. Also, multiple smaller vents will not reduce the pipe resonance problem mentioned earlier.

Vent Cross Section Shape The cross-section shape of the vent affects the end-correction factor and it changes the cross-section dimensions. Three choices are provide: round, rectangle and other. The "other" selection is provided for cross-section shapes that are not round or rectangular. It can only estimate the end correction and the closer the shape is to round, the more accurate it will be.

What is "end correction?" It's an internal parameter that compensates for the fact that a vent usually appears to be slightly longer than it really is. This is because the air immediately outside each end of the vent will vibrate along with the air inside the vent. This is illustrated below:

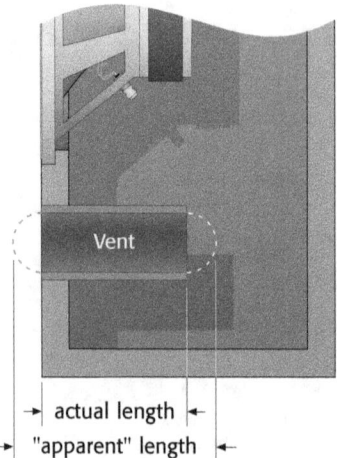

The end correction of a round vent is equal to the internal diameter (Dv) of the vent. For example, a vent with an inside diameter of 2 inches will appear to be 2 inches longer than its actual length (Lv).

The end correction is affected by both the cross-section shape and the end type. BassBox Pro automatically estimates the end correction and uses it for its internal calculations.

Vent End Type The "end type" refers to the way the vent terminates on each end and it affects the end correction described above. Five choices are provided: one flush end, two flush ends, no flush ends, one flared end and two flared ends.

Suggest Minimum Vent Area for Xmax Clicking this button will cause BassBox Pro to estimate the minimum size vent that will avoid turbulence and whistling under the extreme conditions of maximum linear excursion (Xmax). Please refer to the preceding table on page 234 for typical vent diameters for nominal excursions. *Note: This button and feature may be disabled if the required driver parameters are unknown.*

Lock Vent Dimensions This option has a dramatic effect on the way that the vent dimensions and tuning frequency parameter (Fb) are used. Normally, only one of the vent dimensions is "locked". This allows BassBox Pro to control the "unlocked" vent dimension in order to cause the vent to produce the desired resonance frequency. Whenever the internal box volume (Vb) or the "locked" vent dimension is changed, BassBox Pro will automatically recompute the corresponding "unlocked" vent dimension, forcing the vent to produce Fb. **All vent dimensions are locked when the "Lock Vent Dimensions" control is turned on.** This prevents BassBox Pro from changing any vent dimension and it gives you full manual control over them all. It also causes Fb to be recalculated whenever you change a vent dimension. This option can be very helpful when modeling a design that uses a preexisting box and vent.

The current value of Fb is displayed below the "Lock Vent Dimensions" control for your convenience.

Diameter (Dv) The inside diameter of a round vent. This dimension is only available when the cross-section shape is set to "round". Units: mm, cm, m or in.

Height (Hv), Width (Wv) The inside height and width of a rectangular or square vent. These dimensions are only available when the cross-section shape is set to "rectangle". Units: mm, cm, m or in.

Area (Av) The inside cross-sectional area of the vent. This dimension is only available when the cross-section shape is set to "other". Units: sq.mm, sq.cm, sq.m or sq.in.

Length (Lv) The physical length of the vent. If one or more flared ends are present, Lv includes the flared ends in its value. This means you must subtract the flared ends to find the length of the straight tube portion of the vent.

QLv The Q of the vent losses. This parameter only has an effect when the "Include Vent Resonance Peaks" option is turned on in the graphs. See the Chapter 6 for details.

The lower the value of QLv, the more the losses and visa versa. A typical value is 80 and this is what BassBox Pro assumes when QLv has not been entered. A very low value, like 0.0001, will effectively disable the vent and make a vented box behave like a closed box.

Passive Radiator

The fourth tab of the Box Properties window becomes a "Passive Rad" tab when a passive radiator box is selected. Use it to enter the passive radiator parameters.

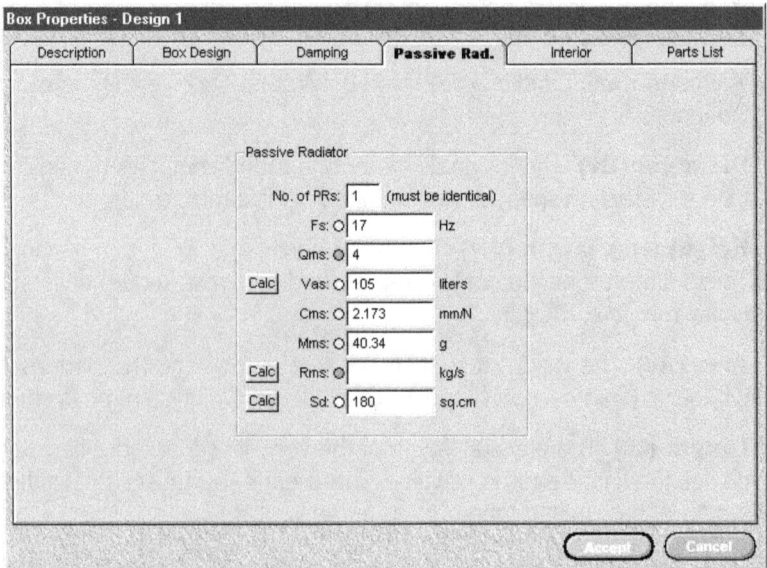

The passive radiator parameters are very similar to the driver parameters. A minimum selection of the parameters must be entered before a passive radiator box can be modeled. Here is a list of some of the combinations that can satisfy the minimum requirement:

        Fs, Qms, Vas, Sd   Fs, Qms, Vas, Cms
        Fs, Qms, Cms, Sd   Cms, Mms, Rms, Sd
        Fs, Qms, Mms, Sd

It is not common to find a full set of parameters for a passive radiator. BassBox Pro includes a built-in procedure to help you measure these parameters. Refer to Chapter 11 for details.

Notice in the illustration above that the "Passive Rad" tab includes two features that are found in the "expert mode" of the "Parameters" tab of the Driver Properties window. These are status indicators and "Calc" buttons.

Status Indicators The status indicators show whether the parameters have passed or failed BassBox Pro's automatic tolerance test. The tolerance settings for the test can be set with the "Driver" tab of the Preferences window using the same settings as are used for the "Parameter" tab's expert mode (see Chapter 13). This self-analyzing feature can quickly reveal "typos" and other mistakes immediately when they are made. It also provides a convenient way to see if a passive radiator's data contains mistakes.

The color code for the parameter status indicators is:

    Grey – there is not enough information available to test the parameter.
    Green – the parameter is probably correct.
    Yellow – the parameter appears to be slightly out of tolerance.
    Red – the parameter is probably incorrect.

When the status indicators of multiple parameters turn red it does not mean that they are all necessarily incorrect. When two or more related parameters fail a test, BassBox Pro can't know which one is incorrect and so it must flag them all.

What are some of the reasons that cause a parameter value to be out of tolerance?

- The wrong units were selected before the parameter was entered.
- The decimal place is incorrect.
- The value has been rounded too much.
- An error was made when the passive radiator was tested.
- An error was made when the parameter was calculated.

Calc The "Calc" buttons appear whenever it is possible to calculate an unknown parameter or a known parameter that does not have a green status indicator.

The remainder of this section will describe each of the passive radiator parameters:

No. of PRs The number of passive radiators. When the "No. of PRs" is set higher than "1", the "net" value of the parameters will be displayed beside the single-PR values as shown below.

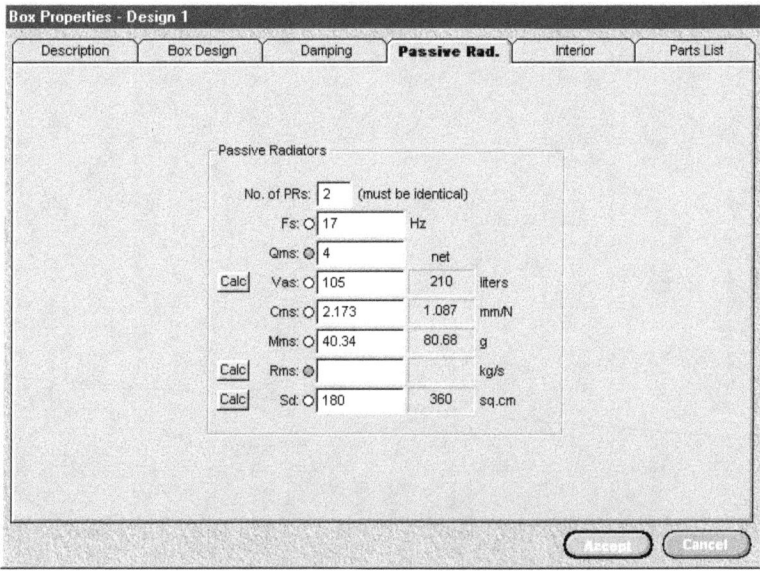

When more than one passive radiator is specified, BassBox Pro assumes that all of them are identical and it will use the "net" values for all internal calculations. You should still enter the parameters (Fs, Qms, Vas, Cms, Mms, Rms and Sd) for just one passive radiator. *Note: When two passive radiators are used instead of one, the resonance of the box will increase. You will need to double the diaphragm mass (Mms) and recalculate Fs and Qms if you want to maintain a similar box resonance as if one passive radiator were used.*

Fs The free air resonance frequency of the passive radiator. Units: Hz. *Note: Fs will change if Mms is changed.*

Qms The mechanical resonance magnification of the passive radiator at Fs. *Note: Qms will change if Fs is changed (usually because Mms was changed first).*

Vas The volume of air equal to the mechanical compliance of the passive radiator's suspension. Units: liters, cu.cm, cu.m, cu.ft or cu.in.

Cms The mechanical compliance of the passive radiator's suspension. Units: μm/N, mm/N, cm/N, m/N or in/lb.

Mms The moving mass of the diaphragm or piston including the air load. Units: g, kg or oz. *Note: Mms is often adjusted when tuning a passive radiator system. When Mms is changed, you will need to recalculate Fs and Qms. This can be done with the "Calc" buttons if Cms, Rms and Sd are known.*

Rms The mechanical resistance of the passive radiator resulting from its suspension losses. Units: kg/s, lb/s or mohms.

Sd The diaphragm or piston area of the passive radiator. Units: sq.mm, sq.cm, sq.m or sq.in.

Reference: 4 Box Properties BassBox 6 Pro

Interior

The "Interior" tab contains information about the volume inside the box that is displaced by internal objects like the driver, vent and braces. The box will need to be enlarged slightly to compensate for these volumes and the "Total" Vb value on the "Box Design" tab reflects this.

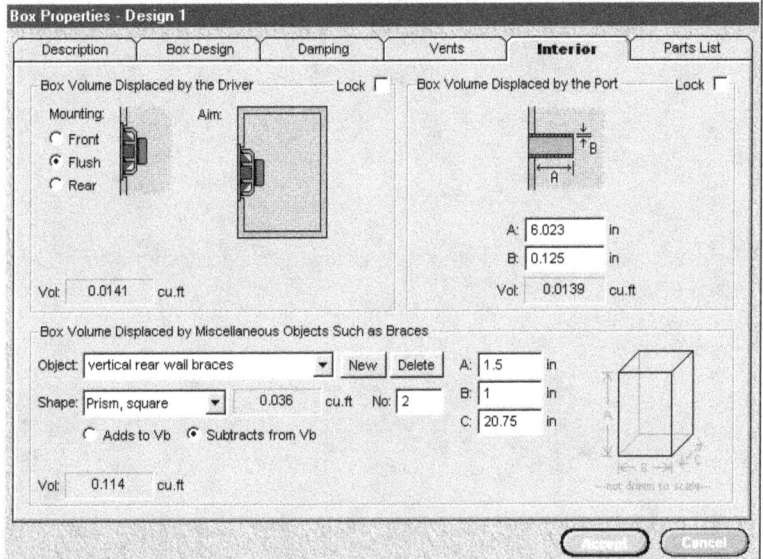

Box Volume Displaced by the Driver

Mounting The way the driver is mounted to the box will affect how much of it protrudes into the interior of the box and therefore, how much box volume it displaces. The three choices are: front, flush and rear. Flush mounting is generally considered the best because it results in the least diffraction.

Front Mounting

Flush Mounting

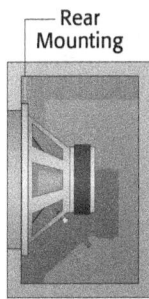
Rear Mounting

BassBox 6 Pro User Manual © D.E.Harris

BassBox 6 Pro

Reference: 4 Box Properties

Aim The direction the driver is facing in the box also affects its displacement volume. Three configurations are covered by this option.

Compound push-pull mechanical configurations:

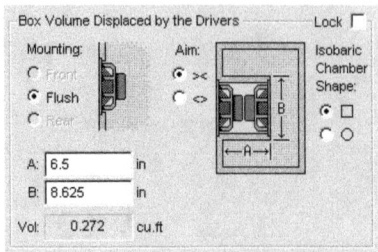 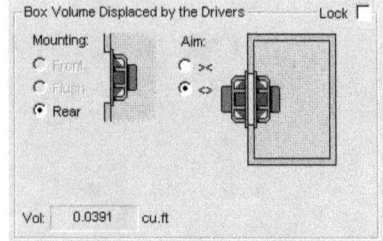

Select "><" when the drivers face away from each other and "<>" when they face toward each other as shown above.

Bandpass boxes with two chambers:

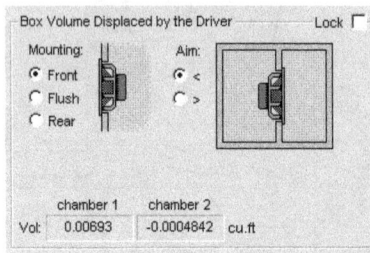 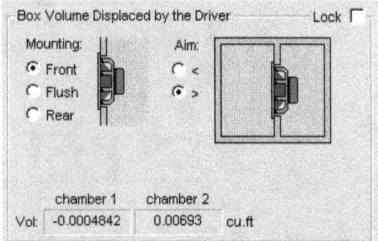

Select "<" or ">" as indicated above. In this configuration, the chamber faced by the driver will usually have an increase in volume which results in a negative "Vol" displacement.

Bandpass boxes with three chambers:

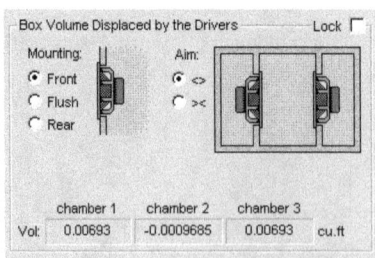 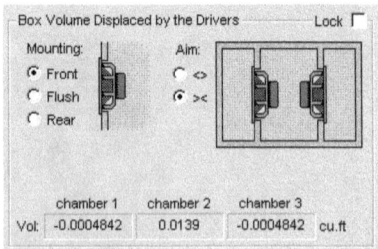

Select "><" when the drivers face away from each other and "<>" when they face toward each other. Chambers with a negative "Vol" will increase rather than decrease in volume.

Reference: 4 Box Properties   BassBox 6 Pro

Isobaric Chamber Shape Compound box designs (except compound bandpass) may use a small isobaric chamber when the push-pull option is used. The dimensions (A and B) of the isobaric chamber will need to be entered so that the volume it displaces inside the box can be calculated. The Isobaric Chamber Shape specifies whether the isobaric chamber is square or round. It is rectangular when an oval or rectangular driver is used.

"B" is a single dimension for most drivers. 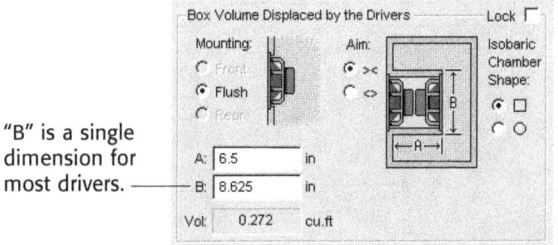 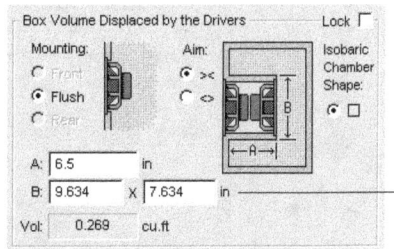 "B" is the height x width when an oval or rectangular driver is used.

    A The depth of the isobaric chamber inside the box. Units: mm, cm, m or in.

    B The definition of this dimension changes with the driver shape and isobaric chamber shape. It is the diameter when the isobaric chamber is round. It is the length of one side when the isobaric chamber is square (shown in the left example above). And it is the height and width when the isobaric chamber is rectangular (shown in the right example above). Units: mm, cm, m or in.

    Vol The total volume displaced by all drivers and, if present, isobaric chambers inside the box. It is usually positive but it can be negative when the driver doesn't subtract but rather adds volume to the box or chamber (displaces negative volume). Units: liters, cu.cm, cu.m, cu.ft or cu.in.

Normally, BassBox Pro will automatically calculate/estimate the volume displaced by each driver if enough information about the driver and, if present, the isobaric chamber is available. However, you can enter the volume manually. If you do, you will need to "lock" it with the "Lock" checkbox to prevent BassBox Pro from changing it when an associated box or driver dimension changes.

    Lock Prevents BassBox Pro from changing the box volume displaced by the driver and, if present, the isobaric chamber.

Box Volume Displaced by the Vent

Notice in the example below that the volume displaced by the vent can be configured separately for each chamber of a bandpass box. With triple-chamber bandpass boxes, the two outer chambers (1 and 3) are always the same.

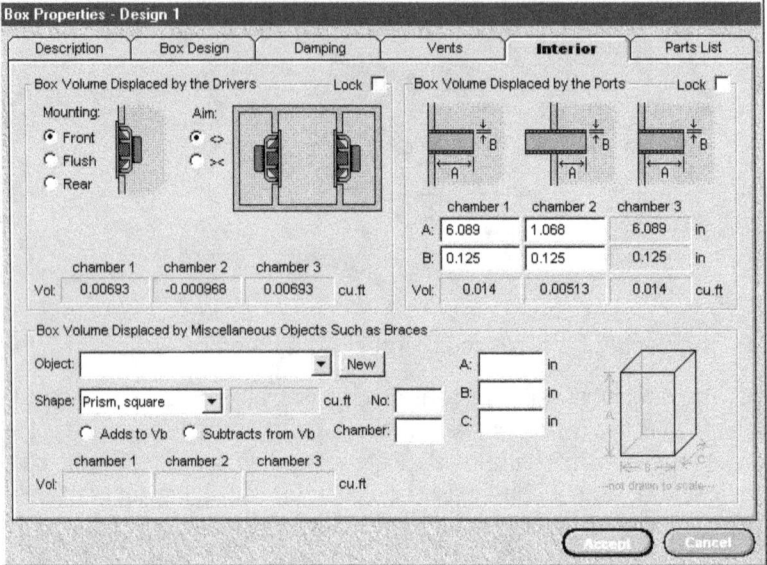

A The depth of the vent inside the box or chamber. BassBox Pro calculates this by subtracting the box wall thickness from the vent length (Lv). Units: mm, cm, m or in.

B The wall thickness of the vent. Units: mm, cm, m or in.

Vol The total volume displaced by all vents inside the box. This "Vol" is always positive. Units: liters, cu.cm, cu.m, cu.ft or cu.in.

Normally, BassBox Pro will automatically calculate/estimate the value of "Vol" if enough information about the vent is available. However, you can enter it manually. If you do, you will need to "lock" it with the "Lock" checkbox to prevent BassBox Pro from automatically recalculating it when a relevant box or driver parameter changes.

Lock Prevents BassBox Pro from changing the box volume displaced by the vent.

Reference: 4 Box Properties   BassBox 6 Pro

Box Volume Displaced by Miscellaneous Objects Such as Braces

This section of the "Interior" tab provides a means for other internal objects like braces to be entered. Up to 30 different object sizes can be entered.

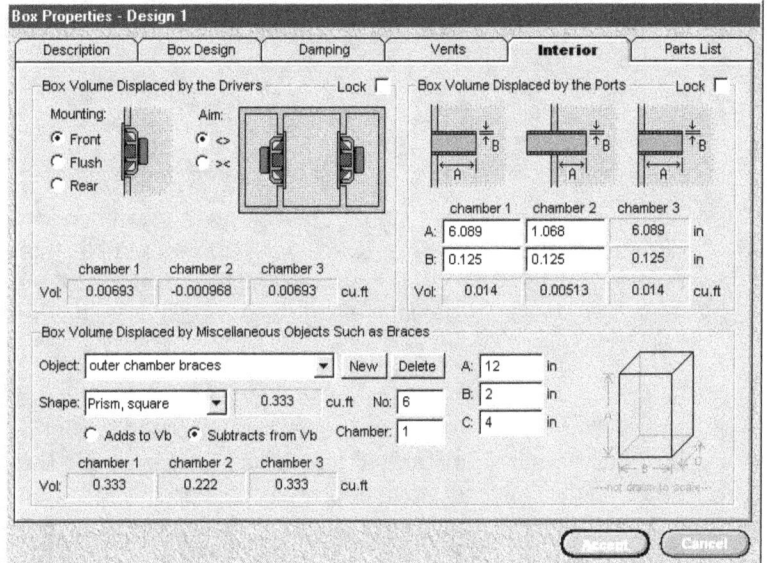

Follow these steps to enter an object:

1. Click the "New" button. The name "object #" will be entered into the object list box (as shown below) where "#" is a number that is temporarily added to the name to prevent more than one object from having the same name.

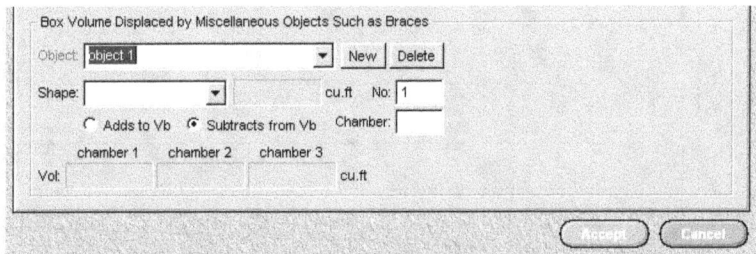

2. Select the object name and enter a name which describes the object. In the example shown on the next page "center chamber braces" was entered.

BassBox 6 Pro

Reference: 4 Box Properties

3. Select the shape from the shape list as shown below:

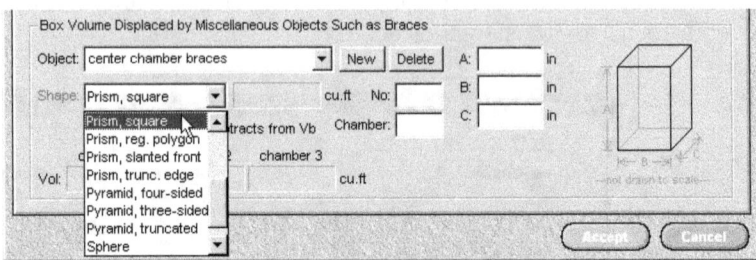

After a shape is selected, input boxes for the required dimensions and a drawing with dimension labels will be displayed. In the above example, three dimensions are required: A, B and C. *Note: The drawing is not drawn to scale—it is provided to identify how the dimensions are labeled. In the example, "A" is the height, "B" is the width and "C" is the depth.*

4. Select the "Adds to Vb" option if the object adds to the internal box volume (Vb). This setting is useful when the box shape is complex and cannot be defined solely with the box shape on the "Box Design" tab. In such cases you can use additional object volumes on the "Interior" tab to add additional portions to the box.

   Select "Subtracts from Vb" if the object reduces the internal box volume. This is the typical setting for objects inside the box such as braces.

5. Enter the requested information for the object. This includes the number of objects (No), the chamber number (if the box is a bandpass design) and the dimensions.

   The volume of the object will be calculated and displayed as shown below.

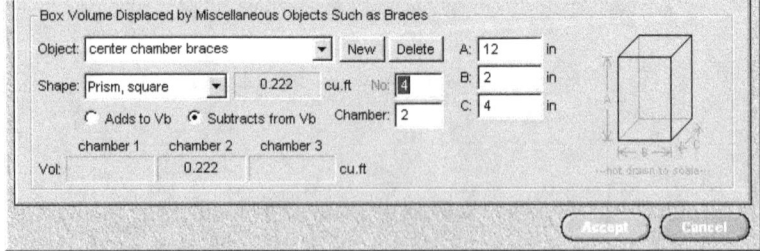

In the above example, the volume of each object is 0.055 cu.ft. Since a quantity (No) of 4 was entered for this object, the total volume for all four is 0.222 cu.ft. Notice that this volume appears two places. First, it appears next to the "Shape" list. Second, it appears in the "chamber 2" volume (Vol) box which keeps a running total of all object volumes entered for chamber 2.

Reference: 4 Box Properties   BassBox 6 Pro

Each object and input is described next:

Object Use the "Object" drop-down list to select an existing object. It also serves as an input box so you can enter or change an object's name.

New The "New" button is used to enter a new object. It will enter a temporary name into the "Object" list. Enter a name that describes the object.

Delete Click the "Delete" button to delete the selected object. It will then be removed from the "Object" list.

Shape The "Shape" drop-down list is used to select an object shape. There are 15 shapes:

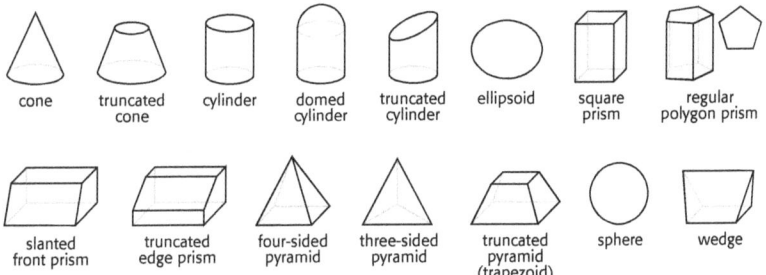

When a shape is selected, the input boxes for the required dimensions are made visible along with a drawing which identifies each dimension.

A 16th shape option, "Known Volume", is also provided. Select it when you want to enter a volume directly.

Object Volume The object volume is displayed to the right of the "Shape" list. If there are more than one of the object, the volume represents the total volume.

No The number or quantity of objects.

Chamber (bandpass boxes only) The chamber number (1 or 2). *Note: Do not enter any objects for Chamber 3 of a triple chamber box. Enter it in Chamber 1 because all objects that are entered in Chamber 1 are mirrored in Chamber 3.*

A, B, C, D, E The dimensions of the object. Units: mm, cm, m, in.

Vol The total volume displaced by all of the miscellaneous objects inside the box. A negative value indicates that the volume adds to the box (displaces negative volume). Units: liters, cu.cm, cu.m, cu.ft or cu.in.

Reference: 4 Box Properties

Parts List

The "Parts List" tab is the last tab of the Box Properties window. It displays the two-dimensional drawings of the box parts and vent(s) plus a parts list. This tab requires no direct input but it does require that the box dimensions be entered in the "Box Design" tab, including the box wall thickness.

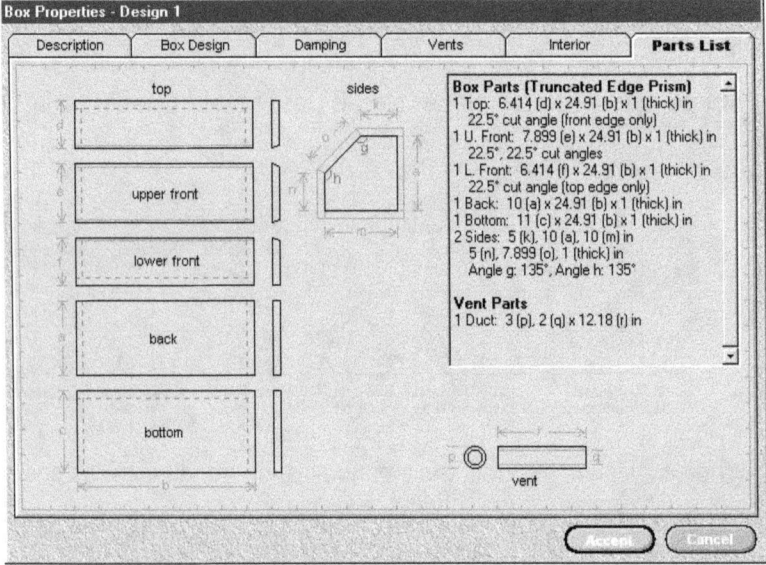

The "Parts List" tab will be empty when some or all of the box dimensions are missing.

Where appropriate, the parts are drawn in the context of adjoining parts so it will be easier to see how they fit together. A solid black line is used for the visible edges of a part. A solid grey line inside the part is used to represent a hidden line such as an edge which is visible from the back side. A solid grey line outside the part is a visible edge of an adjoining point. A dashed grey line is used to show where an adjoining part butts against another part.

When appropriate, cut angles are included with the dimensions in the parts list.

5 Room/Car Acoustic Properties

There are two modes for the acoustic environment: architectural and automotive. To change the acoustic mode, use the "General" tab of the Preferences window. If the architectural (room) mode is selected, the design panels in the main window will have a "Room" button. If the automotive (car) mode is selected, the design panels in the main window will have a "Car" button. Click on the Room/Car button to open the Room/Car Acoustic Properties window. The automotive version is shown below:

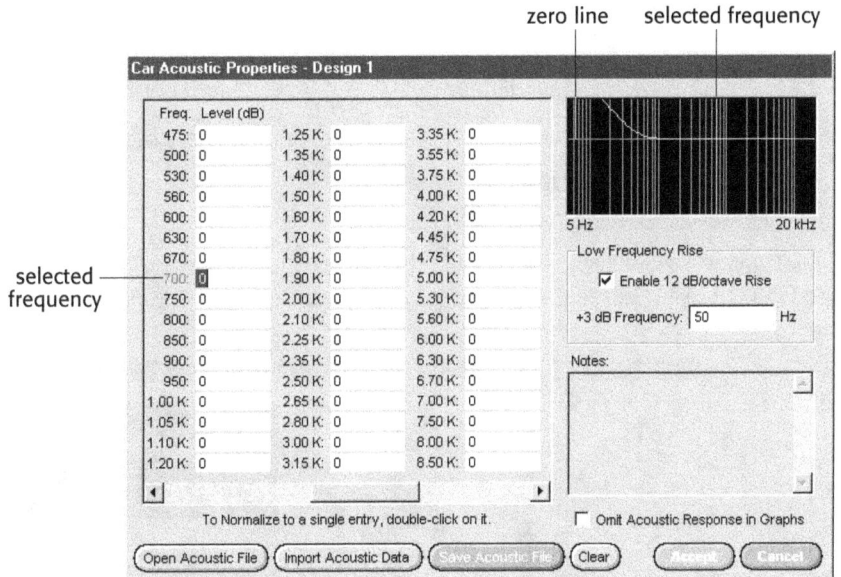

This window is similar to the "Response" tab of the Driver Properties window but it accepts acoustic information about the external environment. This can be very important for some acoustic environments because of the pronounced effect they can have on the sound. For example, most automotive interiors have a 12 dB/octave low-frequency rise that begins at about 50 Hz. That can have a huge effect on the design of a low-frequency enclosure.

There are two ways to enter the acoustic response. It can be manually entered using the individual "Level" input boxes for each data point or it can be imported from one of several measurement systems.

The preview graph in the upper right corner shows the shape of the response while you edit or import data. Notice also that the preview graph includes a "zero line" to help you adjust the overall level of the data so that it is normalized. A red vertical line is also provided to show what part of the response will be changed if the input box of the selected frequency is changed.

Level The normalized amplitude response in dB. The response of the driver can be manually entered or edited with the "Level" input boxes. There are 134 data points from 5 Hz to 20 kHz. Use the horizontal scroll bar to access all of the "Level" settings.

The acoustic data must be normalized to zero dB. If it is not, it can be by selecting one of the levels to use as the zero reference and then double-clicking on its input box. The value of the selected level will be subtracted from all the level settings after BassBox Pro first asks for confirmation.

Low Frequency Rise Several low-frequency rise filters are provide to model the boost in bass response of many interiors. One filter is available in the automotive (car) mode:

> 12 dB/octave rise filter (shown in the illustration on the previous page)

Six low-frequency filters are available in the architectural (room) mode:

> 3, 6, 12 dB/octave rise filters
> 3, 6, 12 dB/octave shelf filters

The illustration below depicts the use of both measured acoustic data and a 3 dB/octave shelf filter. The filter raises the low-frequency response below 50 Hz.

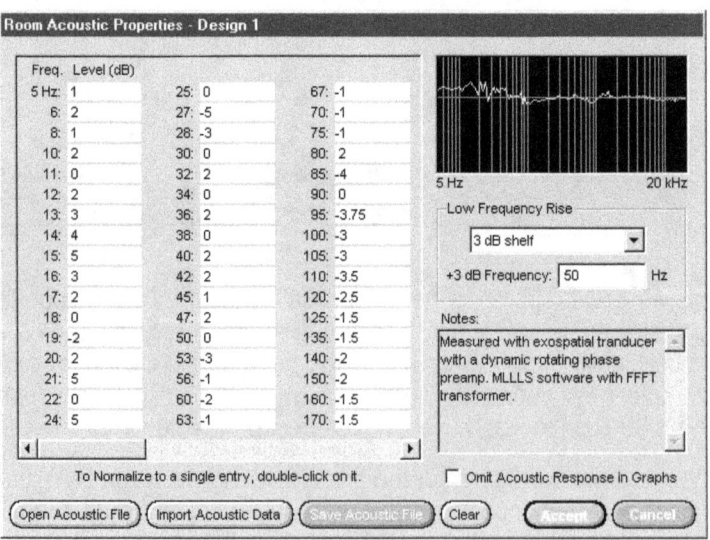

The +3 dB frequency of each filter is adjustable. 50 Hz is a good choice for most automotive interiors of various sizes (cars, trucks, vans) and this is the value shown in both of the preceding illustrations. The +3 dB frequency you choose for an architectural environment must be determined by measurement and is not predicted by BassBox Pro.

Reference: 5 Room / Car Acoustic Properties   BassBox 6 Pro

Notes Use the "Notes" textbox to enter relevant comments about the acoustic data.

Omit Acoustic Response in Graphs Check this checkbox to disable the architectural/automotive acoustic response in all relevant graphs. This can also be controlled with a checkbox in the Graph Properties window and directly from the graph window by right-clicking (🖱) on the graph and using the popup menu.

Open Acoustic File Prompts the user for a file name and path and then opens the selected acoustic data file. These files end in the file name extension "iad".

Import Acoustic Data Frequency-domain data files from several popular measurement systems (including Brüel & Kjaer, CLIO, IMP, LMS, MLSSA, OmniMic, Sample Champion, Smaart, TEF®-20 and TrueRTA) can be imported with the "Import Acoustic Data" button. Clicking it will open the Import Acoustic Response File window so you can select the data file. The file types are identified by the file name extension. Use the "Files of type" drop-down list to select the file type as shown below:

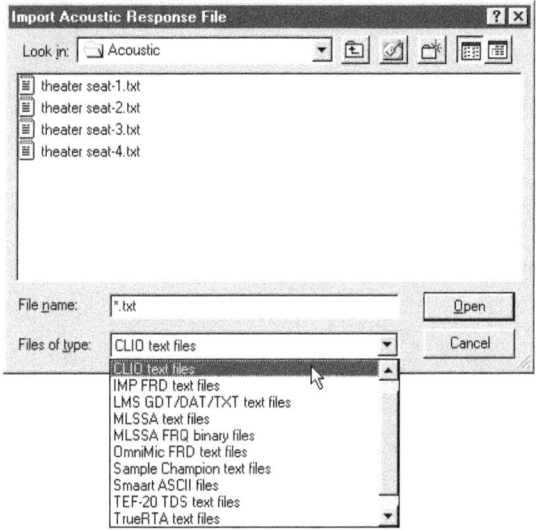

Note: The above Import Acoustic Response File window will vary in appearance based on the version of Windows. For example, the "Files of type" list may not be labelled and may be located to the right of the "File name" input box in Windows 8, 7 and Vista. It is important that the file type be selected with this list so it's type will be correctly identified.

The program will open a Select Data File Type window to prompt the user to select the file type if a file was chosen which the user forgot to identify by the "Files of type" list and BassBox Pro cannot identify it by its file name extension.

The acoustic data import filter works in the following manner:

- The import filter will interpolate between data points if some are missing.
- The import filter will average data points if there are extra ones.
- The import filter will give you the option of extrapolating the beginning or end of the data if it begins at a frequency above 5 Hz or ends below 20 kHz. The slope between the first or last five data points will be used if you choose to extrapolate.
- The import filter for some file types will automatically adjust the overall level of the data so that it will be visible in the preview graph.

Save Acoustic File Saves the acoustic data after first requesting a file name and path. Long file names are supported and the file name must end with the extension ".iad". The "iad" extension will be automatically appended if it is omitted from the end of the name.

Clear Use this button to clear all "Level" settings.

Note: The acoustic response of the environment is optional. It does not have to be entered in order to complete a speaker design.

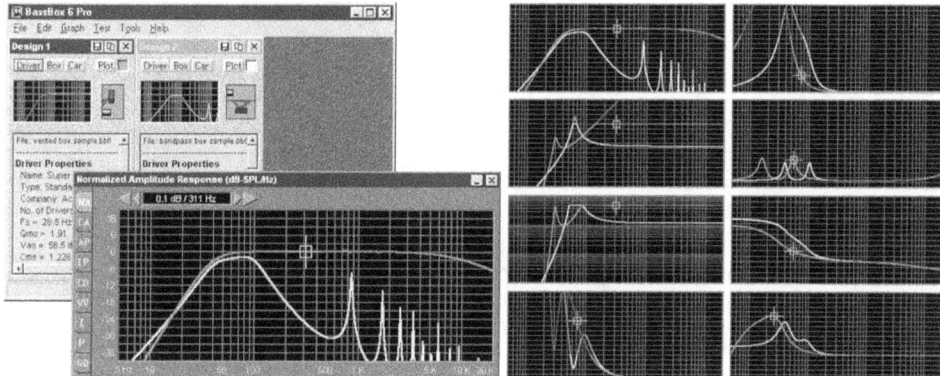

6 Evaluating Performance

A skillful speaker design is a matter of balancing the various compromises that must be made. No speaker yet devised is perfect. In fact, speakers are still considered the weakest "links" in the high-fidelity "chain". Always begin the design process by honestly evaluating the intended purpose and use for the speaker. Remember that purpose when evaluating its performance. This will help you decide which are the most important criteria to achieve.

Every design panel in the main window of BassBox Pro includes a mini preview graph which depicts the normalized amplitude response of the speaker design from 5 Hz to 2 kHz. Because it automatically updates whenever the design is changed, it helps you immediately evaluate the merits of your design decisions. But this mini preview graph only touches the surface of BassBox Pro's powerful graphing capabilities.

This chapter explores the features of BassBox Pro's nine full-size graphs and their use in evaluating the performance of a speaker design.

Graph Modes

There are two graph modes: "single window" mode and "individual windows" mode. The mode can be selected from the Graph menu of the main window (shown below). The default graph mode can be set with the "Graph" tab of the Preferences window.

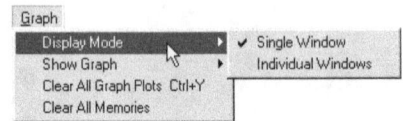

Single Window Mode

The "single window" mode combines all nine graphs into a single window. It is recommended for computers with a low graphics resolution (less than XGA or less than 1024 x 768 pixels) because it economizes space on the Windows desktop. It also uses less system resources which may be helpful if your computer doesn't have much memory and/or you often run other large applications at the same time as BassBox Pro.

"single window" mode

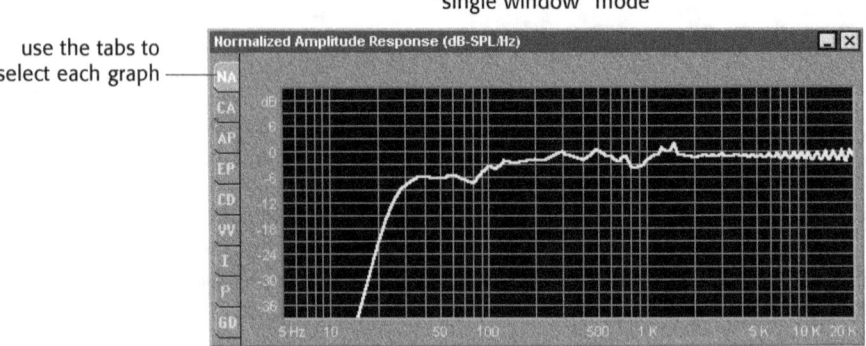

use the tabs to select each graph

A row of vertical tabs along the left side of the window is used to select a graph. Their abbreviations represent:

    NA – normalized amplitude response
    CA – custom amplitude response (requires input power or voltage)
    AP – maximum acoustic power
    EP – maximum electric input power
    CD – cone displacement (requires input power or voltage)
    VV – vent air velocity (requires input power or voltage)
    I   – system impedance
    P   – phase response
    GD – group delay

Individual Windows Mode

The "individual windows" mode displays each graph in its own window, allowing multiple graphs to be viewed simultaneously. This can be a great time-saver when you need to view several things at the same time. However, the graphs can quickly crowd the Windows desktop unless your computer graphics system has a high resolution (like 1600 x 1200 pixels). The "individual windows" mode also provides one other advantage: a large-size option. This allows you to enlarge the size of one or more of the graph windows.

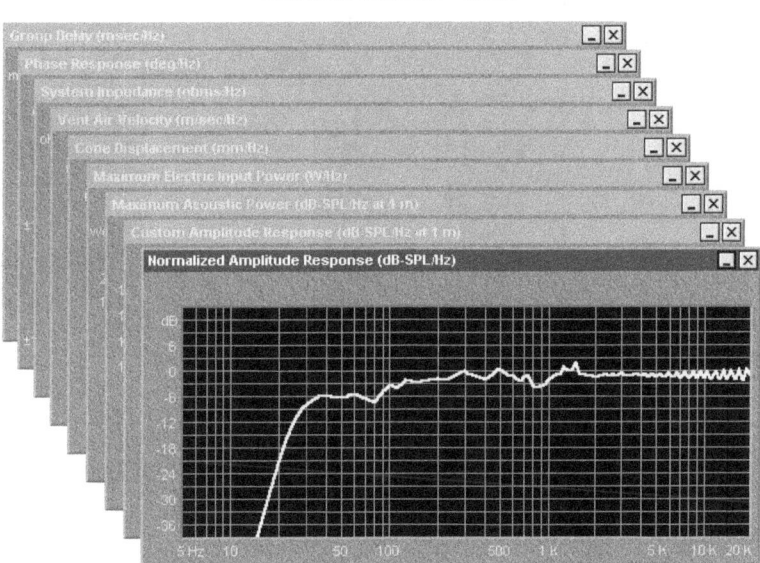

"individual windows" mode

Graph Features

There are many graph features and many of them have default values which can be selected in the Preferences window. The seven options listed below can only be configured with the "Graph" tab of the Preferences window (see pages 340-342):

- the plot line width
- overlay / don't overlay the plot lines
- the grid line darkness
- two custom plot colors
- the shape of the graph cursor
- the linking of the input power/voltage settings
- the units used for some vertical scales

 **BassBox 6 Pro**   *Reference: 6 Evaluating Performance*

The majority of the graph options can be configured from the graph popup menu or the Graph Properties window. These two different methods are provided so you can choose the one that you prefer.

Graph Popup Menu
The graph popup menu (shown below) is accessed by right-clicking (🖱) on the graph. Most of the graph options can be controlled from this menu.

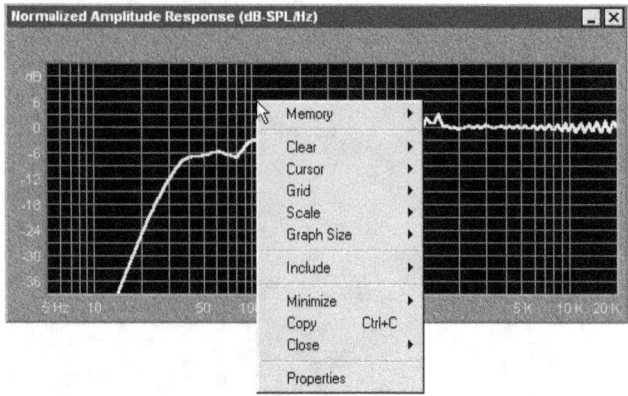

Graph Properties Window
The Graph Properties window (shown at right below) is opened by selecting the "Properties" command from the graph popup menu.

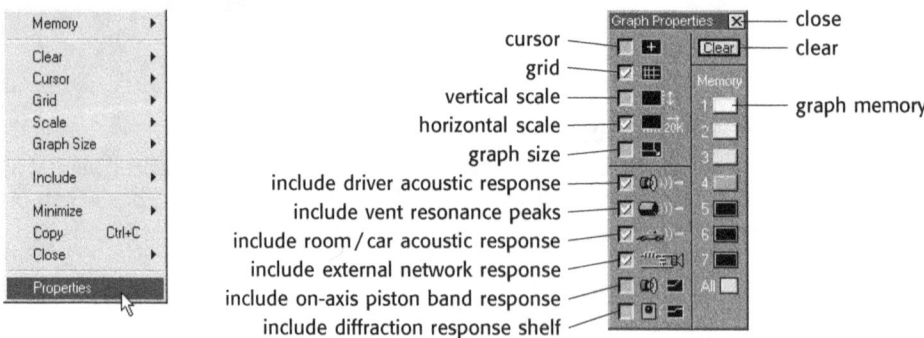

The Graph Properties window contains most of the same options as the graph popup menu. Its advantage is twofold: 1) Its many controls serve as indicators so you can quickly see how a graph is configured and which graph memories have been used; and 2) many options can be changed with a single mouse click (🖱).

Some of the settings in the Graph Properties window are specific to an individual graph. For example, the cursor can be turned on in the Normalized Amplitude Response graph and turned off in the System Impedance graph. When more than one graph is open in the "individual windows" mode, the Graph Properties window will show the settings of the selected graph window. To see the settings of a different graph window, simply click on its title bar to select it.

Each graph option is described next:

Memory Seven graph memories are provided to store a "snapshot" of the design so it can be recalled later and replotted from memory. The design parameters are stored in memory rather than the plot line data points. This enables all open graphs to be replotted when a memory is recalled—even if some of the graphs were closed when the design was originally stored into the graph memory.

To **store** the most recently plotted design into a graph memory, select the desired "Memory > Store" command from the graph popup menu (keyboard shortcut: Shift+F1 to F7). You can also hold down the Shift key and click a memory button in the Graph Properties window. The memory button will be filled with the plot color.

To **recall** a design from the graph memory and replot it in all open graphs, select the desired "Memory > Recall" command from the graph popup menu (keyboard shortcut: Shift+Ctrl+F1 to F8). You can also click on a memory button in the Graph Properties window.

To **clear** an individual graph memory, select the desired "Memory > Clear" command from the graph popup menu. You can also hold down the Alt key and click a memory button in the Graph Properties window. All graph memories can also be cleared with the "Clear All Memories" command of the Graph menu of the main window.

Clear Clears the plot lines from either the selected graph or all open graphs. The graph popup menu provides a separate selection for each. (Keyboard shortcuts: Ctrl+X to clear the selected graph only and Ctrl+Y to clear all graphs.) To clear only the selected graph with the Graph Properties window, click on the "Clear" button. To clear all graphs with the Graph Properties window, hold down the Shift key and click on the "Clear" button.

Cursor Several cursor options are provided in the graph popup menu. The cursor in the selected graph or all graphs can be turned on/off and linked/unlinked. A keyboard shortcut is provided for turning the cursor on and off in the selected graph. Use [Ctrl]+[U] to show the cursor and [Ctrl]+[H] to hide the cursor. The Graph Properties window only allows the cursors to be turned on/off. To turn on/off the cursors for all open graphs with the Graph Properties window, hold down the [Shift] key and click the Cursor button. A sample graph with a cursor is shown below:

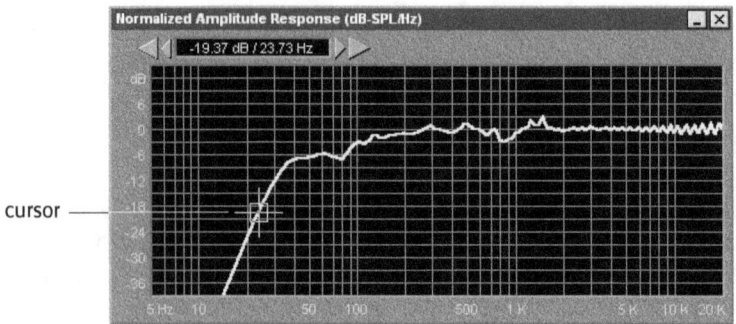

When the cursor is turned on, a set of large-step and a set of small-step cursor controls are available to move the cursor left and right in the graph. The small-step controls move the cursor one pixel at a time. The large-step controls move the cursor 20 pixels at a time. (A pixel is a single dot of light on the computer screen.) The cursor can also be moved with the keyboard using the [←] and [→] keys for small steps and [Shift]+[←] and [Shift]+[→] for large steps. A cursor display is located between the cursor controls to display the value and location of the cursor as it moves.

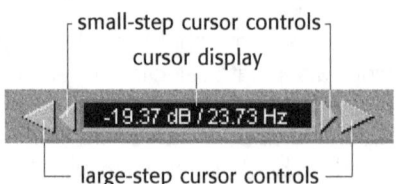

The cursor normally tracks the most recent plot line. You can rotate the cursor through the last ten plot lines by holding down the [Ctrl] key while clicking one of the large-step cursor buttons. This can also be accomplished from the keyboard with the [↑] and [↓] keys.

In the "individual windows" mode the cursors of all the graph windows can be linked. When this is done they will all move together so that clicking on the cursor button of one graph will cause the cursors to move in unison through all the open graphs. The link option is not available for the "single window" mode because it has only one cursor. To link the cursors,

Reference: 6 Evaluating Performance   **BassBox 6 Pro**

right click (🖱) on the graph and use the graph popup menu. The cursors can also be linked with default settings in the "Graph" tab of the Preferences window.

Graph-Specific Cursor Features

Three of the graphs have unique cursor features not shared by the other graphs. These graph-specific cursor features are explained next:

Maximum Acoustic Power graph Most designs will have two plot lines in the Maximum Acoustic Power graph as shown below. The bold plot line depicts the lower of both the displacement limited acoustic power and the thermal limited acoustic power. The dim plot line that continues off the top of the graph depicts the displacement limited acoustic power when it rises above the thermal limit. Normally, the cursor will track the bold plot line as shown below:

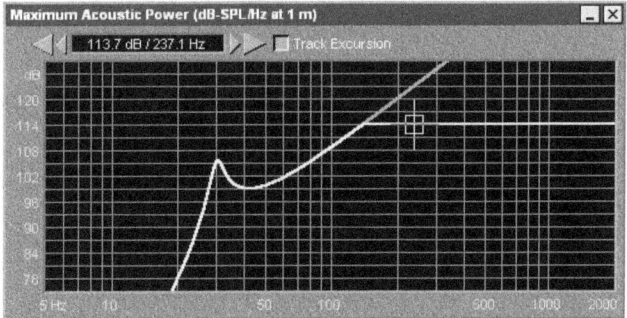

If desired, the cursor can be forced to track the displacement limited plot line only. This is accomplished by clicking on the "Track Excursion" control as shown below:

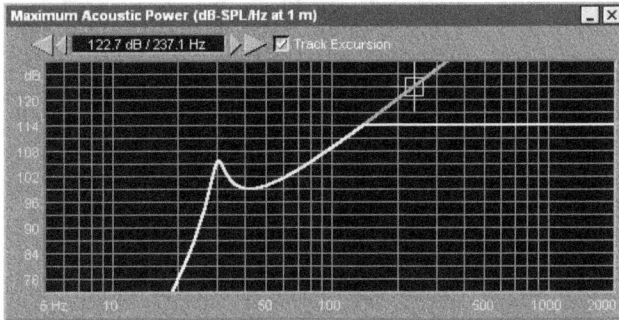

Note: Forcing the cursor to track the displacement limit allows you to read the acoustic power level of the plot line when it rises above the top of the graph.

Cone Displacement graph Passive radiator box designs have two plot lines in the Cone Displacement graph, one for the driver and another for the passive radiator. A "Driver" and "PR" checkbox is provided so you can select which plot line you want the cursor to track as shown below.

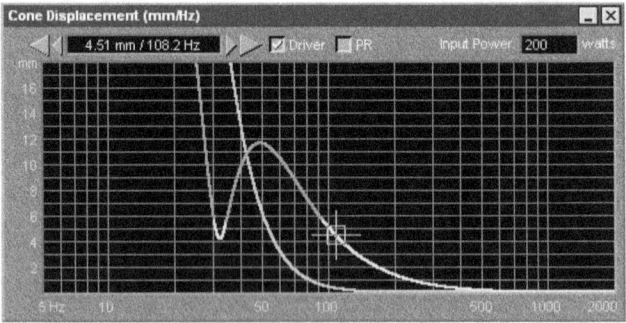

The cursor can also be toggled between the driver and passive radiator plot lines from the keyboard with [Ctrl]+[←] and [Ctrl]+[→].

Vent Air Velocity graph Double-tuned bandpass boxes have two vents, each tuned to a different frequency. This results in two different plot lines, one for the lower frequency (LF) vent and another for the upper frequency (HF) vent. A "LF Vent" and "HF Vent" checkbox is provided so you can select which plot line you want the cursor to track as shown below:

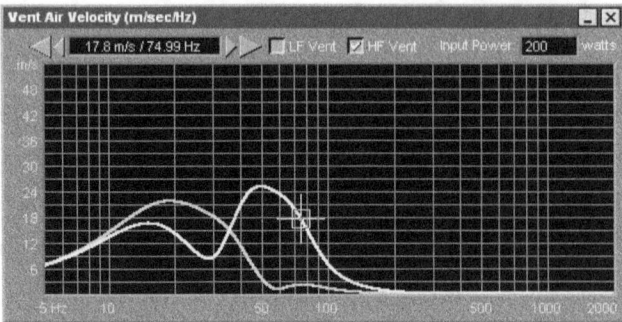

The cursor can also be toggled between the LF vent and HF vent plot lines from the keyboard with [Ctrl]+[←] and [Ctrl]+[→].

Grid If desired, the grid in one or all of the graphs can be turned off. An example with the grid turned off is shown below:

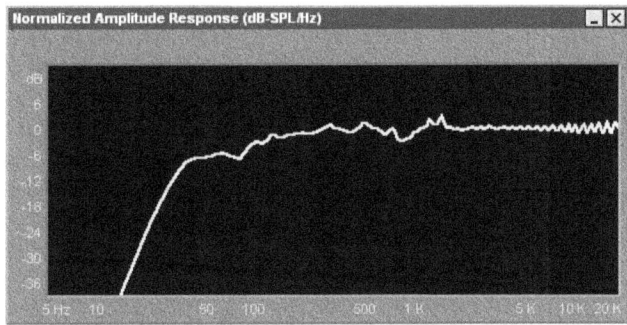

Scale The vertical and horizontal scales are set separately. The vertical scale has a "normal" and "expanded" setting. The actual scale used will depend on the graph. In general, the expanded selection will display a larger vertical range. The horizontal scale can be switched between 5 Hz to 2 kHz and 5 Hz to 20 kHz.

Changing either graph scale will cause the graph to be cleared and then replotted. Only the last ten plots will be replotted because BassBox Pro has a ten-plot graph buffer. If there are more than ten plots, some will be lost. Use the graph memories to store important plots so they can be recalled even if they are not restored after one of the scales is changed.

Graph Size ("individual windows" mode only) The individual graph windows are available in a "normal" and "large" size. All the previous examples in the chapter have used the "normal" size graph windows. A large one is shown below:

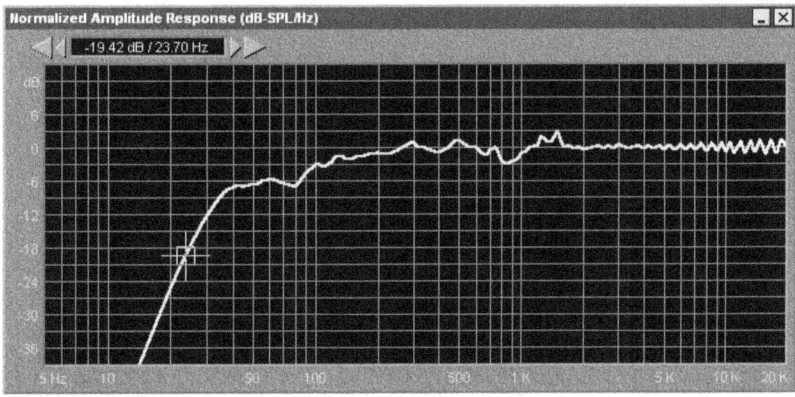

The large-size graph window does not increase the range of either the vertical or horizontal scales but it does add more data points, improving the detail of the plot lines. This is because the plotting functions in BassBox Pro calculate a graph data point for each pixel column in the graph.

Include Six options can be included in the plotting function of several graphs. These options are global in scope. This means that turning on an "include" option will turn it on in all relevant graphs, including the mini preview graph of the main window. However, changing an "include" option will have no effect on existing plot lines—it only affects future plotting. Some of the "include" options will not be available when there is insufficient data or when an option is not relevant to the open designs. Each one is described next:

Driver Acoustic Response
If the acoustic response of the driver has been entered, it will be included in three graphs when this option is turned on. (Keyboard shortcut: Ctrl+I.) The graphs affected by this option are: the Normalized Amplitude Response, Custom Amplitude Response and Maximum Acoustic Power graphs.

Vent Resonance Peaks
The "pipe" resonance peaks created by the vent in vented and bandpass boxes will be included in all nine graphs when this option is turned on. (Keyboard shortcut: Ctrl+V.) See the "Vents" section of Chapter 4 (page 235) for more information about vent "pipe" resonance.

Room/Car Acoustic Response
If the acoustical data for the architectural/automotive environment has been entered, it will be included in three graphs when this option is turned on. The graphs affected by this option are: the Normalized Amplitude Response, Custom Amplitude Response and Maximum Acoustic Power graphs. (Keyboard shortcut: Ctrl+R.)

Note: The Driver Properties window will display a car icon for this setting when the "automotive" acoustic mode is selected and it will display a house icon when the "architectural" acoustic mode is selected. An example of each is shown below:

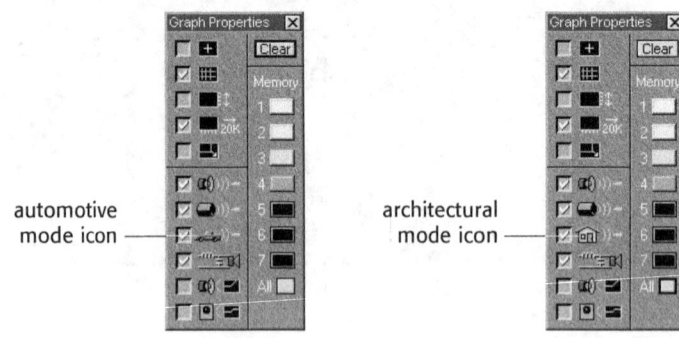

External Network Response

If an external network has been entered, it will be included in seven graphs when this option is turned on. (Keyboard shortcut: Ctrl+E.) All graphs are affected by this option except the Maximum Acoustic Power and Maximum Electric Input Power graphs.

On-Axis Piston Band Response

The "piston band" of a driver is related to the diameter of its diaphragm or piston. It is the frequency band where the driver maintains a constant load versus frequency. This begins at the frequency whose wavelength is equal to the circumference of the diaphragm and it extends upward in frequency. In the piston band, the on-axis response of most drivers will begin to increase with frequency at a rate of 6 dB/octave because the beamwidth or coverage angle of the driver will gradually narrow as the frequency increases. This effect can be included in the Normalized Amplitude Response and Custom Amplitude Response graphs. (Keyboard shortcut: Ctrl+T.)

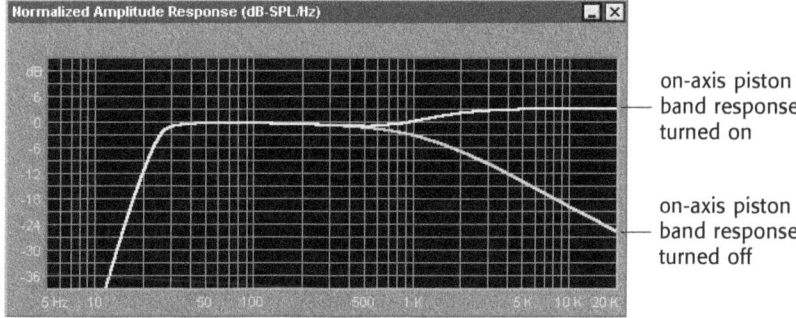

In the example above, the lower plot line shows the response with the on-axis piston band response option turned off. The reason the response takes a "nose dive" at higher frequencies is because the reactance of the voice coil gives it an "apparent" inductance (Le) of 1 mH. This normally isn't a problem for a woofer that is being used to produce low frequencies—not high frequencies. And you would not see much of the phenomenon if the graph's horizontal scale was set to stop at 2 kHz. But with a 20 kHz scale, it looks awful.

The upper plot line shows the same driver with the on-axis piston band response option turned on. This is what the driver should really sound like on-axis. However, as you move off-axis, the high-frequency response will begin to fall.

How important is it to use this option? The answer varies with each driver. If the driver's acoustic response has been entered, then this phenomenon was probably included in the acoustic measurement data. If this is the case, you should turn off this option. In the past, the on-axis piston band response has been ignored because it is somewhat counteracted by the −6 dB/octave roll-off that results from the inductive rise of many voice coils. However, as shown in the preceding illustration the two seldom cancel each other perfectly and you may want to enable this option whenever the Le parameter has been entered and disable it whenever the Le parameter is unknown.

Finally, this option is not as relevant when the graph's horizontal scale is limited to 2 kHz or when a crossover network makes this phenomenon a mute point.

Diffraction Response Shelf

Diffraction is the bending of sound waves as they pass near an edge or corner of a solid object. The box shape and mounting location of the driver can have an enormous effect on how severely the sound diffracts as it emanates from the driver. The "diffraction response shelf" option provides a generalized approximation of some of the diffraction effects for three similar box shapes: cube, square prism and optimum square prism. This option is not available for other box shapes and it assumes that the driver is mounted near the center of the front side. (This is not the best place to mount a driver because it results in the greatest diffraction effects.)

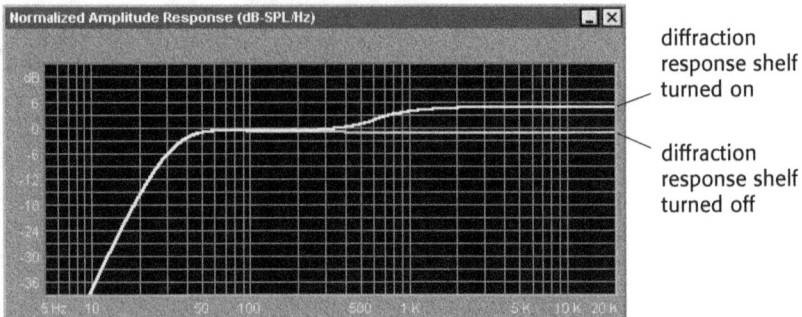

diffraction response shelf turned on

diffraction response shelf turned off

This option is not available when the box dimensions are unknown because it requires the front panel circumference. The graphs affected by this option are: the Normalized Amplitude Response and Custom Amplitude Response graphs. (Keyboard shortcut: [Ctrl]+[F].)

Minimize Hides either the selected graph or all open graphs and places a graph icon on the Windows desktop. To restore the window, simply click on its icon.

Copy Creates a copy of the selected graph in the Windows clipboard so that it can be pasted into another application like a word processor or page layout program. (Keyboard shortcut: Ctrl+C.) It actually "prints" to the clipboard, creating a printable version of the graph with a white background as shown below. The color of the plot lines is preserved, except for white which is switched to black. Be aware, however, that some colors like yellow may be very light.

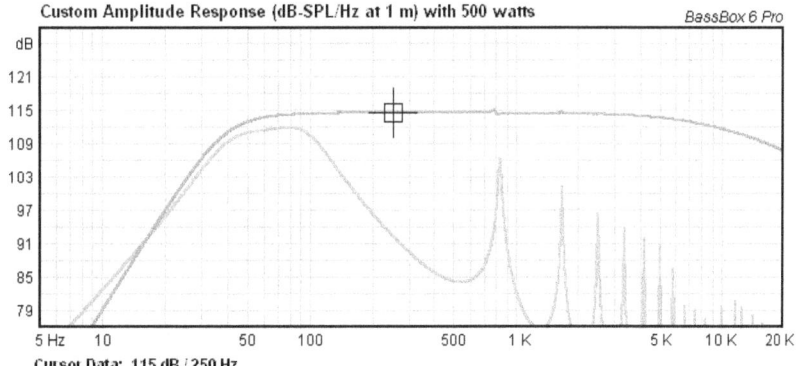

The Copy function includes as many as the ten most recent plot lines. It also includes the cursor, if it is turned on.

Close Closes either the selected graph or all open graphs. The graph popup menu provides a separate command for each. You can also use the Close button in the title bar of the graph. Holding down the Shift key while clicking the Close button of a graph title bar will close all open graphs.

The Graph Properties window will be automatically closed when the last graph is closed.

Properties Opens the Graph Properties window. As long as you never explicitly close the Graph Properties window (that is, you never click on its Close button), BassBox Pro will automatically open and close it whenever the first graph opens and the last graph closes.

That concludes the coverage of all of the graph features in the graph popup menu and Graph Properties window. But there is one more feature that should be mentioned here...

Changing the Plot Color

Each design panel in the main window has a "Plot" button and beside it is a plot color indicator. This is shown below:

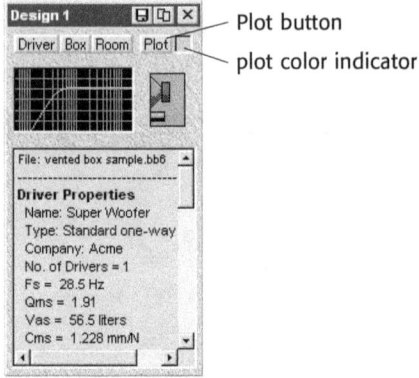

The plot color can be changed by clicking on the plot color indicator. The color will advance one-at-a-time through a 12-color palette with each single click. The first ten colors of the palette are fixed (red, orange, yellow, greenish-yellow, green, cyan, blue, magenta, white and light grey). But the last two colors are "custom" colors and you can make them anything you'd like. You can set the default plot color that is used for ten designs (the maximum number of designs that can be open at the same time) and you can select the custom colors with the "Graph" tab of the Preferences window.

Changing a plot color will not change the color of plot lines that already exist for the selected design. It will only affect future plotting of that design.

Next, let's examine each of the nine graphs. To select a graph use the "Show Graph" command of the Graph menu as shown below:

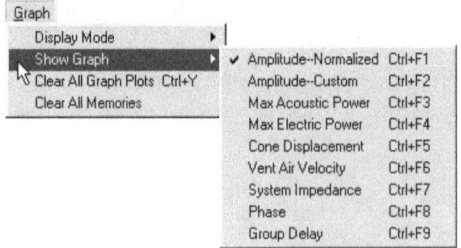

You can also use the keyboard shortcuts: Ctrl + F1 to F9.

Normalized Amplitude Response

Two types of amplitude response graphs are provided by BassBox Pro. This section will describe the "Normalized" Amplitude Response graph. It is "normalized" to zero dB and does not display differences in efficiency between different box types. This makes it easier to compare the shape of the response curves of different boxes. An example is shown below. *Note: Amplitude response graphs are also referred to as "frequency response" or "magnitude response" graphs.*

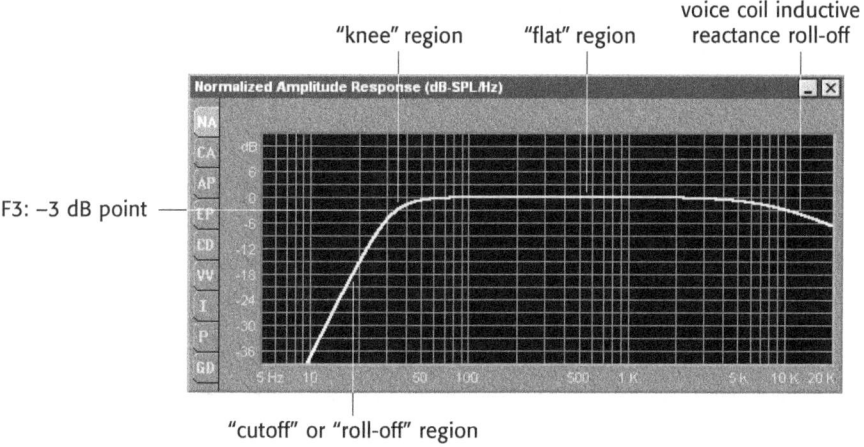

Notice that several characteristics of a typical amplitude response curve are labeled in the above example. They are explained next:

"Cutoff" region Most speakers have a low-frequency response limit—a point beyond which they can no longer reproduce sound waves efficiently. This is visible in an amplitude response graph as the portion of the plot line where the response drops rapidly. This region is known as the "cutoff" or "roll-off" region. Most designers of high-fidelity speakers try to push this cutoff region to the lowest possible frequency in order to get the best possible bass response. The slope of the cutoff region varies with the box type. See Chapter 3 ("How to Choose a Box") in the *Box Designer's Guide* earlier in this manual for more information. *Note: Bandpass speakers have two cutoff regions, a lower frequency cutoff and an upper frequency cutoff.*

F3 The "F3" parameter is considered a very important number because it usually identifies the beginning of the cutoff region. Because of this, F3 is often referred to as the "corner" or "cutoff" frequency and it occurs at approximately 34 Hz in the graph above. It is the frequency where the response has fallen 3 dB. Since a drop in 3 dB corresponds to a halving of the speaker's acoustic power, F3 is also referred to as the "half-power" frequency. F3 is usually located in the "knee" of the response—the portion of the plot line where the cutoff region begins. However, BassBox Pro calculates F3 a bit differently. It tracks the amplitude response in a descending frequency direction until the first −3 dB level is found and identifies this point as F3. This may or may not be in the "knee" of the response curve because some speaker designs have a sagging or rippled bass response. BassBox Pro uses the method it does because it provides an absolute −3 dB frequency rather than a relative one. *Note: Bandpass speakers have two F3s, a lower frequency F3 and an upper frequency F3.*

"Knee" region The "knee" region of the plot line is the area where the response begins to rapidly drop or bend downward. It is also the point where the cutoff region begins. F3 is usually, but not always, located in the "knee" (exceptions were noted earlier).

"Flat" region The "flat" region of the plot line represents the frequencies where the speaker reproduces sound waves with the same loudness. For high-fidelity playback, an ideal speaker should usually have a flat response for its entire operating range or passband. Therefore, most designers try to achieve the flattest possible response. This is also referred to as "maximally flat".

Inductive reactance roll-off All speakers that use a moving coil type motor have inductive reactance because a coil is an inductor. This is represented with the driver parameter "Le". Unfortunately an inductor acts like a 1st-order (6 dB/octave) low-pass filter to limit the high-frequency response of audio signals flowing through it. Generally speaking, larger voice coils have greater inductance and therefore, a greater high-frequency roll-off. This is represented by the falling plot line at the high-frequency end of the graph. *Note: In many speakers, this high-frequency inductive reactance roll-off is balanced by an on-axis 6 dB/octave rise in response due to the piston band narrowing of the driver's coverage angle.*

Is a flat response always ideal? No. Sometimes, a non-flat response is needed to complement the acoustic environment where a speaker will be used. Musical instrument speakers, which are used for sound creation rather than reproduction, often do not have a flat response. However, most of the time a flat response is desired for sound reproduction.

Note: The algorithms in BassBox Pro calculate the industry-standard "half-space" response. This is the response of a speaker radiating into a hemisphere as if the speaker were located next to a wall or a floor. In some listening rooms, a speaker may radiate into a smaller space. For example, placing a speaker next to both a wall and a floor will cause the sound to radiate into a "quarter-space". Placing a speaker into a corner and near a floor will cause it to radiate into an "eighth-space". Reducing the radiating area will cause a rise in low-frequency response because the longer omnidirectional sound waves will be concentrated into a smaller area. A 6 dB increase in bass per halving of area is typical.

Multiple Drivers
Because the sound level in this graph is normalized, the height of the plot line is not affected by the number of drivers in the box.

Custom Amplitude Response

Two types of amplitude response graphs are provided by BassBox Pro. This section will describe the "Custom" Amplitude Response graph. It is "customized" by the input power or voltage level that you choose and it shows the corresponding loudness at a distance of 1 meter (3.3 feet) on-axis from the speaker. If no input level is chosen, the program will use the driver's net power rating (Pe). An example is shown below. *Note: Amplitude response graphs are also referred to as "frequency response" or "magnitude response" graphs.*

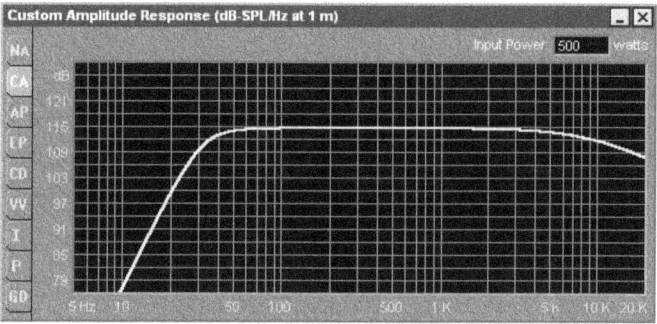

Please refer to the previous "Normalized Amplitude Response" topic for a detailed description of the response plot line and what it's shape represents. The only difference here is that this graph is not normalized to zero dB. Instead, it shows the actual, on-axis direct sound pressure level that you would expect to measure one meter in front of the speaker, in this case 115 dB over most of its passband while being driven with 500 watts.

It is easy to switch between volts and power. Simply click on the input label as noted below.

Click on the input label to toggle between power and voltage.

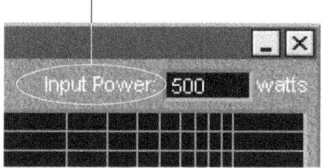 

Internally, BassBox Pro uses only voltage. When an input power level is entered, it converts the level to a voltage based on the net impedance of the driver or drivers.

Multiple Drivers

Since BassBox Pro allows you to design boxes with multiple drivers, it is important to understand how the program interprets the input power and voltage setting of this graph. With one exception, the input level is the total power or voltage driving ALL drivers and the sound pressure level in the graph represents the summed level from ALL drivers. For example, an input power level of 500 watts is used with a speaker that has two woofers. BassBox Pro will interpret the 500 watts as the total for both woofers so it will assume that each woofer receives only 250 watts. If the graph shows a 121 dB sound pressure level, it should be assumed that this is the level that results when the sound waves from both woofers are summed together.

The one exception occurs when the "Separate" electrical configuration is selected on the "Configuration" tab of the Driver Properties window (see Chapter 3). In this case the input level is the individual power or voltage driving just ONE driver. If the "Drivers do NOT add coherently" mechanical setting is turned OFF on the same tab, then the sound pressure level in the graph represents the summed level from ALL drivers. However, if the "Drivers do NOT add coherently" mechanical setting is turned ON then the sound pressure level in the graph represents just ONE driver.

Maximum Acoustic Power

This graph shows the maximum loudness that the speaker is capable of achieving before it reaches either its steady-state linear excursion limit or its thermal limit. An example is shown below:

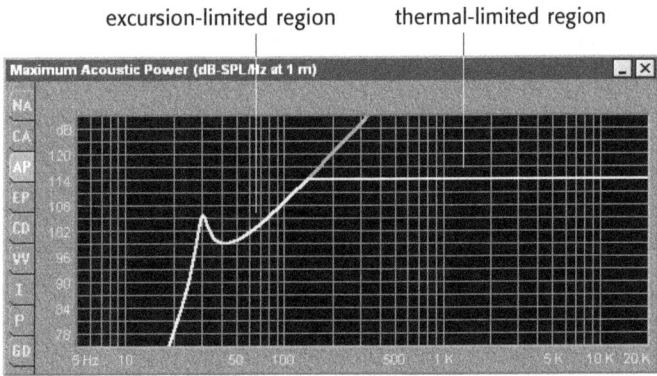

The bold plot line shows the combined excursion-limited and thermal-limited maximum acoustic power. The darker line that continues off the top of the graph shows what the maximum acoustic power level would be if there were no thermal limit. *Note: The thermal limit is a mid-band limit as expressed in the driver parameter Pe.* It is easy to distinguish the linear excursion limit from the thermal limit because the excursion limit changes with frequency and falls rapidly at low frequencies. The thermal limit is a flat horizontal line. The maximum acoustic power is determined by whichever of these limits is lower. Notice in the example above that there is a peak in the excursion-limited region of the plot line around 30 Hz. This peak shows that this is a vented box and the peak occurs because driver excursion is reduced at the resonance frequency of the box.

Note: "Excursion" and "displacement" are synonyms. In other words, the excursion limit is identical to the displacement limit.

One of the challenges in speaker design is to achieve the desired low frequency response <u>and</u> stay within the linear excursion and power-handling limits of the driver. Ideally, you'd like the plot line in this graph to be flat and at the maximum level for the entire passband shown in the amplitude response graphs. It is obvious in our example, that this is not the case. In our example, the amplitude response is flat down to about 50 Hz and yet the maximum acoustic power begins to reach its excursion limits at around 130 Hz and has dropped a whopping 15 dB by the time it gets to 43 Hz. The bottom line: this driver doesn't have enough linear excursion (Xmax is too low) to support its low-frequency response.

Steady-State versus Peak

The excursion-limited plot line represents the driver diaphragm being displaced to its linear excursion limit (Xmax). In this case Xmax is reached only at the peak of the displacement and yet acoustic power must be calculated from an RMS level because this more closely represents the way our hearing perceives loudness. So the maximum acoustic power formulas effectively reduce the excursion level to a "steady-state" or RMS level.

Direct comparisons cannot be made between the Maximum Acoustic Power graph and the Cone Displacement graph because the latter represents only peak displacement. For example, the transition from being thermally-limited to being excursion-limited in the Maximum Acoustic Power graph does not coincide with the point where Xmax is reached on the Cone Displacement graph. In addition, the Cone Displacement graph assumes that a constant input voltage is present while the Maximum Acoustic Power graph allows the input voltage to change as needed to maintain a constant excursion.

The thermal limit is calculated from the driver's power rating (Pe). It can represent either an RMS level or a peak level, depending on which power rating is entered for parameter Pe. A conservative RMS power rating is usually recommended.

Why show acoustic power instead of sound pressure?

The acoustic power is the total power radiating from the speaker in every direction. As such, the acoustic power level includes the on-axis sound waves as well as the off-axis sound waves that radiate into the room at other angles, adding to the total perceived loudness. This is important because most speakers are fairly omnidirectional at low frequencies. So much of their sound energy will radiate off-axis. In contrast, the sound pressure level is valid at only one location, usually on-axis from the speaker, and it does not include the sound that radiates at other angles. Because of this, comparisons of maximum low-frequency loudness between different speaker designs are more accurate when using acoustic power.

What happens when an excursion limit is exceeded?

When a driver is pushed beyond its Xmax limit, it becomes increasingly nonlinear and so its nonlinear distortion rises. If the driver is pushed hard enough it will reach its mechanical excursion limit (Xmech). When Xmech is reached, the driver may suffer physical damage like a torn spider or surround or a damaged coil former. The Cone Displacement graph shows just how far the driver's piston must move for a particular input power or voltage level.

What happens when the thermal limit is exceeded?

If the thermal limit is exceeded for only a very short time, the driver may not be damaged. However, prolonged high power levels that exceed the thermal limit will damage ("burn") the voice coil and destroy the driver. In a worst-case scenario, the voice coil may ignite, creating a life-threatening situation if the fire spreads.

What do you do when the driver can't handle the power?

You can select a different driver or you can try a smaller box that is tuned to a higher frequency. However, the latter choice may result in an amplitude response that is not flat and so the fidelity may be less.

Multiple Drivers

Since BassBox Pro allows you to design boxes with multiple drivers, it is important to understand what this graph represents when multiple drivers are used. With one exception, the maximum acoustic power level in the graph represents the summed level from ALL drivers.

The one exception occurs when the "Separate" electrical configuration AND "Drivers do NOT add coherently" mechanical configuration on the "Configuration" tab of the Driver Properties window are both selected at the same time (see Chapter 3). In this case the maximum acoustic power level in the graph represents just ONE driver.

Maximum Electric Input Power

This graph is derived from the maximum acoustic power level of the previous graph. It shows how much amplifier power the speaker can handle before the driver reaches the first of either its linear excursion or thermal limit. An example is shown below:

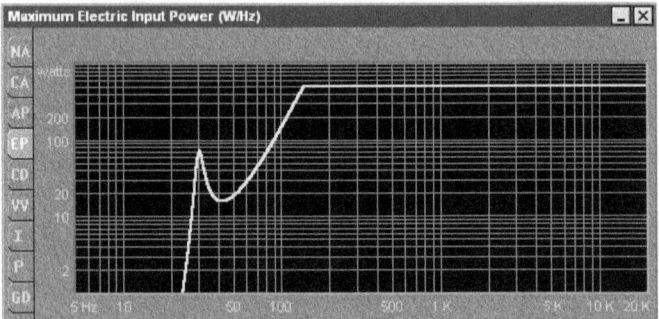

Ideally, the speaker design should to be able to handle a desired amount of power over its entire passband. This is not the case in this example because it reaches its steady-state excursion limit much earlier than the 50 Hz cutoff frequency shown in the Normalized Amplitude Response graph. Depending on how much additional travel the mechanical excursion limit may allow, it might be prudent to add a high-pass filter to this speaker to protect it from full power levels at low frequencies.

The plot lines in this graph may look worse than they really are. This is because there is only a 3 dB change in loudness every time the electrical input power is halved. For example, the 500 watt power level in this graph corresponds to the 115 dB sound pressure level in the Custom Amplitude Response graph. The 250 watt level will be just 3 dB less or a sound pressure level of 112 dB. The 125 watt level will be another 3 dB less or a sound pressure level of 109 dB.

There are situations not considered in this graph that can further reduce the power handling of the speaker. For example, driving the speaker with a heavily distorted or "clipped" signal will reduce the maximum power it can handle. This has surprised many people! It means that a speaker that can handle 200 watts may be damaged by a 100 watt amplifier that is frequently pushed into heavy distortion.

Multiple Drivers

Since BassBox Pro allows you to design boxes with multiple drivers, it is important to understand what this graph represents when multiple drivers are used. With one exception, the maximum electric input power level in the graph represents the total level that ALL drivers in the speaker can handle. For example, a maximum electric input power level of 500 watts for a speaker with two woofers means that the speaker can handle a total of 500 watts but each woofer can only handle 250 watts.

The one exception occurs when the "Separate" electrical configuration setting is selected on the "Configuration" tab of the Driver Properties window (see Chapter 3). In this case the maximum electric input power level in the graph represents just ONE driver.

Cone Displacement

This graph shows how far the driver (or passive radiator) diaphragm or piston must move in one direction for a specified input power or voltage. An example is shown below:

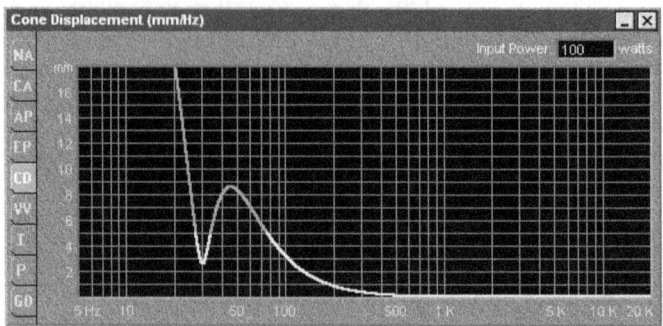

Notice that the plot line has two intensities. The darker part of the plot line indicates that the driver is exceeding its linear excursion limit (Xmax). This graph is very helpful at seeing just how far beyond Xmax the driver's diaphragm may be asked to move. Ideally, the plot line should never go above the Xmax level as long as the input power or voltage is kept within the limits of the driver.

The Cone Displacement graph depicts the peak diaphragm displacement and it assumes that the input voltage is constant. This means that it cannot be compared directly with the Maximum Acoustic Power or Maximum Electric Input Power graphs which both use a steady-state excursion limit and allow the input voltage to vary with frequency.

The units for the vertical scale of this graph can be switched between metric (millimeters) and English (inches) with the "Graph" tab of the Preferences window (see Chapter 13).

Like the Custom Amplitude Response graph, the input for this graph can be switched between power and volts by clicking on the input label.

Multiple Drivers

Since BassBox Pro allows you to design boxes with multiple drivers, it is important to understand how the program interprets the input power and voltage setting of this graph. With one exception, the input level is the total power or voltage driving ALL drivers. For example, an input power level of 100 watts is used with a speaker that has two woofers. BassBox Pro will interpret the 100 watts as the total for both woofers and it will assume that each woofer receives only 50 watts. The one exception occurs when the "Separate" electrical configuration is selected on the "Configuration" tab of the Driver Properties window (see Chapter 3). In this case the input level is the individual power or voltage driving just ONE driver.

The displacement level in the graph always represents just ONE driver.

Vent Air Velocity

This graph shows how fast air will move in and out of the vent for a specified input power or voltage. An example is shown below:

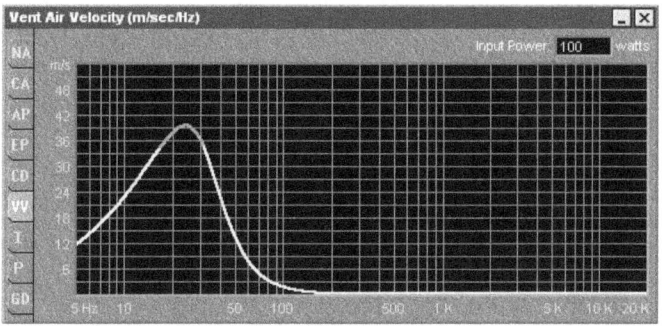

Notice that the plot line in the Vent Air Velocity graph has two intensities. The darker part of the plot line indicates that the air velocity in the vent is exceeding 10% of the velocity of sound in air (approximately 35 m/s or 113 ft/s) and will probably cause audible turbulence or whistling. Ideally, it would be better to keep the air velocity in the vent below 10 m/s or 33 ft/s to be certain that no vent noise is heard but this can be very challenging to do for some designs—especially with low-frequency speakers like subwoofers.

If the air velocity in the vent is too high, it can be reduced by one or both of the following:

- Increase the cross-section area of the vent(s). This can be done by using more vents or by increasing the diameter of the vent(s). *Caution: Increasing the number of vents or their diameter will also require them to be longer.*
- Decrease the amplifier power with which the speaker will be driven.

For more information on vents, please review the "Vents" section in Chapter 4.

The units for the vertical scale can be switched between metric (meters/second) and English (feet/second) with the "Graph" tab of the Preferences window (see Chapter 13).

Like the Custom Amplitude Response graph, the input for this graph can be switched between power and volts by clicking on the input label.

Multiple Drivers

Since BassBox Pro allows you to design boxes with multiple drivers, it is important to understand how the program interprets the input power and voltage setting of this graph. With one exception, the input level is the total power or voltage driving ALL drivers. For example, an input power level of 100 watts is used with a speaker that has two woofers. BassBox Pro will interpret the 100 watts as the total for both woofers and it will assume that each woofer receives only 50 watts. The one exception occurs when the "Separate" electrical configuration is selected on the "Configuration" tab of the Driver Properties window (see Chapter 13). In this case the input level is the individual power or voltage driving just ONE driver.

The vent air velocity in the graph represents the level created by the movement of all driver diaphragms with one exception. When the "Separate" electrical configuration AND the "Drivers do NOT add coherently" mechanical configuration are both selected, the vent air velocity will represent the level created by the movement of just ONE driver diaphragm.

System Impedance Response

This graph shows the impedance response. The interpretation of the graph will depend on the presence of an external network. An example is shown below:

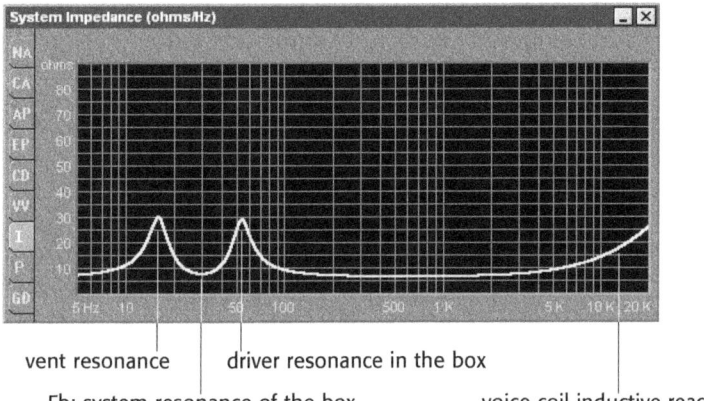

vent resonance | driver resonance in the box
Fb: system resonance of the box | voice coil inductive reactance

The impedance response of a vented box is displayed above. The significant features of the plot line are labeled.

The tuning frequency, Fb, is the system resonance of the box and is located at the minima between the two peaks. Fb is the product of both the box volume and the vent dimensions. The first peak is the vent and the second peak is the driver. The driver resonance is always higher in the box than in free air. **Tip:** The phase of the impedance should pass through zero degrees at the vent resonance peak, box system resonance minima and driver resonance peak. The cursor display shows both the impedance magnitude and phase.

If an external passive network (such as a passive crossover network, impedance equalization network or L-pad) is present between the amplifier and the speaker and its parameters have been entered into BassBox Pro then the impedance plot will represent the net impedance of the speaker (including the network) as seen by the amplifier. If no external passive network is present or if its parameters have not been entered then the impedance plot will represent the impedance of the speaker box and drivers only.

Ideally, the impedance seen by the amplifier or an external network should be as flat as possible and, in the case of an amplifier, should be within its acceptable load range. For example, an amplifier that is designed to drive a speaker with an impedance no lower than 4 ohms may have trouble driving a speaker whose impedance dips significantly below 4 ohms because the speaker may attempt to draw too much current from the amplifier.

Multiple Drivers
Since BassBox Pro allows you to design boxes with multiple drivers, it is important to understand what this graph represents when multiple drivers are used. With one exception, the impedance response in the graph represents the net impedance of ALL the drivers in the speaker. The net impedance is calculated according to the electrical configuration setting on the "Configuration" tab of the Driver Properties window (see Chapter 13).

The one exception occurs when the "Separate" electrical configuration setting is selected. In this case the impedance response in the graph represents just ONE driver.

Phase Response

This graph shows how much the sound waves emanating from the speaker will lag behind the input signal. This delay is expressed as a phase angle in degrees. It is literally the difference between the phase of the input signal and the phase of the output signal. An example is shown below:

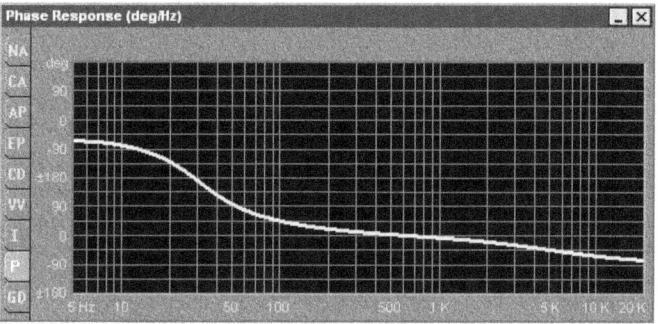

The phase of a perfect sine wave rotates 360° in one complete cycle or wavelength. If this perfect sine wave were used to drive the speaker and the sound that emanated from it had no delay, the phase response would be zero degrees (0°) because the phase of both the input and output would match perfectly. If the sine wave emanating from the speaker were delayed by half a wavelength, the phase response would be 180° at that frequency. In this case, the sine wave emanating from the speaker would be inverted—it would be negative when the input is positive, and visa versa as shown below. See Chapter 2 ("What Sound Is") in the *Box Designer's Guide* earlier in this manual for more information.

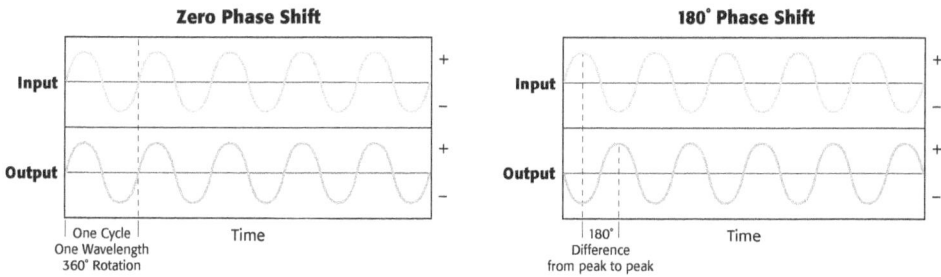

The phase response is most important in the crossover region between two different and adjacent drivers. If the phase response at the crossover frequency of the woofer is 180° different from that of the tweeter in a two-way speaker, then the sound emanating from the two drivers will cancel at the crossover frequency, creating a notch in the amplitude response.

Ideally, there should be no difference in phase from one frequency to the next. This would result in a flat plot line. In practice however, the phase angle changes at both ends of the response. At the low end, the phase angle increases as the frequency decreases because of the natural low-frequency roll-off of all drivers. This is increased by the presence of a box. (Closed boxes typically experience a 180° total phase shift while vented boxes typically experience a 360° total phase shift. Bandpass boxes have an even greater phase shift.) At the high end, the phase angle decreases as the frequency increases because of the inductive reactance of the voice coil. See Chapter 3 in the *Box Designer's Guide* earlier in this manual for more information.

A gradual phase shift is not very audible. However, sudden large shifts in phase are audible and should be avoided. A speaker with minimal phase shift will respond better to transient audio signals than a speaker with sudden shifts in phase. This would be important for a speaker that will be used to reproduce percussive sounds.

Multiple Drivers
BassBox Pro does not do phase summing of multiple drivers and so this graph is not affected by the number of drivers in the box.

Group Delay

This graph is similar to the Phase Response graph except that it expresses the lag of the output audio signal as a delay in milliseconds. The group delay is derived from the slope of the phase response. An example is shown below:

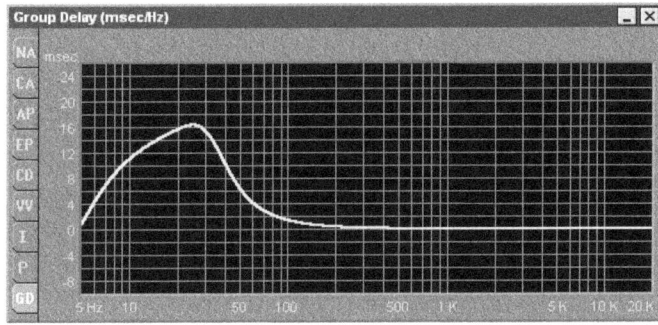

Ideally, there should be no delay from one frequency to the next. However, a uniform delay is usually not a problem as long as it is modest or, if it is significant, the speakers are not used for live sound reinforcement or lipsync to a video image. The most important thing to avoid is a sudden large change in group delay since this can be very audible. An ideal plot line would be flat or have only a modest slope.

Typically, the group delay increases as the frequency decreases. Much of the discussion about the phase response in Chapter 3 of the *Box Designer's Guide* earlier in this manual is also relevant to the group delay. A speaker with a smooth and relatively flat group delay will respond better to transient audio signals than a speaker with a steep, rapidly changing group delay. This would be most important for a speaker that will be used to reproduce percussive sounds.

Multiple Drivers

BassBox Pro does not do phase summing of multiple drivers and so this graph is not affected by the number of drivers in the box.

7 Saving/Opening a Design

While most computer programs run they exist in the "memory" of your computer. This is a temporary or volatile condition because the memory is erased every time the computer is shut off. In fact, BassBox Pro is erased from memory each time that it is shut down with the "Quit" command of its File menu ([Ctrl]+[Q]) or Close button in the title bar.

Without a way to save your work, you would have to re-enter the driver and box parameters each time that you want to resume work on a design. Fortunately, most computer programs provide a way to save and reopen your work and BassBox Pro is no exception.

Saving Your Work

It is considered good practice to frequently save your work. You never know when a power outage or other "glitch" may cause your computer to "hang" or shut down. Fortunately, it is very easy to save a speaker design in BassBox Pro. Simply click on the Save button in the design panel of a design. It is located in the main window as shown below:

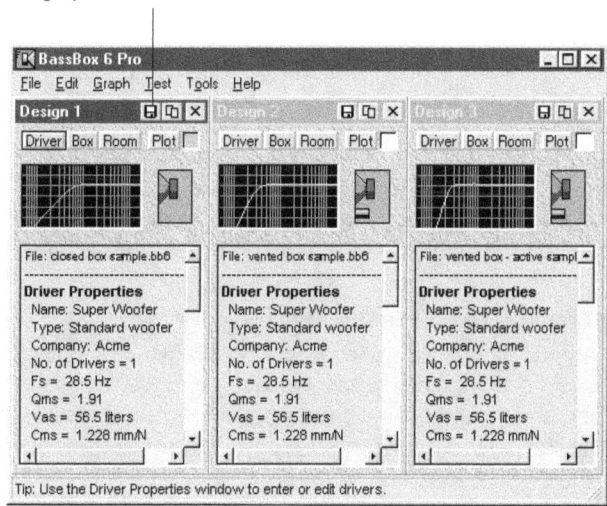

each design panel has its own Save button

Hard Drive Basics

BassBox Pro designs are saved in "files" on the "hard disk" drive of your computer. The hard disk stores the programs and files on your computer—even when the computer is turned off. Hard disks are assigned a drive letter, usually beginning with "C" ("A" and "B" are reserved for floppy disk drives). Because a hard disk can contain millions of files, the hard disk is organized into "folders" or subdirectories. Folders can contain both files and other folders.

For example, BassBox Pro was stored into the "C:\Program Files\HT Audio" folder if the default settings were used during installation. This means that BassBox Pro is located in the "HT Audio" folder inside the "Program Files" folder on hard drive "C". The location of a file or folder is also called its "path". For more information about your computer's hard drive, files and folders, please consult the user manuals of your computer and Microsoft Windows.

Saving a New Design

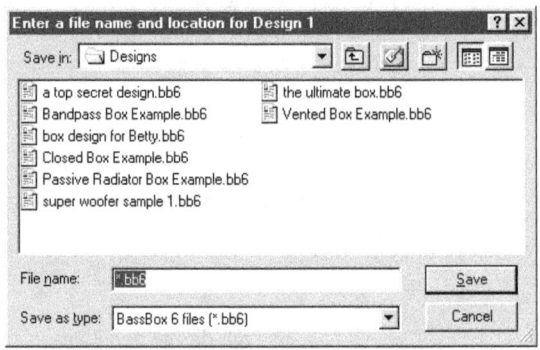

When a new design is saved for the first time, an "Enter a file name and location for Design #" dialog box (shown at right) will open to prompt you for both a location and a file name. BassBox Pro supports long file names but you must always end the name with the extension ".bb6". (This is why BassBox Pro design files are also referred to as "bb6" files.) The ".bb6" extension will be automatically appended if it is omitted from the end of the name.

The "Enter a file name and location for Design #" dialog box is a standard Windows Open/Save dialog box. Its appearance may differ slightly depending on your Windows version. Please consult your Microsoft Windows User Manual or Windows online help if you need assistance with it.

The default location for BassBox Pro design files is "C:\Program Files\HT Audio\Designs" if the default settings were used during the installation of the program.

Saving Changes to an Existing Design

A design that has previously been saved already has a BassBox Pro file on the computer's hard disk. As such, it already has a location and file name. If you decide to make changes to the design and save them, you will <u>not</u> be prompted for a location or file name again. Instead, BassBox Pro will save the design to the same location and file name as before, replacing the original file with the updated one.

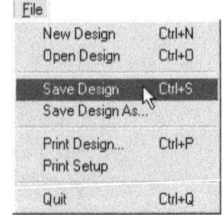

The File menu of the main window (shown at right) has a "Save Design" command. It serves the same function as the Save button in a design panel—it saves the selected speaker design. Notice that it also has a keyboard shortcut: [Ctrl]+[S].

Notice that the File menu contains a second save command. The "Save Design As..." command forces the "Enter a file name and location for Design #" dialog box to appear even when an existing design already has a file name and location. Use this command when you want to save a copy of the existing design with a different file name and/or location. This creates a new copy of the design (the original copy is not changed).

Finally, when you quit BassBox Pro by selecting "Quit" from the File menu ([Ctrl]+[Q]) or clicking on the Close button in the title bar of the main window, the program will first check to see if there are any open designs with changes that have not yet been saved. If there are, BassBox Pro will ask if you want to save them before it shuts down.

Opening a BassBox Design File

After a BassBox Pro design has been saved, how do you open it again later? One way is to use the "Open Design" command in the File menu shown at right. The keyboard shortcut is [Ctrl]+[O].

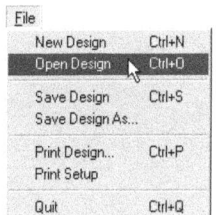

BassBox Pro will open the "Open a BassBox design file for Design #" dialog box to prompt you for a location and a file name. This dialog is a standard Windows Open/Save dialog box and is very similar to the "Enter a file name and location for Design #" dialog box shown on the previous page. Please consult your Microsoft Windows User Manual or Windows online help if assistance is needed.

Opening a Recent Design

BassBox Pro remembers the last four design files that were open and it adds them to the File menu as shown below. This provides a second way to open a BassBox Pro design.

Simply select one of these files from the File menu as shown at right and it will be opened.

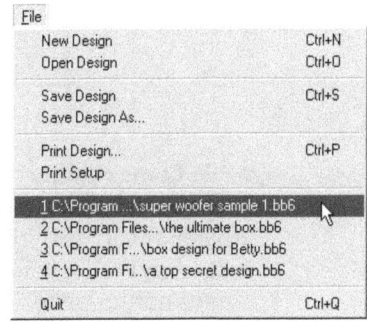

Auto-Starting BassBox Pro by Opening a Design

There is a third way to open a design file. This method works only when BassBox Pro is <u>not</u> running. First, locate the BassBox Pro design file that you want to open. You can do this with the Windows Explorer or you can double-click (🖱🖱) on "My Computer" on the Windows desktop and "drill" down through the folders until you locate the file. If BassBox Pro was installed with the default settings, the default location for speaker design files is "C:\Program Files\HT Audio\Designs". After locating the desired file, double-click (🖱🖱) on its icon. This will cause BassBox Pro to be automatically launched and the selected design file to be automatically opened.

Importing Driver Properties from a BassBox Pro Design File

The driver properties of one BassBox Pro design can be imported into a second BassBox Pro design. Why would you want to do this? Answer: It provides a convenient way to copy driver information from one box design into a different box design without having to re-enter the information. Use the "Import" button on the "Parameters" tab of the Driver Properties window (shown below) to access this feature.

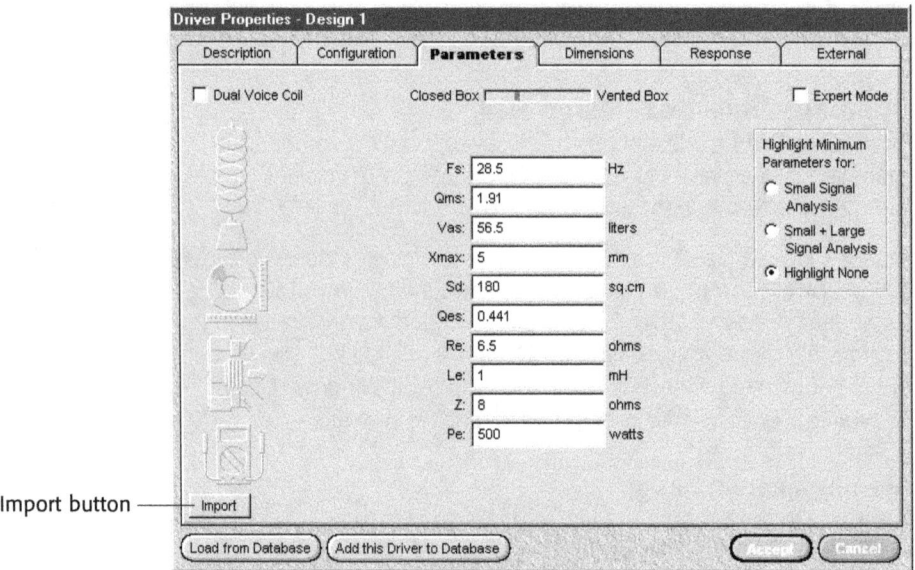

Import button

File Compatibility

BassBox Pro and BassBox Lite "bb6" design files are interchangeable. You can open any BassBox Lite design file with BassBox Pro. Conversely BassBox Lite can usually open BassBox Pro files. Exceptions occur when a BassBox Pro file uses box shapes or certain features not supported by the Lite version.

BassBox Pro can also open files from earlier versions of BassBox including BassBox 5.1, 5.0, 4.0 and 3.0. *Note: To import driver information as described above from an older BassBox file, first open it with the "Open Design" command described on the previous page and save it as a BassBox Pro (bb6) file. Then import the data from the new BassBox Pro file.*

Lastly, X•over Pro (Harris Tech's passive crossover network design program) can also open and import driver and box information from BassBox Pro design files.

Since BassBox Pro design files contain primarily data they are not compatible with third-party programs such as word processors or page layout programs.

Reference: 8 Printing a Design   BassBox 6 Pro

8 Printing a Design

To print a design, select the "Print Design..." command from the File menu of the main window as shown at left below. The keyboard shortcut is Ctrl+P.

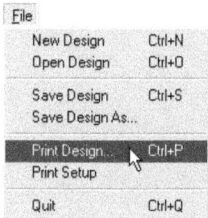 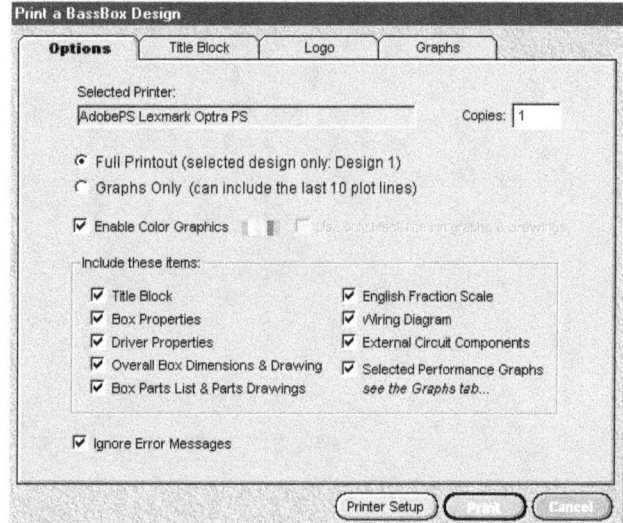

The "Print a BassBox Design" window shown at right above will open. It provides a plethora of printout options which are divided between four tabs. Many of these options can be assigned default values in the Preferences window so that your choices will be selected each time BassBox Pro is launched (see Chapter 13).

Before examining each tab in detail let's discuss the buttons at the bottom of the window:

Printer Setup Click the "Printer Setup" button to open the standard Windows Print Setup window so you can choose another printer and/or change the settings of the printer driver.
Important: Make sure the printer is set to print on either a letter-size or A4-size page in portrait mode. The Print Setup window can also be opened with the "Print Setup" command in the File menu.

Print Click the "Print" button to begin printing using the selected settings.

Cancel Click the "Cancel" button to close the print window and restore the previous settings. If printing has already been started, clicking the "Cancel" button will abort printing and prevent pages that have not yet printed from being printed.

The options of each tab are described next, followed by sample printouts.

Options

The "Options" tab allows you to control the major features of a printout and identifies the selected printer.

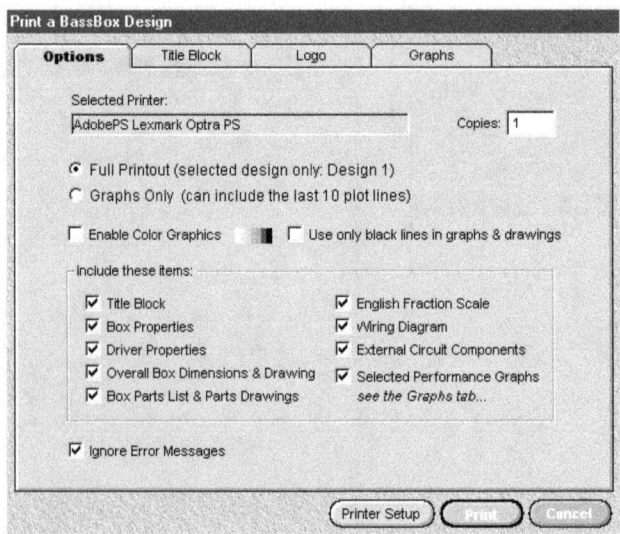

Selected Printer Displays the currently selected printer. The selected printer can be configured or changed by clicking on the "Printer Setup" button. You can also configure or change the printer with the "Print Setup" command in the File menu of the main window.

Copies The number of copies to be printed.

Full Printout (selected design only: Design #) This option makes it possible to include a full description of a design in the printout. However, the printout will only include information about the selected design—it will ignore other designs which may also be open. This also means that the graphs will include only the plot line of the selected design. Other plot lines will be ignored.

Graphs Only (can include the last 10 plot lines) This option limits the printout to including just the graphs so that the printout is <u>not</u> tied to a specific design. This makes it possible to include multiple plot lines of different designs in the graphs when the "Include multiple plot lines" option of the "Graphs" tab is turned on.

Only those plot lines which are visible in the main graphs can be included in the printout. Open designs which have not been plotted will be ignored. This is because the multiple plot line feature uses the graph redraw buffer to gather information for the printout. If an open design has not been plotted, its information will not be present in the graph redraw buffer. *Note: The graph redraw buffer holds only the last 10 plots in memory. This limits the multiple plot line feature to just the last ten plots.*

Since the "Graphs Only" option limits the printout to just the graphs, most of the print options of the "Options" tab will be disabled.

Enable Color Graphics This option enables the graphs to print with color plot lines and the box drawing and box parts drawings to print with color dimension lines. This option will have no effect on the logo picture. If a color logo picture is selected and a color printer is used, the logo will still print in color even if the "Enable Color Graphics" option is turned off.

Use only black lines in graphs & drawings This option controls the way some lines will be printed when the "Enable Color Graphics" option is turned off. It has no effect when the "Enable Color Graphics" option is turned on.

Normally, the "Use only black lines in graphs & drawings" option should be turned off. This will allow grey plot lines to be substituted for the color graph plot lines of the "Graphs Only" mode. The shade of grey of each plot line will correspond to their original color in the on-screen graphs. It will also allow grey lines to be used to represent hidden lines in the box and box parts drawings.

Turn on the "Use only black lines in graphs & drawings" option if your printer has trouble printing grey lines. Only black lines will be used.

Include These Items There are nine items that can be included in a printout. Check the ones that you want to include. They are summarized below:

   Title Block Includes the Title line, Designer line, Company line, Address lines, Tel/Fax line, Notes lines and the logo picture. It is printed on page one.

   Box Properties Includes the general box information such as the box type, description, volume, tuning and (if applicable) vent information.

   Driver Properties Includes the general driver information such as the driver type, description, configuration and parameters.

   Overall Box Dimensions & Drawing Includes a list of both external and internal dimensions. It also includes a scaled external three-dimensional wireframe drawing of the box with dimension labels.

   Box Parts List & Parts Drawing Drawings Include a list of all box parts, vent parts (if applicable) and miscellaneous objects like braces. It also includes a scaled two-dimensional drawing of each box part and the vents.

   English Fraction Scale Includes a small table to convert between the decimal values used in BassBox Pro and the fractions commonly used in the U.S.A.

   Wiring Diagram Includes a configuration wiring diagram similar to the one depicted on the "Configuration" tab of the Driver Properties window. Its primary purpose is to show how to wire a multi-driver design.

External Circuit Components Includes a list of all passive external components included in the design (filter, impedance equalization and/or L-pad). It also adds a schematic of the external network to the wiring diagram.

Selected Performance Graphs Includes the graphs selected on the "Graphs" tab.

Ignore Error Messages Normally this option should be enabled. If you are having print problems, it can be unchecked to enable error messages to aid with troubleshooting.

Title Block

The "Title Block" tab allows you to enter information that will be printed at the top of the first page. Use it to customize the printout. For example, you can add comments for a customer or instructions for a worker.

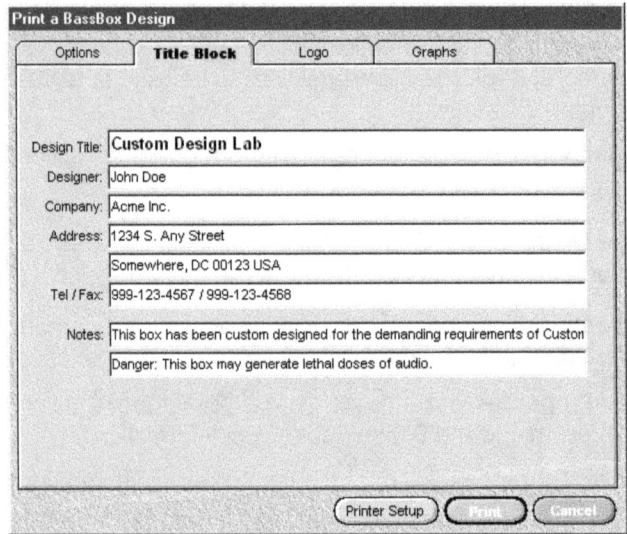

Design Title, Designer, Company, Address, Tel/Fax, Notes These lines are printed at the top of page one in the title block when the Title Block option is enabled on the "Options" tab. *Note: The Designer line will always be preceded with "By ".*

Logo

BassBox Pro allows you to add a custom logo to your printouts. The logo prints in the title block of the first page. A smaller version also prints in the header of successive pages. The logo can be your company logo or a small picture of the box. Three different versions of the BassBox Pro icon are included as default logos.

Notice below that a picture of the logo is displayed in the middle of the "Logo" tab. Whenever the mouse pointer is paused over the logo picture, the path and file name of the logo will be displayed in the balloon help.

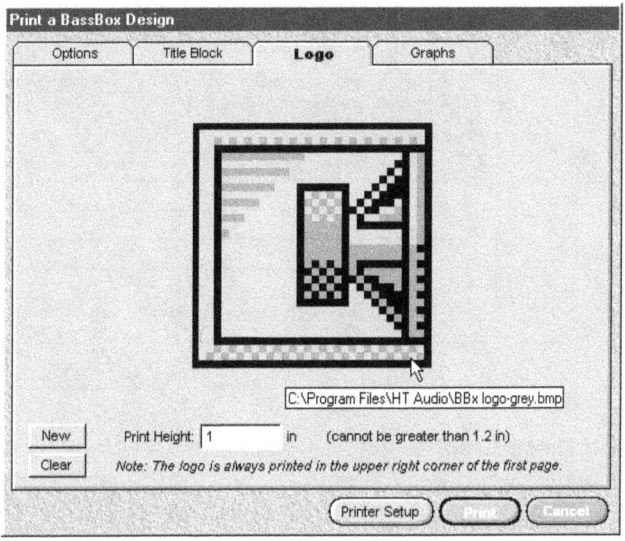

New Click this button to select a different picture for the logo. Several picture file formats are supported including: bmp, dib, gif, jpg, wmf, emf, ico and cur. Three versions of the BassBox Pro icon are included with the program. They are:

BBx logo-color.bmp   a color version of the BassBox Pro icon
BBx logo-grey.bmp   a greyscale version of the BassBox Pro icon
BBx logo-mono.bmp   a black & white version of the BassBox Pro icon

These files are located in the same folder as BassBox Pro ("C:\Program Files\HT Audio").

Clear Clears the logo picture so that no logo will print.

Print Height Sets the height of the logo. The maximum allowable height is 1.2 inches or 30 mm. The logo picture will be proportionally scaled to this height when it is printed. Click the units label to change the units.

Graphs

The "Graphs" tab allows you to control which of BassBox Pro's nine performance graphs will be included in the printout when the "Selected Performance Graphs" option of the "Options" tab is turned on.

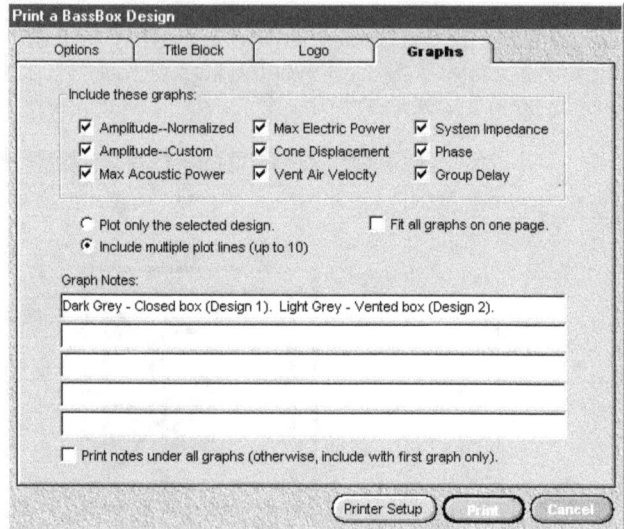

Include These Graphs The nine graph checkboxes in this section select the graphs that will be printed when the "Selected Performance Graphs" option is checked under the "Include these items" section of the "Options" tab.

Some of the nine graphs may be disabled if BassBox Pro is unable to include them. For example, closed box designs do not have vents and so the Vent Air Velocity graph will not be available for a closed box design. Also, some graphs may not be available if there is not sufficient driver and/or box information.

Plot Only the Selected Design Causes the graphs to include the plot lines of the selected design only. This is the only plot line setting that is available when the "Full Printout" option of the "Options" tab is turned on.

Include Multiple Plot Lines (up to 10) This feature is only available when the "Graphs Only" option of the "Options" tab is turned on. It causes the graphs to include the plot lines of multiple designs in the printout.

Only those plot lines which are visible in the main graphs can be included in the printout. Open designs which have not been plotted will be ignored. This is because the multiple plot line feature uses the graph redraw buffer to gather information for the printout. If an open design has not been plotted, its information will not be present in the graph redraw buffer. *Note: The graph redraw buffer holds only the last 10 plots in memory. This limits the multiple plot line feature to just the last ten plots.*

Fit All Graphs on One Page Up to nine graphs can be printed. The total number depends on the type of box and on the amount of driver and box parameters that are available in the design. Normally a page can hold only three graphs. So three pages will be needed to print all nine full-size graphs. This option causes a smaller graph to be printed so that all nine graphs will fit on one page. *Note: Lengthy graph notes may cause even the small graphs to print on more than one page.*

Graph Notes These lines are printed below the graph(s). They provide a convenient way to label graphs—especially when multiple plot lines are included. Each line is limited to 100 characters and blank lines will be ignored. To include a blank line between two lines, enter a space on the blank line so that BassBox Pro will not "think" it is empty.

Print Notes Under All Graphs (otherwise, include with first graph only) When turned on, this option causes the graph notes to be repeated at the bottom of each graph. When turned off, the graph notes will only be printed under the first graph in the printout. This option will be disabled when all graph note lines are empty.

BassBox 6 Pro

Reference: 8 Printing a Design

Sample

A sample printout follows. It includes labels to identify the various parts of each page. The first four pages were created with the "Full Printout" option turned on.

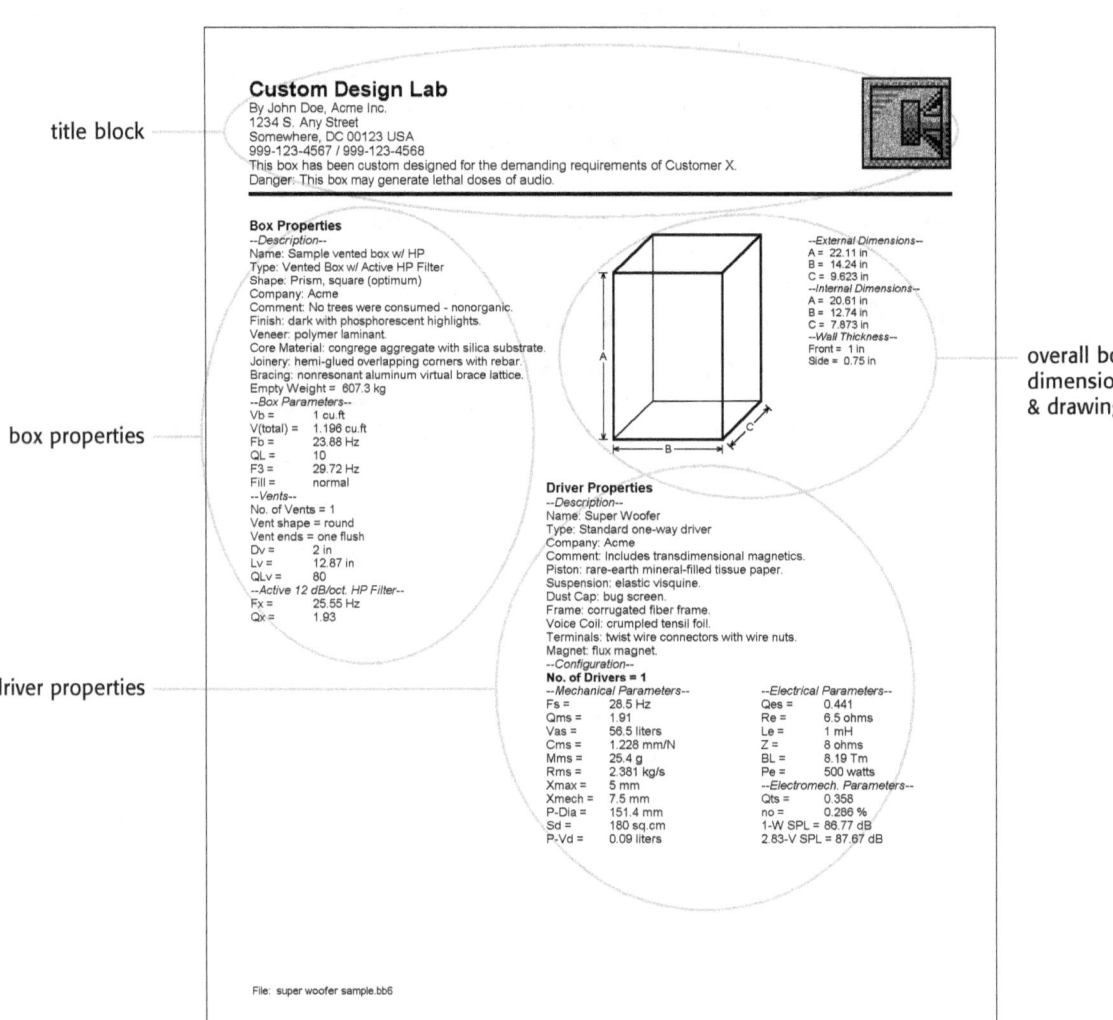

Reference: 8 Printing a Design — BassBox 6 Pro

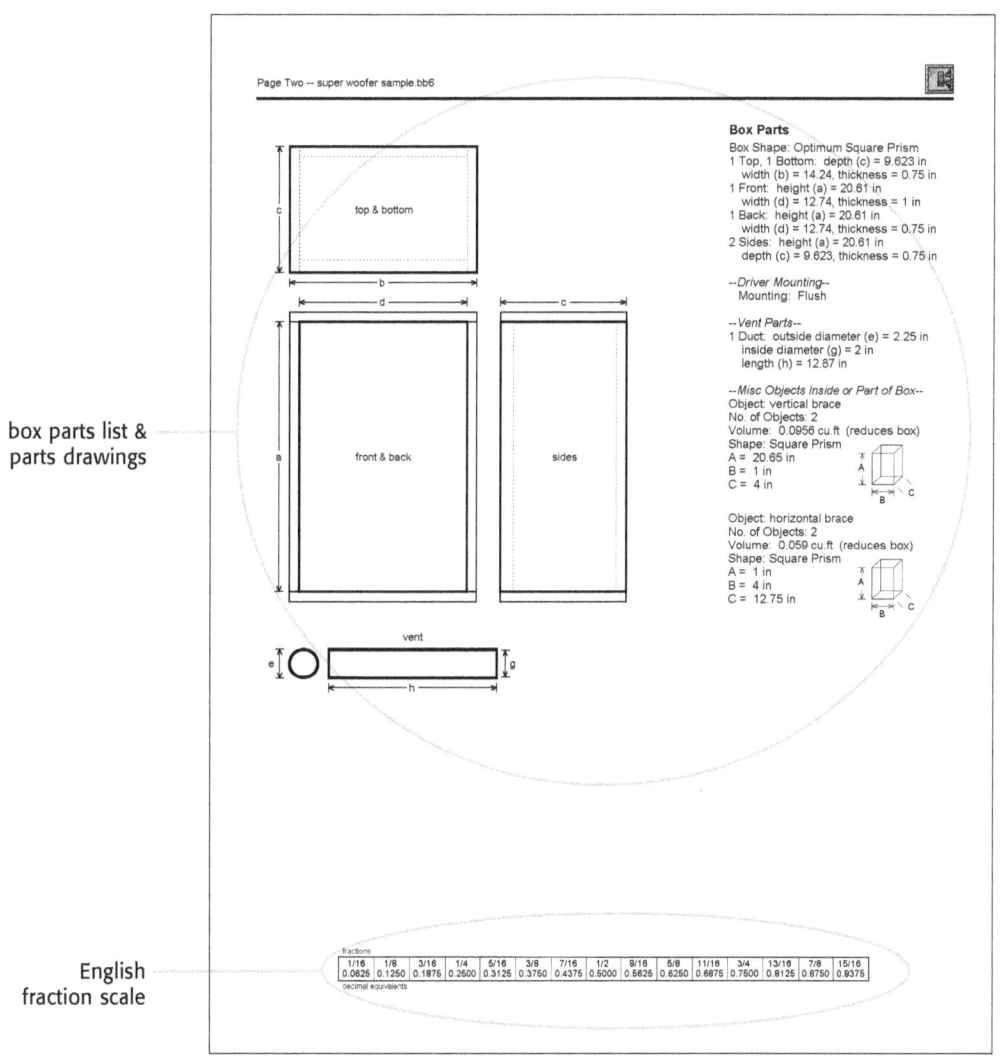

box parts list & parts drawings

English fraction scale

Reference: 8 Printing a Design

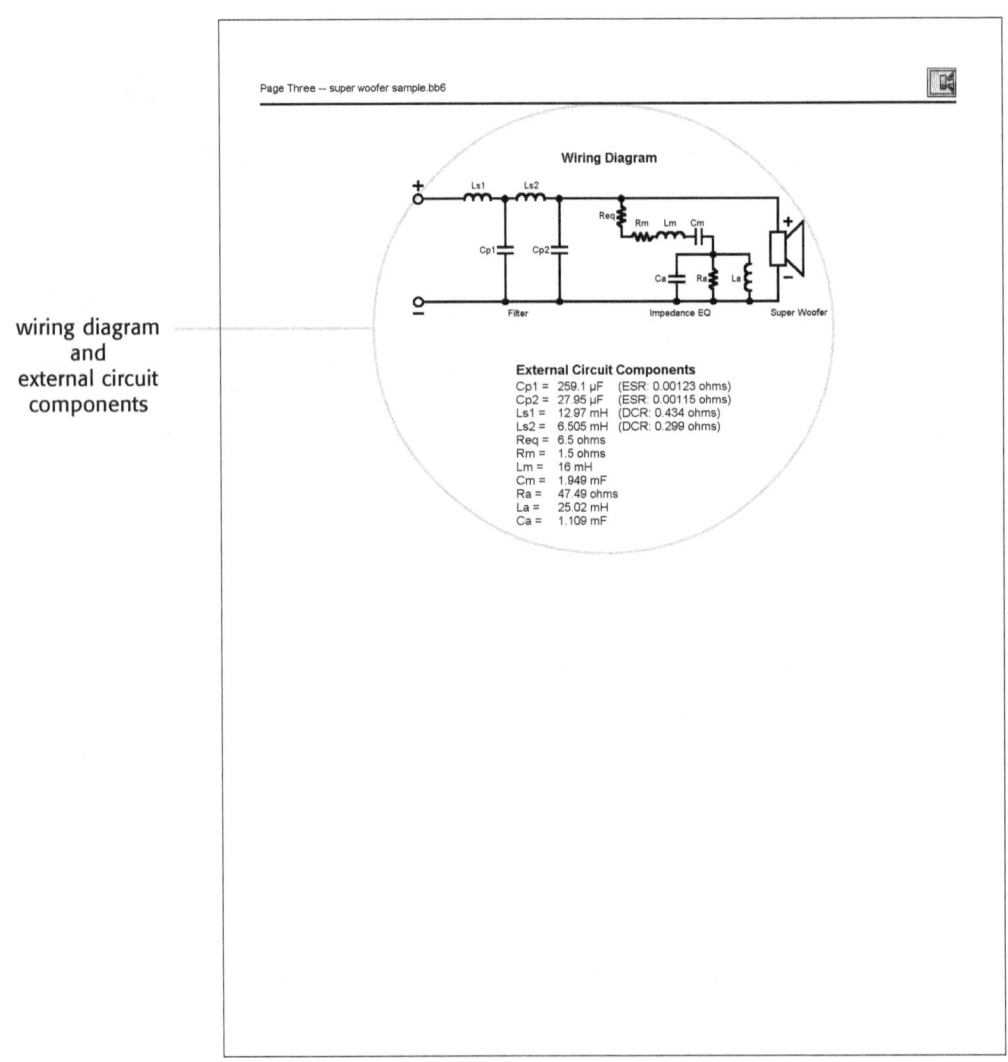

wiring diagram and external circuit components

Notice below that the full-size graphs are used. Only three full-size graphs will fit per page. To print all nine graphs would require two more pages.

full-size graph

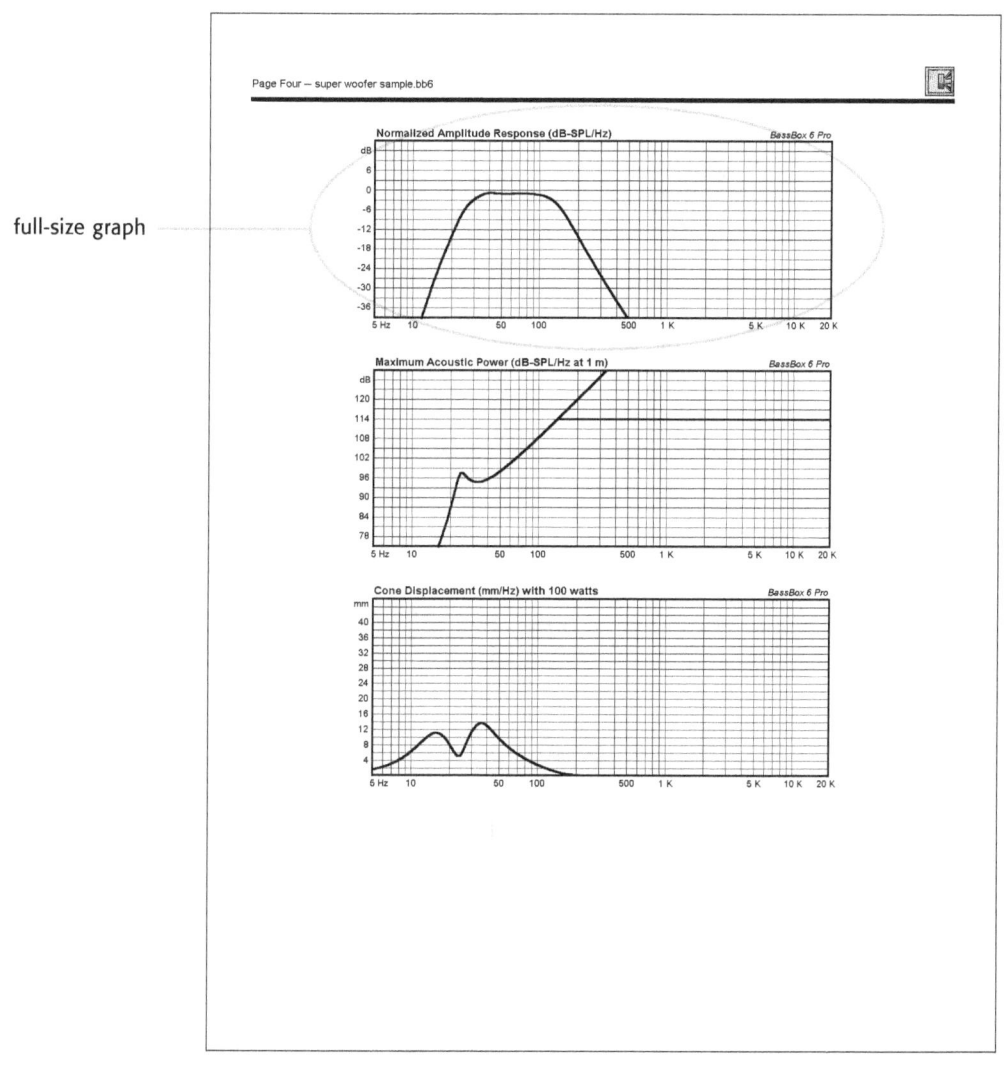

BassBox 6 Pro

Reference: 8 Printing a Design

The sample below shows an alternate Page 4. It shows the graphs when the "Fit all graphs on one page" option is turned on.

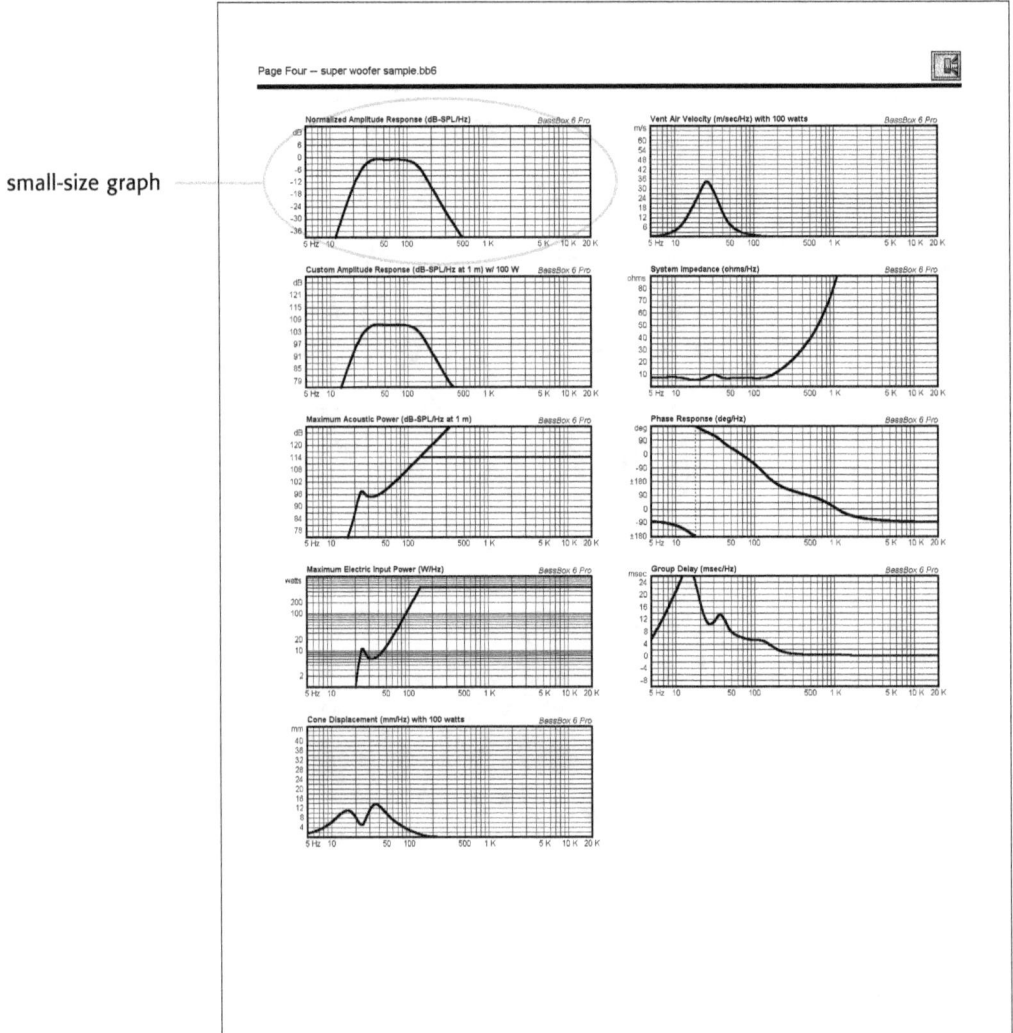

small-size graph

The last sample page was created with both the "Graphs Only" option turned on and the "Include multiple plot lines" option turned on.

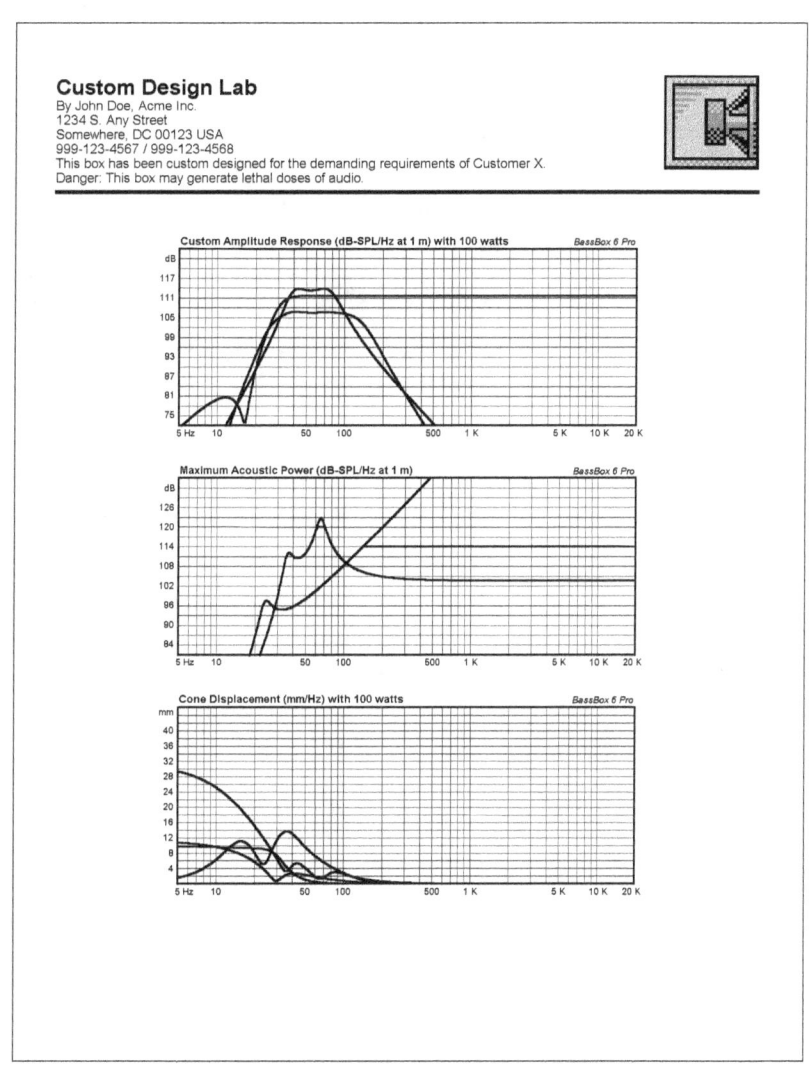

Exporting & Printing a Graph

There is one more way to print a graph. It can be copied to the Windows clipboard and pasted into another application (like a word processor or page layout program) and then printed from the application. This is accomplished by selecting the "Copy" command ([Ctrl]+[C]) from the graph popup menu as shown below:

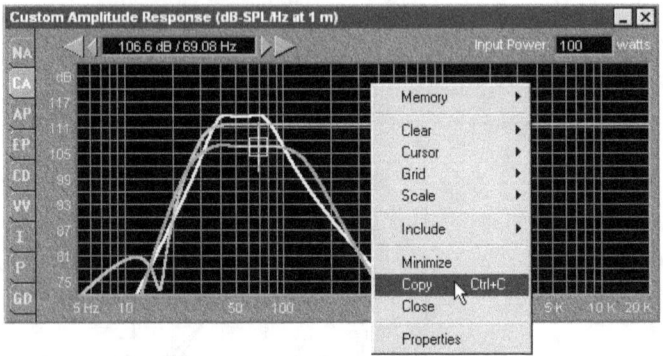

To access the graph popup menu, right click (🖱) on a graph.

When the "Copy" command is executed, the graph is actually "printed" to the clipboard with a white background. The color of the plot lines are preserved, except for white plot lines which are changed to black. Be aware that some colors like yellow may be very light.

A graph that is copied to the clipboard will include the plot lines of the ten most recently plotted designs much the same as a printed graph using the "Include multiple plot lines" option. In addition, a graph that is copied to the clipboard will also include the cursor if the cursor is turned on. The sample below was copied to the clipboard:

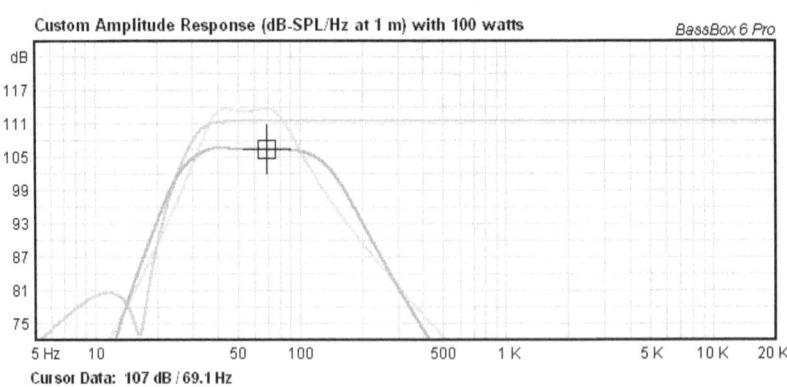

Reference: 9 Clearing/Closing a Design   BassBox 6 Pro

9 Clearing/Closing a Design

Clearing a Design
A speaker design can be cleared without closing it. This can be helpful when you want to start over with a design. Select the "Clear Selected Design" command from the Edit menu of the main window to clear the one design or select the "Clear All Designs" command to clear all open designs.

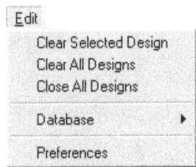

Closing a Design
A single design can be closed by clicking on its Close button in the title bar of its design panel in the main window. BassBox Pro will check for unsaved changes before it executes the command.

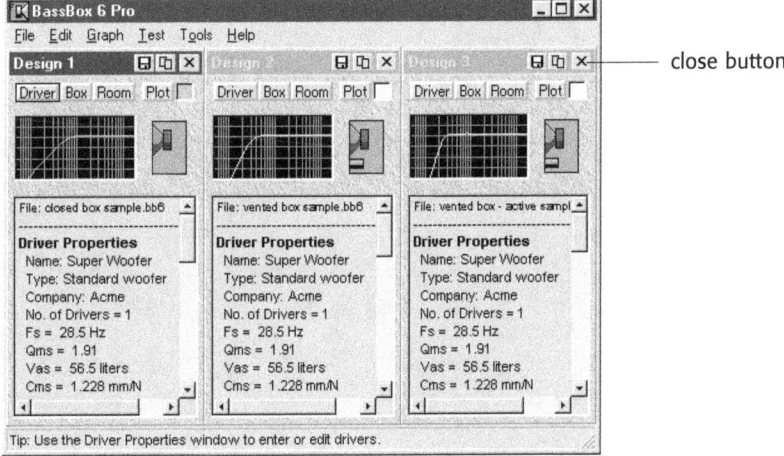

Important: BassBox Pro does not allow a design number to be skipped. When a design is closed, all designs after it will be renumbered. For example, suppose that Designs 1, 2, 3 and 4 are open and you decide to close Design 2. After it has been closed, Designs 3 and 4 will be renumbered 2 and 3, respectively.

It is also possible to close all open designs by selecting the "Close All Designs" command from the Edit menu (shown above).

Reference: 10 Editing the Driver Database BassBox 6 Pro

10 Editing the Driver Database

Chapter 3 (pages 173-182) discussed how to search the driver database to locate a driver and load its parameters into a speaker design. Chapter 3 (page 172) also mentioned the "Add this Driver to Database" button of the Driver Properties window which allows the user to add a new driver to the database. This chapter will show the best way to add a driver to the database as well as how to edit and delete the ones that are already there.

To edit the driver database, select the "Database > Edit Driver Data" command in the Edit menu of the main window as shown below. The keyboard shortcut is [Ctrl]+[W].

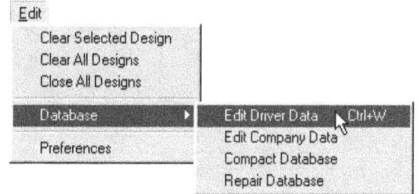

Two windows will open (shown below): the Edit Database Driver Data window and the Driver Locator window. The Edit Database Driver Data window is used to edit the driver data while the Driver Locator is used to navigate the database.

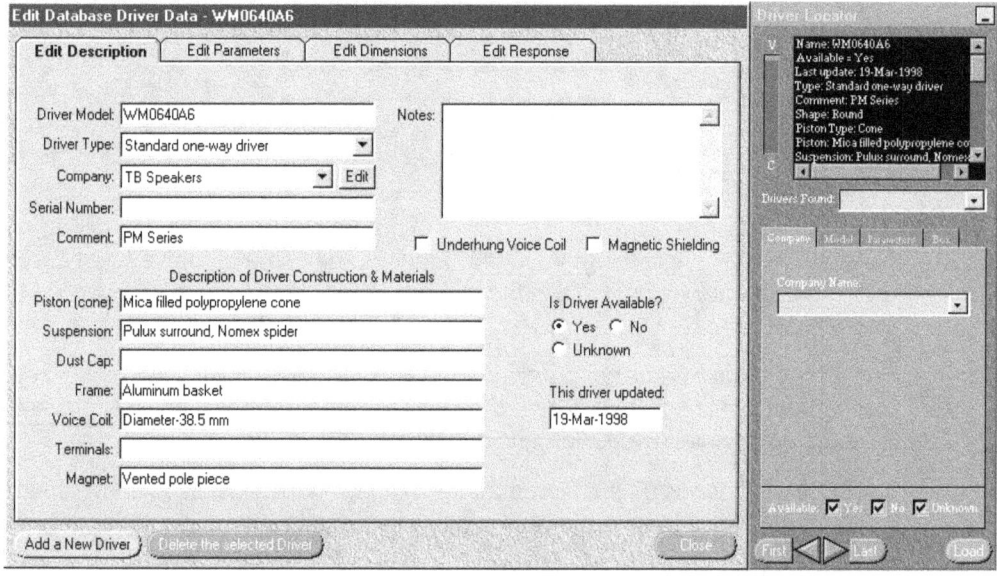

BassBox 6 Pro User Manual © D.E.Harris   307

Since the operation of the Driver Locator window was covered in Chapter 3, only a couple of quick points will be provided here. First, the Driver Locator will not have a Close button in its title bar because it is automatically closed when the Edit Database Driver Data window is closed. Instead it will have a Minimize button which, in addition to serving its normal function, can also be used to switch the size of the Driver Locator window. This is done by holding down the ⇧Shift key when you click on the Minimize button.

Second, the default company which is selected when the Driver Locator first opens is set in the Preferences window (see Chapter 13). If no default company is selected in the Preferences window, the Driver Locator will simply open to the first driver in the database.

Now on to the Edit Database Driver Data window...

Let's notice a couple things about the window. The name of the selected driver is added to the end of the window's title in the title bar. In the above example, driver "WM0640A6" is selected.

Number values are never rounded in the Edit Database Driver Data window even if the number rounding feature is turned on for the rest of the program. This prevents rounding errors when entering data into the database.

Notice the buttons on the bottom of the window in the illustration above and on the previous page. Which buttons are available is determined by the state of the data. Their functions are described next:

Add a New Driver This button is always available to clear the current driver and begin a new one. It allows some of the information to be repeated from the previous driver. This is described in the "Adding a New Driver" section below.

Delete the Selected Driver This button is only available when there are no unsaved changes. It disappears as soon as a driver is edited and reappears when its changes are saved. Its function is to delete a single driver from the database

Save This button is only available when there are unsaved changes. It saves the changes to the database record occupied by the selected driver.

Close This button is always available. It closes both the Edit Database Driver Data window and the Driver Locator window.

Important: BassBox Pro does not consider the editing of the driver database to be a casual affair. It will ask you to confirm the commands that, if executed, will change the database. Deleting a driver, saving a new driver and saving changes to an existing driver are three of the actions that will always require confirmation.

Adding a New Driver
To add a new driver follow these steps:

1. Click on the "Add a New Driver" button. If you don't begin by clicking this button, the program will assume that you are editing the selected driver. After clicking the "Add a New Driver" button, the Add New Driver window shown below will open.

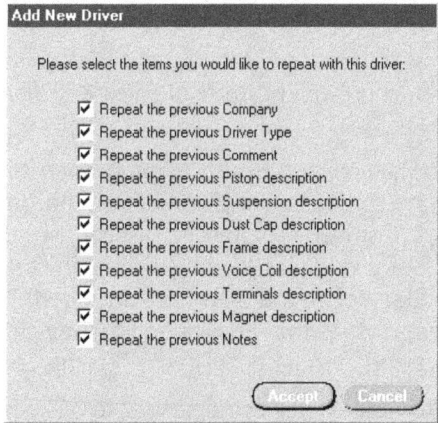

This window allows you to repeat the company of the previous driver and many of the descriptions. This can save time when entering several similar drivers.

2 Enter the requested information in the "Enter Description" tab. Much of this information can be repeated form the previous driver as described in Step 1. The required Description parameters are listed below.

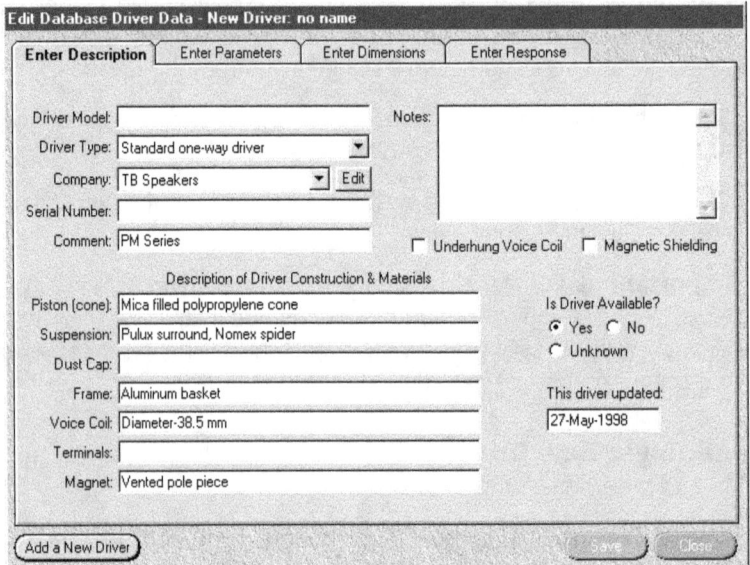

Driver Model The model name can be up to 30 characters long. A company cannot have more than one driver with the same model name.

Driver Type Select one of the four choices: standard one-way driver, two-way coaxial driver, two-way coincident driver or three-way driver. See Chapter 3 (page 183) for a description of each.

Company Each driver must be linked to an existing company in the database. This should be the company that is considered the "manufacturer" of the driver—not the distributor. The "Company" drop-down list contains the names of all these companies. If the new driver is manufactured by a company that is not in the database, the company must be added before the driver can be added. To enter a new company (or edit an existing one) click on the "Edit" button to the right of the "Company" drop-down list. This is described in more detail in the "Editing Companies" section later in this chapter.

Is Driver Available? Select the answer to this question. If you do not know, select "unknown".

This driver updated The current date will be added to this field automatically. However, it is best to enter the date of the driver information. From time to time, manufacturers make changes ("improvements") to their drivers. By entering the date of the data, you will know when newer information is available. Some manufacturer's spec sheets will have a date on them. If not, try contacting the manufacturer and find out the date for the specs.

Please enter the date in the international form: dd-mmm-yyyy. For example enter July 27, 2004 as "27-Jul-2004".

This concludes the <u>required</u> information for the "Enter Description" tab. The remaining descriptions and notes of this tab are optional but highly recommended. See Chapter 3 (pages 183-185) if you'd like more information.

3 Using the "Enter Parameters" tab enter the Thiele-Small and electromechanical parameters for the driver.

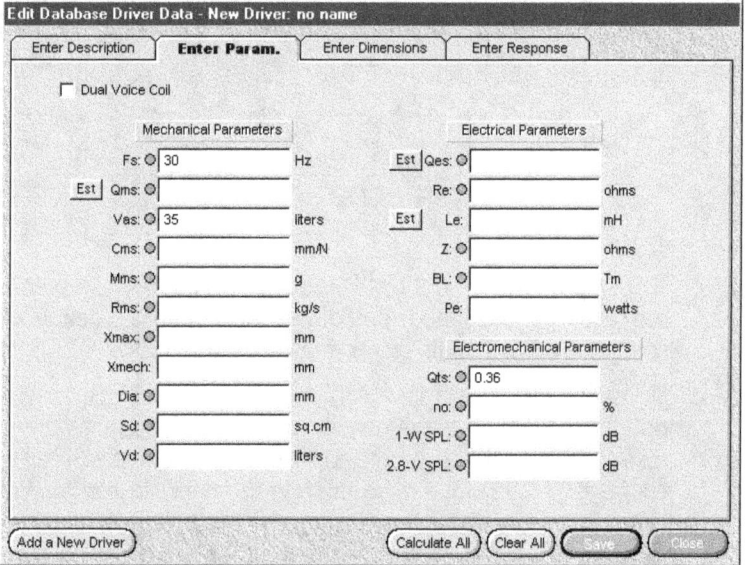

Unlike the "Parameters" tab in the Driver Properties window which has two modes (normal and expert), this "Enter Parameters" tab is always in "expert mode" to help identify data problems before the parameters are saved to the database.

A minimum set of parameters must be entered or the driver cannot be added to the database. These parameters include Fs, Vas and either Qms, Qes or Qts. A model name and valid company name are also required. See Chapter 3 (pages 198-200) for a description of each parameter. Before entering a parameter, remember to select the desired units by clicking on its units label.

4 Using the "Enter Dimensions" tab select an outer shape and piston type for the driver and then enter its dimensions.

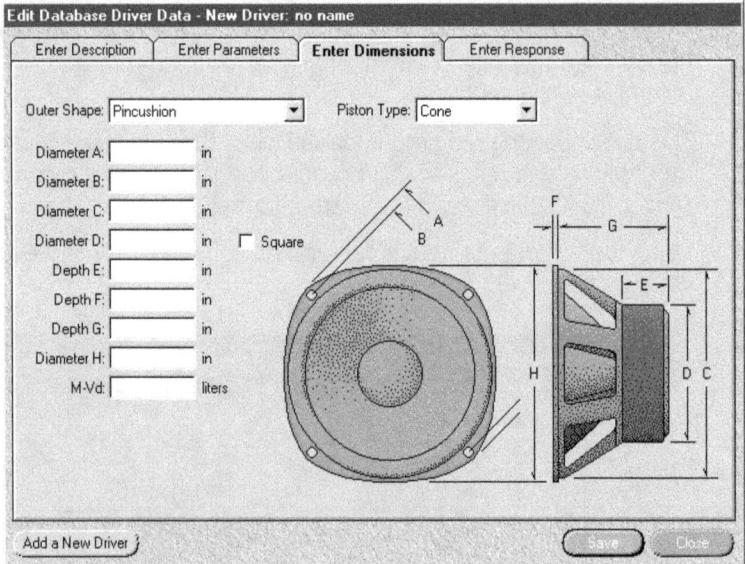

This step is optional but highly recommended. See Chapter 3 (pages 201-202) for a description of the shapes and dimensions. Before entering a dimension, remember to select the desired units by clicking on its units label.

5 Using the "Enter Response" tab enter the acoustic response of the driver. This step is optional and requires the measured acoustic response data to be normalized to the response curve that is predicted by the driver's Thiele-Small parameters. See Chapter 3 (pages 203-209) for a description of the normalization process. **Important:** Before the acoustic data can be normalized to the predicted T-S response, the driver's parameters must be entered in Step 3 above and then the driver must be "loaded" into a design. If the driver was measured on a test box, the box information must also be entered into the design (with the "Box Design" tab of the Box Properties window) in order to duplicate the conditions used for the acoustical measurement. This will enable the program to generate the predicted response curve.

6 Add the new driver to the driver database by clicking on the "Save" button. Before the new driver is added to the database, BassBox Pro will check the company name to make sure that it is a valid company in the database and it will check its model name to make sure that it is not a duplicate driver for the company. Then it will check to see if the minimum parameters have been entered. If the data checks out, the driver will be added to the end of the database.

Changing an Existing Driver

Changing an existing driver is easy. First, use the Driver Locator window to find it. Then use the "Description", "Parameters", "Dimensions" and "Response" tabs of the Edit Database Driver Data window to make the desired changes. Then click on the "Save" button.

Deleting an Existing Driver

First, use the Driver Locator window to find the unfortunate driver. Then, with firm resolution, click on the "Delete the Selected Driver" button. BassBox Pro will confirm your request before executing it.

Editing Companies

The Edit Companies window is used to add a new company or edit or delete an existing one. To open the Edit Companies window, click on the "Edit" button to the right of the "Companies" drop-down list on the "Description" tab of the Edit Database Driver Data window. You can also edit the companies directly from the main window by selecting "Database > Edit Company Data" from the Edit menu as shown below:

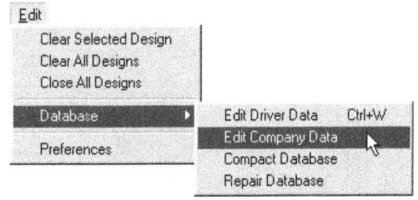

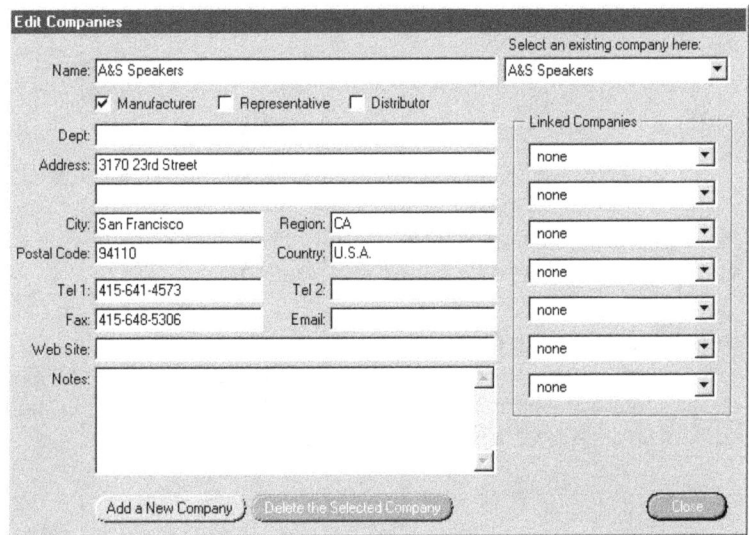

Adding a New Company

To add a new company, click on the "Add a New Company" button and enter its name and other information. A "Save" button will appear when you begin to enter the company. After you have finished, click on the "Save" button to enter the company into the database. *Note: Two or more companies cannot have the same name.*

Important: Drivers should only be linked to companies that are listed as "manufacturers". Otherwise they may not appear in the database. The "representative" and "distributor" settings are discussed on the next page.

Editing an Existing Company

Begin by selecting the company with the "Select an existing company here" drop-down list as shown in the sample below:

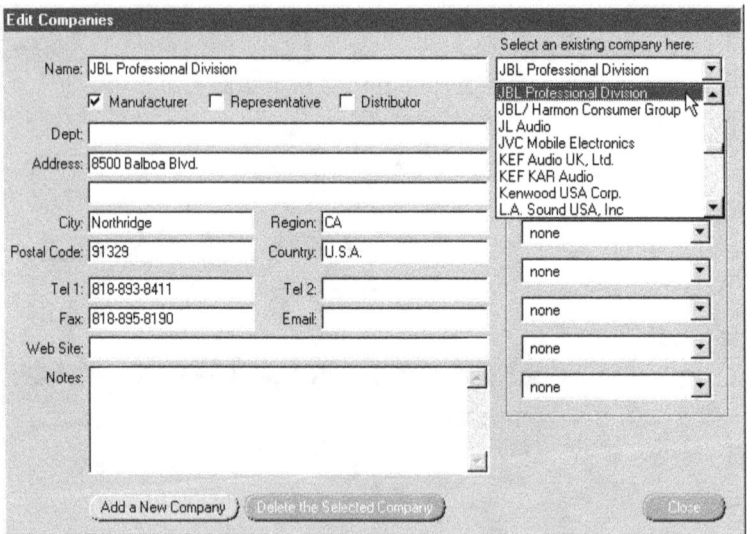

Make the desired changes. The "Save" button will appear when the first change is made. After you have finished, click the "Save" button.

Deleting an Existing Company

Select the unfortunate company with the "Select an existing company here" drop-down list as shown above. Then click the "Delete the Selected Company" button. BassBox Pro will confirm your request before executing it.

Linking Representatives & Distributors

If the manufacturer of a driver does not allow direct communication, you may need to enter the company information for one or more representatives and/or distributors. A representative or distributor is entered just like a manufacturer but the "Representative" and/or "Distributor" checkboxes are checked instead of the "Manufacturer" checkbox.

After the representatives and/or distributors have been entered, return to the company that is the manufacturer. Then use the drop-down lists under "Linked Companies" to select the representatives and/or distributors.

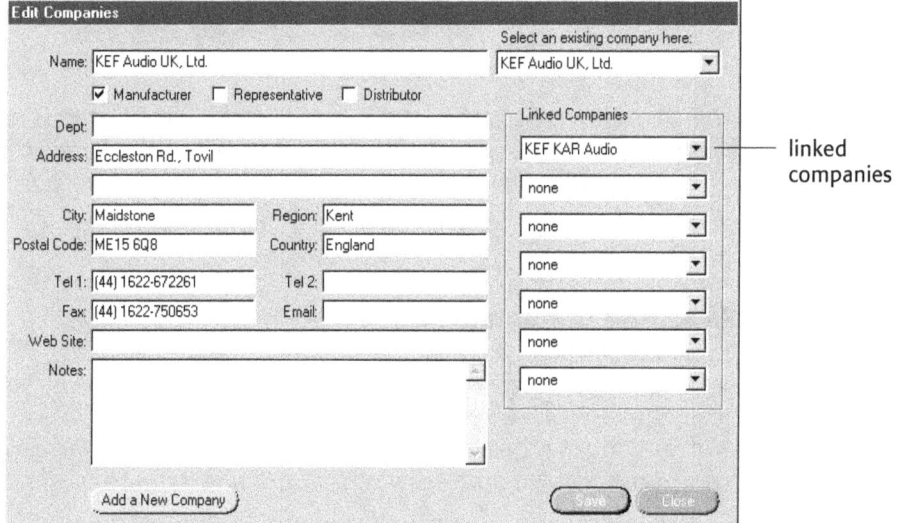

Up to seven representative or distributor companies can be linked to a manufacturer. The linked companies will always appear after the manufacturer in the information box of the Driver Locator.

Important: Do not link representatives and distributors to each other. They should only be linked to a manufacturer.

Note: It is okay for a manufacturer to also be a representative and/or distributor. This situation, although rare, occurs when a company both manufactures its own line of drivers and at the same time distributes another company's line of drivers. In such cases, it is okay to link one manufacturer to another manufacturer.

Compacting the Database

Deleting drivers and/or companies from the database does not decrease its size because that would require the entire database to be resaved, greatly slowing its operation. To compact the database after one or more drivers or companies have been deleted from it, select the "Database > Compact Database" command from the Edit menu of the main window.

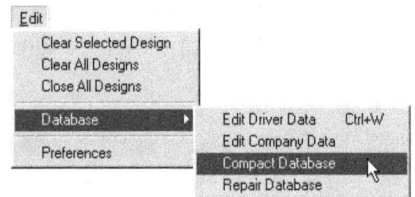

Note: A backup copy of the uncompacted database will be made before the database is compacted. Its name is "htaudio.bak". To restore it, delete "htaudio.mdb" and rename the backup to "htaudio.mdb". The database should be in the same location as the BassBox Pro program. This is "C:\Program Files\HT Audio" if the default settings were used when BassBox Pro was installed.

Repairing the Database

If your computer looses power or "crashes" while the database is being edited, the database may become "corrupt". If, when you attempt to open the database, you ever receive an error message saying that the database is unrecognizable or is not a Microsoft Access database, you may be able to repair it with the "Database > Repair Database" command in the Edit menu of the main window.

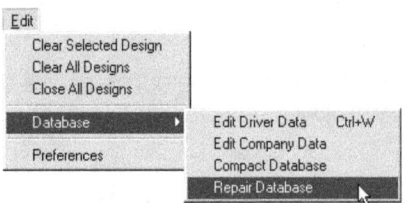

Important: Some data in corrupted portions of the database may be lost when the database is repaired. If this happens, you can reinstall a fresh copy of the database with the BassBox Pro installation CD. However, any changes that you made will be lost.

Note: A backup copy of the unrepaired database will be made before the repair is attempted. Its name is "htaudio.bak". To restore it, delete "htaudio.mdb" and rename the backup to "htaudio.mdb". The database should be in the same location as the BassBox Pro program. This is "C:\Program Files\HT Audio" if the default settings were used when BassBox Pro was installed.

11 Testing Drivers & Passive Radiators

BassBox Pro includes a test procedure to help you measure many of the Thiele-Small and electromechanical parameters of a driver and a passive radiator. These procedures require some basic test equipment. See the list below. *Note: The procedures in BassBox Pro will only assist with the measurement of "small-signal" parameters. "Large-signal" parameters like Xmax and Pe are beyond the scope of the included procedures.*

Test Equipment
- Audio signal generator capable of producing low-frequency sine waves.
- Accurate frequency counter for measuring the precise signal frequency.
- High-quality power amplifier with at least 100 watts continuous output and a flat power response below 20 Hz.
- 1000 ohm resistor with at least a 2 watt capacity.
- Voltmeter that is accurate at low voltages (only one is required but two are recommended).
- Accurate ohmmeter.
- Hanging apparatus for suspending the driver in mid air away from all external surfaces or obstructions.
- Test box of an appropriate size and capable of making a completely airtight seal to the driver. If a passive radiator will be tested, the test box must also have an airtight mounting hole for the passive radiator. And a plug will be needed to seal shut the passive radiator mounting hole during the part of the procedure when the driver is measured without the passive radiator.

  The following list provides some suggested test box volumes for different driver sizes:

Loudspeaker Diameter	Suggested Test Box Volume
5 in (130 mm)	0.125 cubic feet (3.5 liters)
8 in (200 mm)	0.5 cubic feet (15 liters)
10 in (250 mm)	1.0 cubic foot (30 liters)
12 in (300 mm)	2.0 cubic feet (60 liters)
15 in (380 mm)	4.0 cubic feet (120 liters)
18 in (460 mm)	6.0 cubic feet (170 liters)

- A driver capable of producing low frequencies will be required for the passive radiator procedure. It should have a similar piston area (Dia or Sd) and compliance (Vas or Cms) as the passive radiator.

BassBox 6 Pro

Reference: 11 Testing Drivers & Passive Radiators

Testing Drivers

It is often desirable to test drivers to find their precise parameters since manufacturing tolerances can have a very significant effect on design accuracy of vented, bandpass and passive radiator boxes. Sometimes it is necessary to measure the drivers because their parameters are unknown. When accurate data is available, drivers can be paired according to the similarity of their parameters for improved stereo imaging.

The driver test procedure is illustrated to make it easier to perform. Select "Driver" from the Test menu of the main window to open the Driver Test Procedure window:

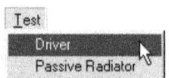

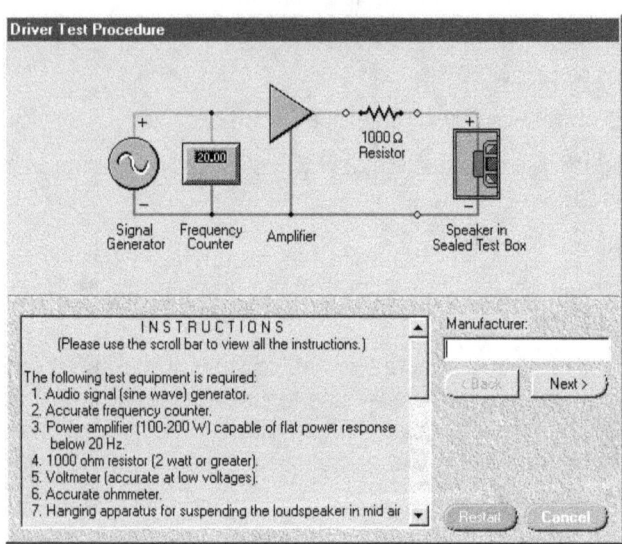

The upper half of the Driver Test Procedure window is used to illustrate the procedure. The basic test circuit diagram is visible when the window first opens. Other illustrations will be used at appropriate times during the procedure. An instruction box is located on the left side of the bottom half of the window. It provides the step-by-step instructions for the procedure. To the right of the instruction box is the input box, where requested information is entered. Beneath it are the "Back" and "Next" buttons which help you navigate through the procedure. In the bottom right corner of the window are the "Restart" and "Cancel" buttons. Clicking the "Restart" button will cause the procedure to start over. The "Cancel" button terminates the procedure and closes the window.

Reference: 11 Testing Drivers & Passive Radiators BassBox 6 Pro

The driver test procedure contains two parts. First, the driver is suspended in the air for the "free-air" measurements. Second, the driver is mounted on the closed test box for the remaining measurements.

Driver Test Setup
The circuit diagram in the Driver Test Procedure window on the previous page shows how to wire the test setup. The illustration below shows the same setup with the test points. The signal generator should be plugged into an input of the amplifier. The frequency counter should be connected across the output of the signal generator so it can measure the frequency of the signal. The output of the amplifier should be connected to the driver with the 1000 ohm resistor in series with the positive lead.

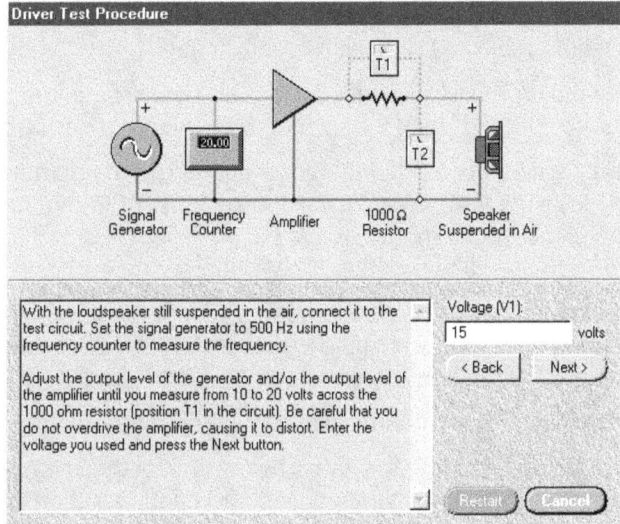

The positive terminal of the driver can be easily found if it is not marked. To do so, connect the negative end of a 1.5 volt penlight battery (size AA) to one side of the driver with a short piece of wire. While carefully observing the driver cone, <u>momentarily</u> touch the positive end of the battery to the remaining driver terminal. The terminal you touched is positive if the cone jumps outward (away from the magnet). If the cone jumps inward (toward the magnet) the terminal is negative. *Caution: Allowing direct current to flow through the driver from the battery for too long may damage the voice coil.*

Driver Test Procedure
The following procedure is used by the Driver Test Procedure window.

1. Enter the name of the driver's manufacturer and click the "Next" button.
2. Enter the model name of the driver and click the "Next" button.
3. Optional: Enter the serial number of the driver and click the "Next" button.
4. Measure the series resistor, enter its exact resistance (R) and press the "Next" button.
5. Select the units for the piston diameter by clicking on the units label. Then enter the piston diameter of the driver (Dia) and click the "Next" button. The piston diameter is usually the diameter of the cone and half the surround.
6. Suspend the driver in the air away from other surfaces or objects. Measure its DC voice coil resistance (Re) using an ohmmeter. Remember to subtract any test lead resistance. Enter the value and click the "Next" button.
7. With the driver still suspended in the air, connect it to the test circuit. Set the signal generator to 500 Hz using the frequency counter to measure the frequency.

    Adjust the output level of the generator and/or the output level of the amplifier until you measure from 10 to 20 volts across the 1000 ohm resistor at test location T1. Be careful that you do not overdrive the amplifier, causing it to distort. Enter the voltage (V1) you used and click the "Next" button.

8. Find the free-air resonance frequency (Fs) of the driver by sweeping the signal generator downward slowly until the maximum voltage is reached at test location T2 in the circuit. Enter the frequency and click the "Next" button.
9. Next, enter the maximum voltage (V2) you measured at Fs and click the "Next" button.
10. Sweep the generator downward below Fs until the voltmeter reads a voltage calculated by the program. Enter the frequency (F1) and click the "Next" button.
11. Sweep the generator upward above Fs until the voltmeter again reads a voltage calculated by the program. Enter the frequency (F2) and click the "Next" button.

    After F2 is entered, the program will check Fs, F1 and F2 to make sure Fs was properly measured. The square root of F1 times F2 should be within ±1 Hz of Fs. If this is not the case, the program will offer you the opportunity to return to Step 7.

12. Carefully mount the driver on the test box. Check to be sure there will be no air leakage from either the driver or the box. It is often more convenient to mount the driver backwards in the box, so that the front of the driver is pointing toward the inside of the box. This prevents you from having to run wires through the side of the box to the driver.

    Select the units for the test box volume by clicking on the units label. Then enter the volume of the test box (Vb) and click the "Next" button. Vb should be the internal volume of the box considering the volume displaced by the driver.

13. With the driver again connected to the test circuit, find the system resonant frequency (Fc) by sweeping the generator from 500 Hz down until you the read maximum voltage across the driver. Enter Fc and click the "Next" button.

14. Next, enter the maximum voltage (V2) you measured at Fc and click the "Next" button.

15. Sweep the generator downward below Fc until the voltmeter reads a voltage calculated by the program. Enter the frequency (F1) and click the "Next" button.

16. Sweep the generator upward above Fc until the voltmeter again reads a voltage calculated by the program. Enter the frequency (F2) and click the "Next" button.

After F2 is entered, the program will check Fc, F1 and F2 to make sure that Fc was properly measured. The square root of F1 times F2 should be within ±1 Hz of Fc. If this is not the case, the program will offer you the opportunity to go back and start over from Step 11.

Results

After successfully completing the above procedure the following driver and system parameters are computed: Fs, Fc, Qms, Qmc, Vas, Cms, Mms, Rms, Sd, Qes, Qec, Re, BL, Qts, Qtc, η_o, and 1-W SPL. (Consult the Glossary of Terms for descriptions of these parameters.)

Finally, a "Load into a New Design" button will appear. If a speaker design is already open, a "Load into Selected Design" button will also appear. Click one of these buttons to load the driver parameters and close the Driver Test Procedure window.

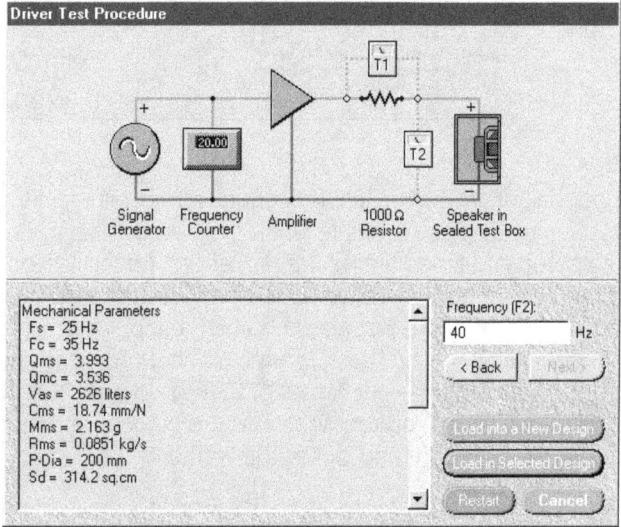

Testing Passive Radiators

One of the more difficult aspects of designing a box with a passive radiator is obtaining a suitable passive radiator along with accurate data for it. Fortunately, it is not difficult to measure a passive radiator to determine the required parameters. This section explains how.

The passive radiator test procedure is illustrated to make it easier to perform. Select "Passive Radiator" from the Test menu of the main window to open the Passive Radiator Test Procedure window:

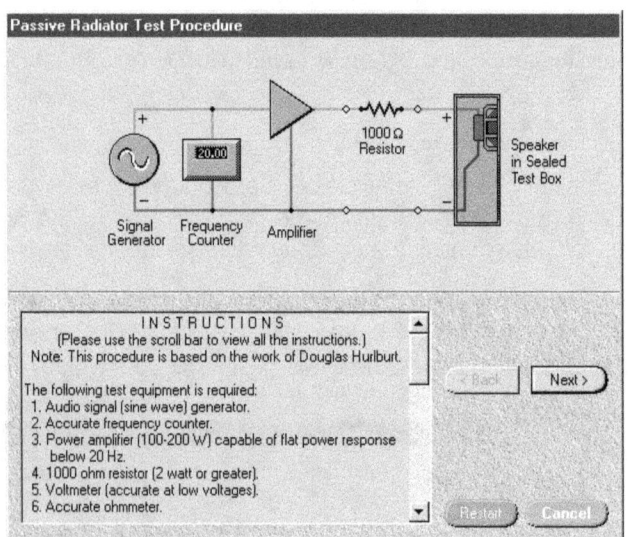

The upper half of the Driver Test Procedure window is used to illustrate the procedure. The basic test circuit diagram is visible when the window first opens. Other illustrations will be used at appropriate times during the procedure. An instruction box is located on the left side of the bottom half of the window. It provides the step-by-step instructions for the procedure. To the right of the instruction box is the input box, where requested information is entered. Beneath it are the "Back" and "Next" buttons which help you navigate through the procedure. In the bottom right corner of the window are the "Restart" and "Cancel" buttons. Clicking on the "Restart" button will cause the procedure to start over. The "Cancel" button terminates the procedure and closes the window.

The passive radiator test procedure in BassBox Pro is based on the work of Douglas

Hurlburt (see page 357). It contains two parts. First, the driver and test box are measured without the passive radiator. Second, the passive radiator is installed and measured.

Passive Radiator Test Setup
The circuit diagram in the Passive Radiator Test Procedure window on the previous page shows how to wire the test setup. The illustration below shows the same setup with the test points. The signal generator should be plugged into an input of the amplifier. The frequency counter should be connected across the output of the signal generator so it can measure the frequency of the signal. The output of the amplifier should be connected to the driver with the 1000 ohm resistor in series with the positive lead.

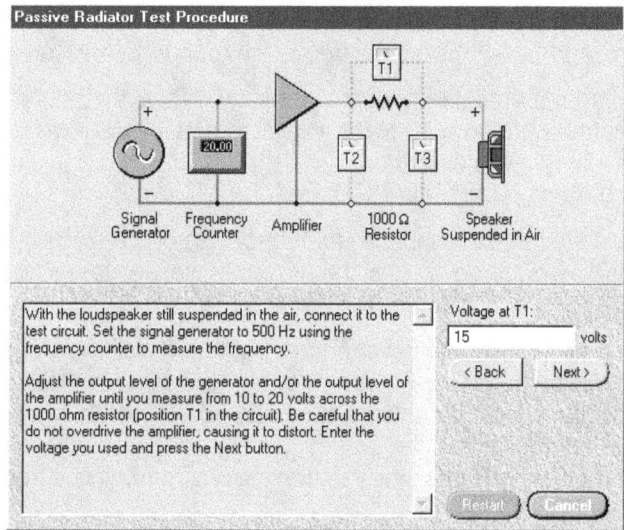

The positive terminal of the driver can be easily found if it is not marked. To do so, connect the negative end of a 1.5 volt penlight battery (size AA) to one side of the driver with a short piece of wire. While carefully observing the driver cone, <u>momentarily</u> touch the positive end of the battery to the remaining driver terminal. The terminal you touched is positive if the cone jumps outward (away from the magnet). If the cone jumps inward (toward the magnet) the terminal is negative. *Caution: Allowing direct current to flow through the driver from the battery for too long may damage the voice coil.*

Passive Radiator Test Procedure

The following procedure is used by the Passive Radiator Test Procedure window. After opening the window press the "Continue" button to begin.

1. Measure the series resistor, enter its exact resistance (R) and press the "Next" button.

2. Select the units for the piston diameter by clicking on the units label. Then enter the piston diameter of the passive radiator (Dia) and click the "Next" button. The piston diameter is usually the diameter of the cone and half the surround.

3. Suspend the driver in the air away from other surfaces or objects. Measure its DC voice coil resistance (Re) using an ohmmeter. Remember to subtract any test lead resistance. Enter the value and click the "Next" button.

4. With the driver still suspended in the air, connect it to the test circuit. Set the signal generator to 500 Hz using the frequency counter to measure the frequency.

    Adjust the output level of the generator and/or the output level of the amplifier until you measure from 10 to 20 volts across the 1000 ohm resistor at test location T1. Be careful that you do not overdrive the amplifier, causing it to distort. Enter the voltage you used and click the "Next" button.

5. Find the free-air resonance frequency (Fs) of the driver by sweeping the signal generator downward slowly until the maximum voltage is reached at test location T3. Enter the frequency and click the "Next" button.

6. Carefully mount the driver on the test box. Install the plug in the passive radiator mounting hole. Check to be sure there will be no air leakage from either the driver, the passive radiator mounting hole or the box. It is often more convenient to mount the driver backwards in the box, so that the front of the driver is pointing toward the inside of the box. This prevents you from having to run wires through the side of the box to the driver.

    Select the units for the test box volume by clicking on the units label. Then enter the volume of the test box and click the "Next" button. The value you enter should be the internal volume of the box considering the volume displaced by the driver.

7. Beginning with the generator set to 500 Hz, find the resonance frequency (Fc) of the system by sweeping the signal generator downward slowly until the maximum voltage is reached at test location T3. Enter the frequency and click the "Next" button.

8. With the frequency still set to Fc, enter the voltage at test location T3 and click the "Next" button.

9. With the frequency still set to Fc, enter the voltage at test location T2 and click the "Next" button.

10. Sweep the generator downward below Fc until the voltmeter reads a voltage calculated

by the program at test location T3. Enter the frequency (F1) and click the "Next" button.

11 Sweep the generator upward above Fc until the voltmeter reads a voltage calculated by the program at test location T3. Enter the frequency (F2) and click the "Next" button.

12 Remove the plug from the passive radiator mounting hole in the box and mount the passive radiator to the box. Make sure that there is no air leakage around the passive radiator.

Beginning at 500 Hz, slowly sweep the generator downward until the first voltage peak is reached at test location T3. Enter the frequency at the peak (FH) and click the "Next" button.

13 Continue to slowly sweep the generator downward until the second voltage peak is reached at test location T3. Enter the frequency at the peak (FL) and click the "Next" button.

14 Set the generator to a frequency calculated by the program (Fb), enter the voltage at test location T2 and click the "Next" button.

15 With the generator still set at Fb, measure the voltage at test location T3 and click the "Next" button.

Results

After successfully completing the above procedure the following passive radiator parameters will be calculated: Fs, Qms, Vas, Cms, Mms, Rms and Sd. (Consult the Glossary of Terms for descriptions of these parameters.)

Finally, a "Load into a New Design" button will appear. If a speaker design is already open, a "Load into Selected Design" button will also appear. Click one of these buttons to load the passive radiator parameters and close the Passive Radiator Test Procedure window.

Undocumented Feature

An alternate passive radiator measurement procedure is included in BassBox Pro. It is based on impedance response measurements made with a Woofer Tester. However, any measurement system capable of making impedance response measurements will work. To use this procedure you'll need to obtain a copy of the June 2002 issue of *AudioXpress* magazine. The article "Measuring Passive Radiators" by David Harris begins on page 50 and provides a detailed explanation of the procedure including the math. To use this alternate procedure in BassBox Pro and have the program perform the calculations for you, select "Passive Radiator" from the Test menu as described earlier. Then hold down the [Shift]+[Ctrl]+[Alt] keys and click on the test picture in the upper half of the Passive Radiator Test Procedure window. The phrase "IMPEDANCE VERSION" will appear in the title bar of the window. To return to the original procedure, hold down [Shift]+[Ctrl]+[Alt] again and click on the test picture.

12 Tools

BassBox Pro provides access to several helpful tools in its Tools menu as shown below:

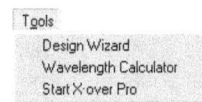

Design Wizard
The Design Wizard serves as a smart assistant to help the user design a speaker. It opens relevant property windows of BassBox Pro as it prompts the user for input and provides helpful information. The Design Wizard can also be launched with the "Run Design Wizard" button of BassBox Pro's title window when the program starts.

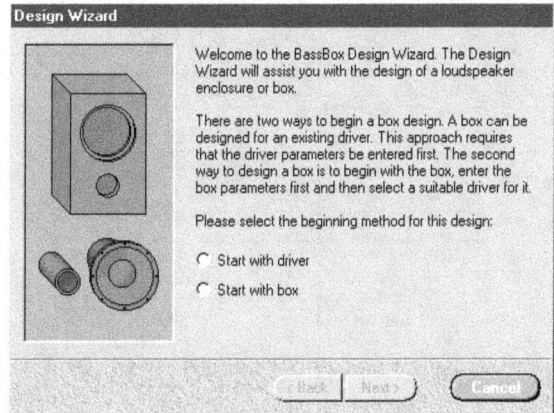

With its on-screen messages, the Design Wizard's operation is self-explanatory. If desired, see Chapter 2 for a brief introduction to the Design Wizard.

Wavelength Calculator
Ever wanted to calculate the frequency of a standing wave inside a box? Or wanted to compare the piston diameter to the longest wavelength in the driver's passband? These kinds of comparisons require conversion between frequency and wavelength. BassBox Pro's Wavelength Calculator makes this very easy.

Simply enter a frequency in Hz and its wavelength will be calculated. Or, enter a wavelength and its frequency will be calculated. If you want to change the units of length, click on the wavelength units label.

Start X•over Pro

The optional "Start X•over Pro" command of the Tools menu launches the X•over Pro program, if available. This command is enabled if you have purchased an X•over Pro license and the program is installed on your computer.

X•over Pro (shown below) is a passive network designer from Harris Tech. Using it you can design passive crossover networks or filters, impedance equalization networks and L-pads.

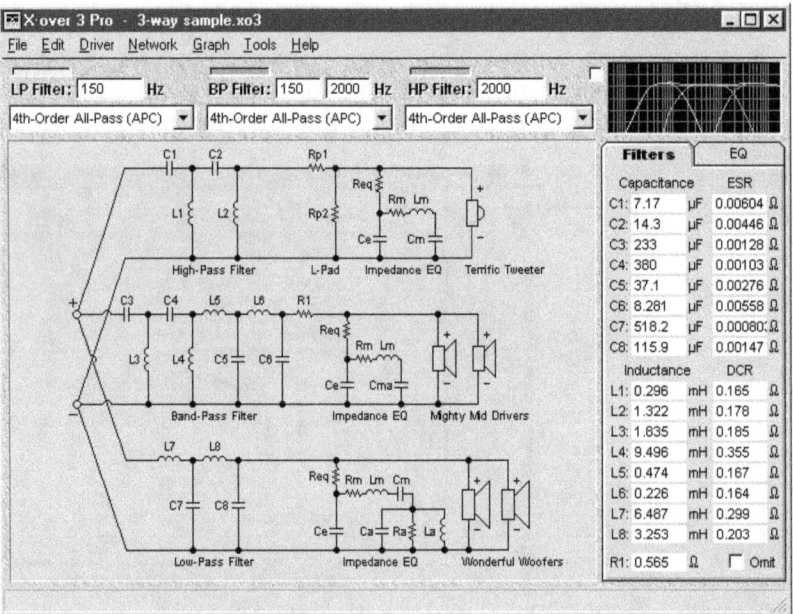

X•over Pro can import driver and box information from BassBox Pro, making it easy to design a crossover network for an existing speaker design. And since X•over Pro can also display the response of sealed back drivers (such as tweeters) you can view the response of the complete speaker in its four performance graphs.

If the "Start X•over Pro" command is not enabled in the Tools menu and yet the program is properly installed on your computer, then its path entry in the Preferences window is probably blank. This is easy to correct by manually entering the location of X•over Pro on the "General" tab of the Preferences window. See Chapter 13 (pages 329 and 331) for details.

To obtain information about X•over Pro and how to purchase a software license for it, please visit Harris Tech's website at www.ht–audio.com or contact them directly via email at sales@ht–audio.com.

13 Preferences

BassBox Pro offers an extensive set of default parameters to tailor the program to your preferences. These "preference" settings are available in the Preference window. To open it, select "Preferences" from the Edit menu of the BassBox Pro main window as shown below:

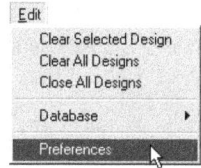

The Preferences window is divided into 6 tabs which will be discussed in detail next. When the window first opens, it always opens to the first "General" tab.

General
The "General" tab contains mostly global settings that affect multiple areas of BassBox Pro.

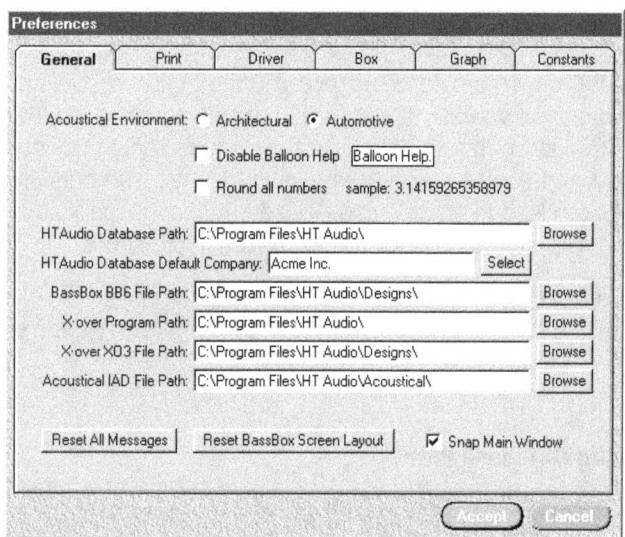

Acoustical Environment
BassBox Pro has the ability to incorporate the acoustical response of the environment into some of the performance graphs in order to provide a more complete "picture" of the sound. Two common environments are the interior of a room and the interior of an automobile. The acoustical environment setting lets you select the preferred environment for the

program. The settings have the following affects:

Architectural Updates relevant buttons and icons in the program to reflect the selection. Configures the Room/Car Acoustical Properties window to include two optional filters for modelling a variety of acoustical phenomenon: a 3, 6 or 12 dB/octave low-frequency boost filter and a 3, 6 or 12 dB/octave low-frequency shelf filter.

Automotive Updates relevant buttons and icons in the program to reflect the selection. Configures the Room/Car Acoustical Properties window to include a 12 dB/octave boost filter to model the bass rise that occurs in most automotive interiors.

Balloon Help

Extensive balloon help is provided in BassBox Pro. It can be turned off with the "Disable Balloon Help" checkbox.

Number Rounding

Many users do not like to see huge numbers with a "gazillion" decimal places. After all, BassBox Pro "models" speaker boxes and even the best models will not match reality perfectly. So those many decimal places are not really necessary. Or are they? BassBox Pro gives you the choice of whether to round numbers before displaying them or whether to show their full double-precision value.

Why is this option provided? To give you the option of avoiding many of the rounding errors that can creep into data. The most common source of rounding errors is in units conversion. All values in BassBox Pro are stored internally in MKS metric units. If the numbers are rounded, some information will be lost when the numbers are displayed in the program and this can lead to small errors. These errors are usually small enough not to cause any serious inaccuracies but they can be unsettling to some.

If you hate huge numbers with lots of decimal places and can live with occasional subtle rounding errors, then check the "Round all numbers" checkbox. If, on the other hand, you want to avoid rounding errors and you can live with lots of decimal places then make sure that the "Round all numbers" is not checked.

HT Audio Database Path

A "path" is like a street address. It tells your computer where to go to find something. The HT Audio Database Path is the location of the driver database file on your computer's hard disk drive and this path setting was set during the installation of BassBox Pro. You shouldn't normally need to change it. If the default settings were used during installation, the path will be "C:\Program Files\HT Audio\". The database file name is "htaudio.mdb". If you ever decide to move it, or want to keep multiple copies in different folders, you can use this path setting to tell BassBox Pro where to find it. A "Browse" button is provided to help you select a new path, if desired.

HT Audio Database Default Company
BassBox Pro allows you to set a default company for the driver database. Each time the database is opened, it will open with the default company selected and its drivers will have been added to the "Drivers Found" list. There are two ways to enter a company name. It can be manually entered directly into the "HT Audio Database Default Company" input box or it can be selected from the driver database using the "Select" button. The latter method is preferred because the company name must be spelled exactly as it is in the database.

BassBox BB6 File Path
When you save a speaker design, BassBox Pro creates a file with the extension "bb6". This path setting tells BassBox Pro where to find these files. If the default settings were used during installation, the path will be "C:\Program Files\HT Audio\Designs\". A "Browse" button is provided to help you select a new path, if desired.

X•over Program Path
X•over Pro is a passive crossover design program from Harris Technologies. It can be launched from BassBox Pro and this path setting tells BassBox Pro where to find it. If the default settings were used during installation, the path will be "C:\Program Files\HT Audio\". A "Browse" button is provided to help you select a new path, if desired.

X•over XO3 File Path
The components of a passive network can be imported into a BassBox Pro design from X•over Pro. This enables BassBox Pro to show the net response of a box design with its crossover network, impedance equalization network and/or L-pad. X•over Pro design files end with the extension "xo3" and this path setting tells BassBox Pro where to find them. If the default settings were used during installation, the path will be "C:\Program Files\HT Audio\Designs\". A "Browse" button is provided to help you select a new path, if desired.

Acoustical IAD File Path
BassBox Pro allows acoustical data from an architectural or automotive environment to be imported from various measurement systems as well as manually entered. Once the acoustical data is in BassBox Pro, it can be saved to a special BassBox Pro acoustic file so that it can be recalled later for use in other designs. These files end with the extension "iad" and this path setting tells BassBox Pro where to find them. If the default settings were used during installation, the path will be "C:\Program Files\HT Audio\Acoustic\". A "Browse" button is provided to help you select a new path, if desired.

Reset All Messages
Some of BassBox Pro's message windows can be disabled by clicking on a "Don't show this message in the future" checkbox in the message window. Clicking on the "Reset All Messages" button will uncheck all these settings so that all message windows will appear again.

 *Reference: 13 Preferences*

Reset BassBox Screen Layout
BassBox Pro remembers the last location of many of its various windows so that they will appear in their previous locations on the screen each time that they are opened. What would happen if the windows were positioned around the screen while it was at a very high resolution such as 1600 x 1200 pixels and then the screen was later reduced to a much smaller resolution such as 640 x 480 pixels? In such cases, some of BassBox Pro's windows may no longer be visible on the screen. Clicking the "Reset BassBox Screen Layout" button will restore all BassBox Pro windows to their default positions so they will be visible.

Snap Main Window
The main window of BassBox Pro can be resized. However, only the height of the design panels (not their width) can be resized in it. When the Snap Main Window option is enabled, it forces the width of the main window to be a multiple of the design panel width. This prevents a design panel from being only partially visible.

Print
Most of the print options have a "preference" setting so that a default printout can be created. These settings are available in the "Print" tab of the Preferences window.

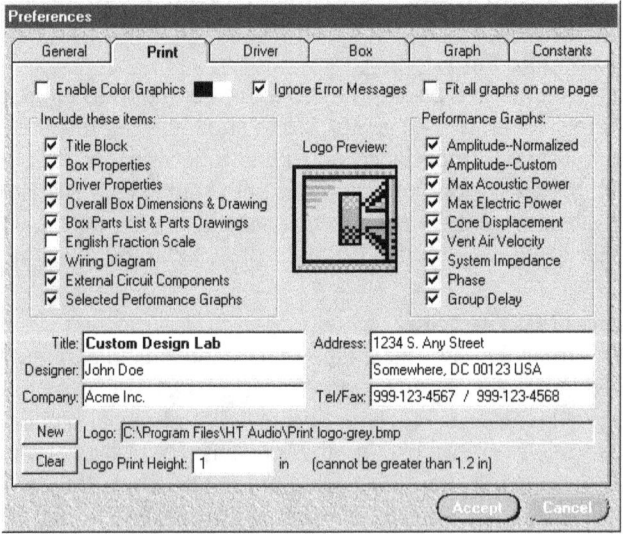

Enable Color Graphics
This option enables the graphs to print with color plot lines and the box drawing and box parts drawings to print with color dimension lines. This option will have no effect on the logo picture. If a color logo picture is selected and a color printer is used, the logo will still print in color even if the "Enable Color Graphics" option is turned off.

BassBox 6 Pro

Ignore Error Messages
Normally this option should be enabled. If you are having print problems, it can be unchecked to enable error messages to aid with troubleshooting.

Fit All Graphs on One Page
Up to nine graphs can be printed. The total number depends on the type of box and on the amount of driver and box parameters that are available in the design. Normally a page can hold only three graphs, so three pages will be needed to print all nine full-size graphs. This option causes a smaller graph to be printed so that all nine graphs will fit on one page.

Include These Items
There are nine items that can be included in a printout. Check the ones that you want to include. They are summarized below:

Title Block Includes the Title line, Designer line, Company line, Address lines, Tel/Fax line, Notes lines and the logo picture. It is printed on page one.

Box Properties Includes the general box information such as the box type, description, volume, tuning and (if applicable) vent information.

Driver Properties Includes the general driver information such as the driver type, description, configuration and parameters.

Overall Box Dimensions & Drawing Includes a list of both external and internal dimensions. It also includes a scaled external three-dimensional wireframe drawing of the box with dimension labels.

Box Parts List & Parts Drawings Includes a list of all box parts, vent parts (if applicable) and other miscellaneous objects such as braces (if entered). It also includes a scaled two-dimensional drawing of each box part and the vents.

English Fraction Scale Includes a small table to convert between the decimal values used in BassBox Pro and the fractions commonly used in the U.S.A.

Wiring Diagram Includes a configuration wiring diagram similar to the one depicted on the "Configuration" tab of the Driver Properties window. Its primary purpose is to show how to wire a multi-driver design.

External Circuit Components Includes a list of all passive external components included in the design. It also adds a schematic of the external network to the wiring diagram.

Selected Performance Graphs Includes the graphs which are selected in the "Performance Graphs" section.

Logo Preview
The logo preview shows the picture that is presently selected for the logo. The logo prints in the title block of the first page. A smaller version also prints in the header of successive pages. Use the "New" button to select a different picture for the logo. Use the "Clear" button to clear the logo so none will print.

Performance Graphs
Select the graphs that you want to print when the "Selected Performance Graphs" checkbox is checked in the "Include these items" section.

Title, Designer, Company, Address, Tel/Fax
These lines are printed at the top of page one in the title block when the Title Block option is enabled under the "Include these items" section.

New (logo)
Click this button to select a different picture for the logo. Several picture file formats are supported including: bmp, dib, gif, jpg, wmf, emf, ico and cur. Three versions of the BassBox Pro icon are included with the program. They are:

> BBx logo-color.bmp   a color version of the BassBox Pro icon
> BBx logo-grey.bmp   a greyscale version of the BassBox Pro icon
> BBx logo-mono.bmp   a black & white version of the BassBox Pro icon

These files are located in the same folder as BassBox Pro ("C:\Program Files\HT Audio").

Clear (logo)
Clears the logo picture so that no logo will print.

Logo Print Height
Sets the height of the logo. The maximum allowable height is 1.2 inches or 30 mm. The logo picture will be proportionally scaled to this height when it is printed.

Driver

There are several "preference" settings for drivers. These are depicted below in the "Driver" tab of the Preferences window.

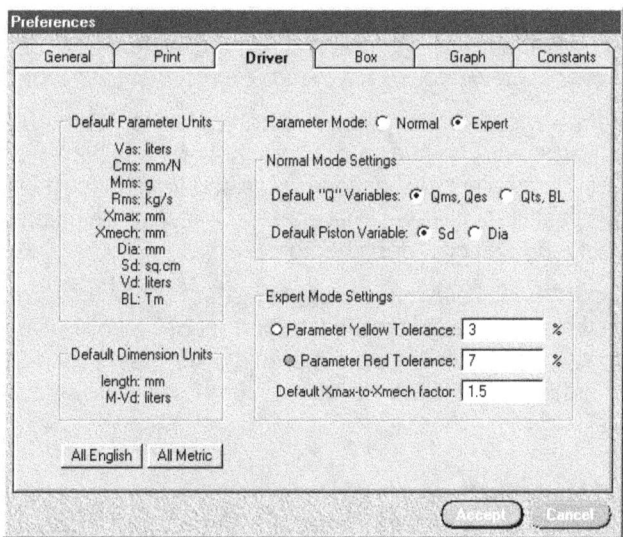

Default Parameter Units
Sets the default units for the driver parameters throughout the program. This affects the main window, Driver Properties window, Edit Database Driver Data window, Driver Locator window and the printouts.

To change the units, click once (🖰) on the desired units label. Each time you click, it will advance to the next available setting. *Note: The default values here can often be overridden in the program by clicking on the individual units labels in the various windows.*

Default Dimension Units
Sets the default units for the driver dimensions throughout the program. This affects the Driver Properties window, Edit Database Driver Data window and the printouts.

To change the units, click once (🖰) on the desired units label. Each time you click, it will advance to the next available setting. *Note: The default values here can often be overridden in the program by clicking on the individual units labels in the various windows.*

All English
Click this button to quickly change all the default parameter and dimension units to English.

All Metric
Click this button to quickly change all the default parameter and dimension units to metric.

Parameter Mode
There are two operating modes for the "Parameters" tab sheet of the Driver Properties window. These modes are "normal" and "expert". *Note: The "Parameters" tab in the Edit Database Driver Data window is always in "expert mode" to provide status indicators.*

The "normal mode" presents a less complex version of the "Parameters" tab with fewer driver parameters. The minimum parameters for a full analysis are included and a "Highlight Minimum Parameters for:" option is provided to identify subgroups of these parameters for a small-signal and/or large-signal analysis. A few of the parameters are selectable with the "Normal Mode Settings" described below.

The "expert mode" presents a complete set of Thiele-Small and electromechanical parameters in the "Parameters" tab. The "expert mode" automatically tests many of the parameters to see if any of them are out of tolerance. A status indicator shows whether they passed or failed. The tolerance settings for the tests are selectable with the "Expert Mode Settings".

Normal Mode Settings
Two sets of parameters are selectable in the "normal mode". One set selects between Qms/Qes and Qts/BL. The second set selects between Sd and Dia. The ones that are selected in the "Normal Mode Settings" will be the ones displayed when the "Parameters" tab of the Driver Properties window is first selected. *Note: The user can override the default values by clicking (🖱) on the appropriate parameter labels in the Driver Properties window.* Select the defaults which best fit the driver data you anticipate using.

Expert Mode Settings
The "Expert Mode Settings" configure the automatic self-analyzing feature.

> **Parameter Yellow Tolerance** Sets the ± percent variation that is allowed before a status indicator is turned yellow.
>
> **Parameter Red Tolerance** Sets the ± percent variation that is allowed before a status indicator is turned red.
>
> **Default Xmax-to-Xmech factor** Is multiplied times Xmax to estimate Xmech and visa versa. Xmax is the maximum linear excursion measured from rest in one direction. Xmech is the maximum mechanical excursion measured from rest in one direction.

Box

There are several "preference" settings for boxes. These are depicted below in the "Box" tab of the Preferences window.

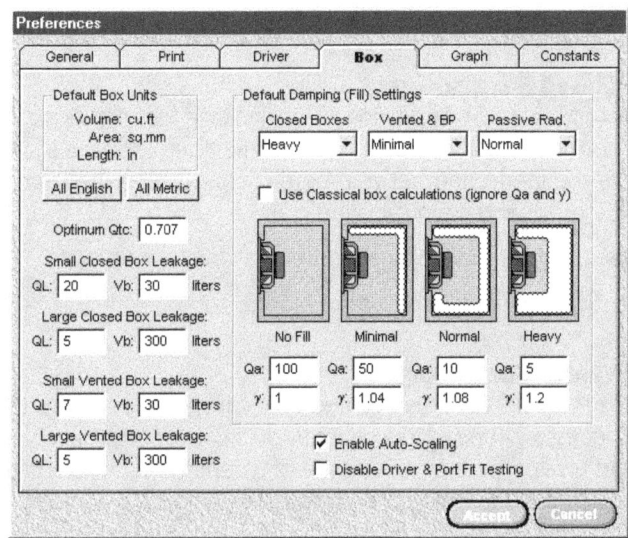

Default Box Units

Sets the default units for the box parameters and box dimensions throughout the program. This affects the main window, Box Properties window and the printouts.

To change the units, click once (🖱) on the desired units label. Each time you click, it will advance to the next available setting. *Note: The default values here can often be overridden in the program by clicking on the individual unit labels in the various windows.*

All English

Click this button to quickly change all the default box units to English.

All Metric

Click this button to quickly change all the default box units to metric.

Optimum Closed Box Qtc

When a "high fidelity" closed box is calculated, the value of Qtc entered here will be used.

Small Closed Box Leakage

BassBox Pro uses the small closed box leakage loss settings to estimate the value of QL when necessary. It does this by interpolating between the small box and large box values to estimate a value for the net volume of a closed box. The small box value represents the highest value of QL and a typical value is 20.

    QL The Q of the leakage losses of a small closed box.

    Vb The net small closed box volume.

Large Closed Box Leakage

BassBox Pro uses the large closed box leakage loss settings to estimate the value of QL when necessary. It does this by interpolating between the small box and large box values to estimate a value for the net volume of a closed box. The large box value represents the lowest value of QL and a typical value is 5.

    QL The Q of the leakage losses of a large closed box.

    Vb The net large closed box volume.

Small Vented Box Leakage

BassBox Pro uses the small vented box leakage loss settings to estimate the value of QL when necessary. It does this by interpolating between the small box and large box values to estimate a value for the net volume of a vented box. The small box value represents the highest value of QL and a typical value is 7.

    QL The Q of the leakage losses of a small vented box.

    Vb The net small vented box volume.

Large Vented Box Leakage

BassBox Pro uses the large vented box leakage loss settings to estimate the value of QL when necessary. It does this by interpolating between the small box and large box values to estimate a value for the net volume of a vented box. The large box value represents the lowest value of QL and a typical value is 5.

    QL The Q of the leakage losses of a large vented box.

    Vb The net large vented box volume.

Default Damping (Fill) Settings

The "Default Damping Settings" control the way BassBox Pro models the addition of acoustic absorption material to the interior of the box.

    None, Minimal, Normal, Heavy Selects the amount of damping material that will be automatically entered for a new speaker design. There are separate settings for the different box types. *Note: The vented and bandpass boxes share the same default damping setting because they both have vents.* If desired, the user can override these default settings later with the "Damping" tab of the Box Properties window.

Use Classical box calculations (ignore Qa and γ) When enabled, BassBox Pro will ignore the damping settings and use "classical" box calculations. This means that Qa and γ will be ignored in the box calculations. The result is that the box volumes (Vb) calculated by BassBox Pro will be similar to many other programs. Normally, this control should be turned off so that BassBox Pro is allowed to adjust the box calculations to compensate for the amount of acoustic fill in them. One of the results of including Qa and γ in the box calculations is that the program will recommend smaller box volumes for a maximally flat response.

Qa Sets the values of the Q of the absorption losses (Qa) for the different damping settings. The factory settings for BassBox Pro are 100, 50, 10 and 5 for none, minimal, normal and heavy damping settings, respectively.

γ (gamma) Sets the γ values for the different damping settings. Gamma is the ratio of heat at constant pressure to that at constant temperature for the air in the box with the specified amount of fill. BassBox Pro uses a delta (ratio of change) value that represents the difference from normal air (which is around 1.4). You can think of gamma as the ratio of the physical box volume to that of the "apparent" box volume. Adding absorption (damping) to the interior of a box lowers the total Q of the speaker as if the box were bigger than it really is. The factory settings for BassBox Pro are 1.00, 1.04, 1.08 and 1.20 for none, minimal, normal and heavy damping settings, respectively. For example, a "heavy" amount of damping will make the box look approximately 1.2 times larger than if it had no damping.

Enable Auto-Scaling
When enabled, the box will be automatically drawn to scale in the "Box Design" tab of the Box Properties window whenever a dimension is changed. When disabled, the box will no longer be drawn to scale. Disabling this setting may be helpful for slow computers.

Disable Driver & Port Fit Testing
BassBox Pro can automatically check the diameter of a driver and the length of a vent to make sure that they will fit in the box. If they don't fit, the program can warn the user. In some circumstances you may want to defeat this feature by unchecking this checkbox.

Graph

There are many "preference" settings for graphs. These are depicted below in the "Graph" tab of the Preferences window.

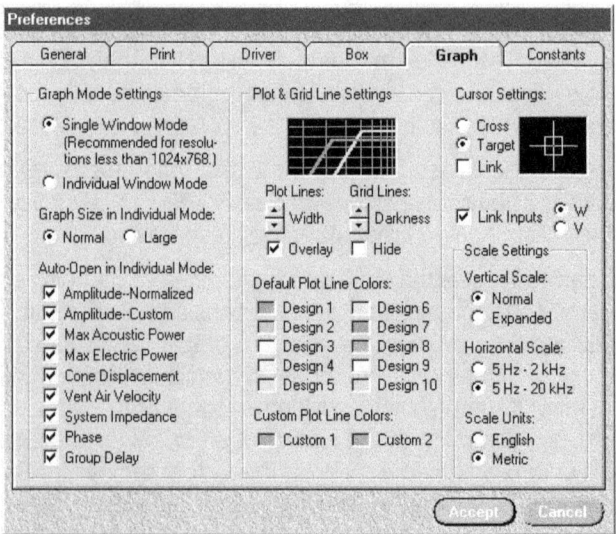

Graph Mode Settings

Mode Selects the default graph mode. Two graph modes are available. The "Single Window Mode" displays all graphs one-at-a-time in a single window. This is recommended for computers with video resolutions less than 1024 x 768 pixels (XGA). The "Individual Window Mode" displays each graph in its own individual window, allowing multiple graphs to be viewed simultaneously.

Graph Size in Individual Mode Sets the default size of the graph windows in the "individual window" mode.

Auto-Open in Individual Mode Sets which graphs will automatically open when <u>no</u> graphs are open and the first "Plot" button is clicked. These settings apply only to the "individual window" graph mode.

Plot & Grid Line Settings

Plot Line Width Use the up/down buttons to select the width of the plot lines.

Plot Line Overlay Check this checkbox to prevent the graphs from being erased before each new plot.

Grid Line Darkness Use the up/down button to select the desired darkness of the grid lines.

Hide Grid Lines Check this checkbox to hide the grid lines. The graphs will then be plotted on a solid black background.

Default Plot Line Colors Sets the plot line color given to each design when it is first opened. Click (🖱) on a color box to change the color. Each click will cause the color to advance through the 12 possible colors, including the two custom colors.

Custom Plot Line Colors Enables you to select custom colors to be used for the plot lines of any designs. Click (🖱) on one of the two custom color boxes and the windows Color window shown below will open so you select a custom color.

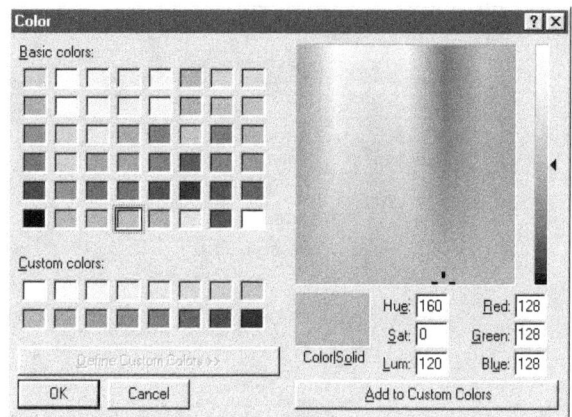

Note: The available color range and color quality will depend upon the capabilities and settings of the video system in the computer.

Cursor Settings
The shape of the cursor can be selected with the "Cross" and "Target" options. The cursors in all graphs can be linked with the "Link" option. When linked, all cursors will move in unison, making it easy to control the cursors in multiple individual graphs.

Link Inputs
Three graphs require the user to provide either an input power or input voltage. These are the Custom Amplitude Response, Cone Displacement and Vent Air Velocity graphs. Check the "Link Inputs" checkbox to link all of their inputs so a change to one input power/voltage is mirrored in the others. Use the "W" and "V" options to select whether you want the default graph input value for these three graphs to be watts (W) or volts (V).

Scale Settings
Vertical Scale Selects one of two sizes as the default. The range of the "normal" and "expanded" scales vary depending on the graph.

Horizontal Scale Selects either a 5 Hz to 2 kHz or a 5 Hz to 20 kHz range for the default horizontal frequency scale for the graphs.

Scale Units Selects either English or metric units for the Cone Displacement and Vent Air Velocity graphs. When set to "English", the Cone Displacement graph uses inches for the vertical scale and the Vent Air Velocity graph uses feet per second for the vertical scale. When set to "Metric", the Cone Displacement graph uses millimeters and the Vent Air Velocity graph uses meters per second.

Constants

There are two "constants" used by BassBox Pro which pervade the calculations of the program. These are the density of air and the velocity of sound in air. They can be changed using the "Constants" tab of the Preferences window.

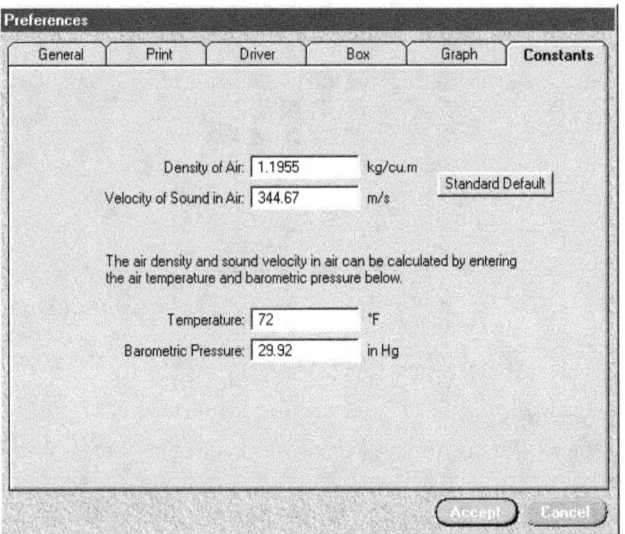

There are two ways to change the values of these constants. A new air density or sound velocity value can be manually entered into its respective input box. Or, the temperature and barometric pressure can be entered and the air density and the sound velocity will be automatically calculated from them. Before making changes, remember to click on the units labels to select the desired units.

Finally, to restore the factory values for the constants, click on the "Standard Default" button.

Appendix A: Command Shortcuts

The following keyboard shortcuts can be used to swiftly access many commonly-used BassBox Pro commands. A "+" (plus) sign means to hold down the first key(s) while pressing the last key. For example, [Ctrl]+[P] means to press and hold the [Ctrl] key and then press and release the [P] key. Release the [Ctrl] key after the [P] key has been released.

General

[F1]	Open the on-screen manual.
[Ctrl]+[P]	Print the graphs or a selected design.
[Ctrl]+[Q]	Close BassBox Pro.

Files

[Ctrl]+[O]	Open an existing speaker design from disk.
[Ctrl]+[S]	Save design changes to disk in a BassBox Pro bb6 design file.

Design

[Ctrl]+[N]	Begin a new speaker design.
[Ctrl]+[D]	Open the Driver Properties window.
[Ctrl]+[B]	Open the Box Properties window.
[Ctrl]+[N]	Open the Room / Car Acoustic Properties window.

Graphs

[Ctrl]+[F1]	Open or select the Normalized Amplitude Response graph.
[Ctrl]+[F2]	Open or select the Custom Amplitude Response graph.
[Ctrl]+[F3]	Open or select the Maximum Acoustic Power graph.
[Ctrl]+[F4]	Open or select the Maximum Electric Input Power graph.
[Ctrl]+[F5]	Open or select the Cone Displacement graph.
[Ctrl]+[F6]	Open or select the Vent Air Velocity graph.
[Ctrl]+[F7]	Open or select the System Impedance graph.
[Ctrl]+[F8]	Open or select the Phase Response graph.
[Ctrl]+[F9]	Open or select the Group Delay graph.
[Ctrl]+[0] to [9]	Plot the specified design in all open graphs. For example, use [Ctrl]+[1] to plot Design 1, [Ctrl]+[2] to plot Design 2, etc. Use [Ctrl]+[0] (zero) to plot Design 10.

BassBox 6 Pro

Appendix A: Command Shortcuts

Ctrl+Alt+0 to 9	("Single window" graph mode only). Plot the specified design in the selected (visible) graph only. For example, use Ctrl+Alt+1 to plot Design 1 in the selected graph in the combination graph window. Use Ctrl+Alt+0 (zero) for Design 10.
Shift+F1 to F7	Store the last plotted design into graph memory 1 to 7.
Shift+Ctrl+F1 to F8	Recall and replot memory 1 to 7. Use Shift+Ctrl+F8 to recall and replot all stored graph memories.
Ctrl+X	Clear the selected graph.
Ctrl+Y	Clear all open graphs.
Ctrl+U	Activate the cursor in the selected graph.
Ctrl+H	Hide the cursor in the selected graph.
Ctrl+I	Include or ignore the driver acoustic response.
Ctrl+V	Include or ignore the vent resonance peaks.
Ctrl+R	Include or ignore the room/car acoustic response.
Ctrl+E	Include or ignore the external network response.
Ctrl+T	Include or ignore the on-axis piston band response.
Ctrl+F	Include or ignore the diffraction response shelf.
Ctrl+C	Copy the selected graph to the Windows clipboard so it can be pasted into another Windows application as a picture (bitmap).
←	Move the graph cursor left one pixel.
→	Move the graph cursor right one pixel.
Shift+←	Move the graph cursor left 20 pixels.
Shift+→	Move the graph cursor right 20 pixels.
Ctrl+←	Switch the cursor to the unselected plot line when a graph creates two plot lines for a design. This is possible in only the Cone Displacement graph (passive radiator boxes) and Vent Air Velocity graph (dual-tuned bandpass boxes). For example, use Ctrl+← to move the cursor from the driver plot line to the passive radiator plot line in the Cone Displacement graph.
Ctrl+→	Same as above.
↑	Move the graph cursor up through the last 10 design plots.
↓	Move the graph cursor down through the last 10 design plots.

Driver Database

Ctrl+W	Edit the driver information in the database.

Appendix B: Glossary of Terms

1-W SPL	The reference sensitivity of a driver with a 1 watt signal and measured at 1 meter as a sound pressure level (dBSPL).
2.8-V SPL	The reference sensitivity of a driver with a 2.83 volt signal and measured at 1 meter as a sound pressure level (dBSPL).
AC	Alternating current. Audio signals are AC signals.
audible	Able to be heard. A sound that cannot be heard is inaudible. What makes a sound audible? Answer: 1) The frequency of the sound waves are within the range of human hearing. 2) The sound waves have sufficient amplitude (loudness) to be heard. 3) The sound waves are not "masked" by a louder sound or noise.
Av	The inside cross-section area of a vent.
BL	The motor strength of a driver.
Cms	The mechanical compliance of the suspension of a driver or passive radiator.
DC	Direct current. DC flows in a steady direction. The direction depends on the polarity of the energy source. A battery produces a DC signal.
DCR	The DC resistance of an electric device such as an inductor.
DF	The system damping factor. This can be the system damping factor in the case of a single-tuned bandpass box or the damping factor of an amplifier output.
Dia	The diaphragm or piston diameter of a driver or passive radiator.
diffraction	The bending of sound waves as they pass near an edge or corner of a solid object.
distortion	When a signal flowing through a circuit is compared at two different points, any change except for magnitude is distortion.
Dv	The inside diameter of a vent.
E-M	The electro-mechanical loudspeaker parameters.
EBP	The efficiency bandwidth product (Fs/Qes). *Note: When Qes is unknown EBP is sometimes estimated with Fs/Qts.*
ESR	The equivalent series resistance of an electric device such as a capacitor.
F3	The half-power (–3 dB) frequency of a system. F3 is sometimes referred to as the "corner" or "cutoff" frequency. However, this is not always true for BassBox Pro as explained in the note below.
	Note: BassBox Pro calculates F3 by tracking the system response in a descending frequency direction until the first –3 dB level is found. This may or may not be below the "knee" of the response curve. This is a little different than other methods which calculate F3 as the –3 dB point relative to the "knee" of the response. BassBox Pro uses the method it does because it shows the absolute –3 dB frequency rather than a relative one.
Fb	The system resonant frequency of a speaker. It is also called the "tuning frequency" of a box with a vent. Double-tuned bandpass boxes have two Fb values.

Fc	The resonant frequency of a driver in a closed box.
Fill	The acoustic absorption or damping material added inside a box to suppress unwanted resonances (and sometimes to increase the apparent box volume).
frequency	The number of sound waves per second. It is expressed in hertz (Hz).
Fs	The free-air resonant frequency of a driver or passive radiator.
γ	(gamma) The ratio of heat at constant pressure to that at constant temperature for the air inside a box. It characterizes the change in the "springiness" of the air inside the box after acoustic absorption or "fill" is added inside. *Note: BassBox Pro uses a delta (ratio of change) value for γ which equals 1 (one) for a box with no fill and increases as the amount of fill inside the box increases.*
Hv	The inside height of a square or rectangular vent.
impedance	Any opposition to the flow of electricity. This opposition is from all sources, both resistive and reactive.
isobaric	Constant pressure. A "compound" pair of drivers are mounted on either end of a small, sealed isobaric chamber. The pressure in the chamber is kept constant because both drivers are fed the same signal and their diaphragms move in the same direction.
Le	The inductance that a driver appears to have at upper frequencies because of the inductive reactance of its voice coil.
Lv	The length or depth of a vent (including, if present, the flared ends).
max flat	(maximally flat) An amplitude response curve with the least possible ripple in the passband.
midrange	A driver designed to produce frequencies in the middle of the audible spectrum (500 to 4000 Hz).
Mms	The mechanical mass of a driver or passive radiator diaphragm and voice coil assembly including the air load.
η_0	(eta zero) The reference efficiency of a driver with a half-space acoustical load.
octave	A two-to-one change in frequency. For example, 100 to 200 Hz is one octave and 5000 to 10000 Hz is also one octave.
on-axis	When a speaker or driver is pointing directly at something it is "on-axis" to it. Most speaker measurements are made on-axis.
overhung	An overhung voice coil is taller than the height of the magnet gap.
parallel	Components that are connected side-by-side are wired in parallel.
passband	The operating frequency band of a driver. The passband is usually defined by F3.
Pe	The maximum electrical power that a driver can handle before it is damaged, usually when the voice coil burns. Also called the "thermal power limit".
phase shift	When a signal flowing through a circuit is compared at two different points, any delay in the signal can be observed as a phase shift. It is expressed as an angle of rotation.

Appendix B: Glossary of Terms

piston band	The frequency band where a driver maintains a constant load versus frequency.
pixel	The smallest dot of light that a computer can turn on and off on the video monitor.
polarity	A driver is said to be "in" polarity when its diaphragm moves outward in response to a positive signal. It is said to be invented or "out" of polarity when its diaphragm moves inward in response to a positive signal.
push-pull	When two drivers are mounted in opposite directions and wired with opposite polarity with respect to each other, they are said to be in a "push-pull" configuration. This is because when the diaphragm of one driver moves away from its magnet, the other driver's diaphragm moves toward its magnet. This results in a reduction in even-order distortion because many nonlinearities are cancelled.
Q	The resonance magnification of a system.
Q't	(Q prime T) The total Q of a driver's suspension with the load of the rear chamber in a single-tuned bandpass box.
Qa	The Q of a box resulting from all absorption losses. Absorption losses usually result from the addition of an acoustical absorber or "fill" to the box interior.
Qec	The Q of a driver in a closed box at Fc considering only its electrical (non-mechanical) resistance.
Qes	The Q of a driver at Fs considering only its electrical (non-mechanical) resistance.
QL	The Q of a box resulting from all leakage losses. Sources of leakage loss include box wall vibration, poor box construction, poor driver gasket seal, a porous driver dust cap and a "lossy" driver surround.
QLv	The Q of a vent resulting from all of its losses. One source of vent loss is the viscosity friction of air moving through the vent.
Qmc	The Q of a driver in a closed box at Fc considering only its mechanical (non-electrical) resistance.
Qms	The Q of a driver or passive radiator at Fs considering only its mechanical (non-electrical) resistance.
Qtc	The total Q of a driver in a closed box at Fc considering both electrical and mechanical resistance.
Qts	The total Q of a driver at Fs considering both electrical and mechanical resistance.
Re	The DC resistance of a driver's voice coil.
reactance	An opposition to the flow of electricity because of capacitive and inductive characteristics. Purely resistive characteristics are not included in reactance.
resistance	An opposition to the flow of electricity without capacitive and inductive characteristics.
resonance	The frequency of peak response of a device that results from the balance of its capacitive and inductive characteristics.
Rms	The mechanical resistance of a driver or passive radiator suspension losses.
RMS	The <u>r</u>oot <u>m</u>ean <u>s</u>quare level. For example, voltage is often measured as an RMS

voltage because this is the "effective" or "steady-state" level that determines how much work or heat can be produced. It is calculated from a peak level by dividing the peak level by the square root of 2.

Sd — The diaphragm or piston area of a driver or passive radiator.

series — Components that are connected end-to-end are wired in series.

SPL — The sound pressure level. It is usually expressed as a decibel (dB) ratio.

stopband — The area outside of the operating frequency band (passband) of a driver.

subsonic — Sound waves with such a low frequency that they cannot be heard and are therefore beyond the range of human hearing.

subwoofer — A driver designed to produce ultra-low frequencies (below 100 Hz).

supertweeter — A driver designed to produce ultra-high frequencies (above 5000 Hz).

T-S — The Thiele-Small driver or passive radiator parameters, named after A.N. Thiele and R.H. Small who popularized the "lumped sum" method of enclosure analysis used by many in the audio industry. Many of their key papers are included in the Audio Engineering Society *Loudspeaker Anthologies*, listed in Appendix E.

tweeter — A driver designed to produce high frequencies (2000 to 20000 Hz).

ultrasonic — Sound waves with such a high frequency that they cannot be heard and are therefore beyond the range of human hearing.

underhung — An underhung voice coil is shorter than the height of the magnet gap.

Vas — The volume of air having the same compliance or "springiness" as the suspension of a driver or passive radiator.

Vb — The net internal volume of a box.

Vd — The diaphragm or piston displacement volume of a driver or passive radiator. For a driver this is usually the volume displaced at Xmax. For a passive radiator this is the volume displaced at Xmech.

woofer — A driver designed to produce low frequencies (20 to 2000 Hz).

Wv — The inside width of a square or rectangular vent.

Xmax — The maximum linear excursion of the driver or passive radiator. It should be measured in one direction from a resting position.

Xmech — The maximum mechanical excursion of the driver or passive radiator. With some drivers Xmech is reached when the driver's diaphragm has moved as far as the suspension will allow. In other drivers Xmech is reached when the voice coil former hits the back plate of the magnet structure. Xmech should be measured in one direction from a resting position.

Z — The nominal electromagnetic impedance of a driver.

Appendix C: The Driver Shapes in BassBox Pro

Shape: **Round**

Shape: **Round with square sides**

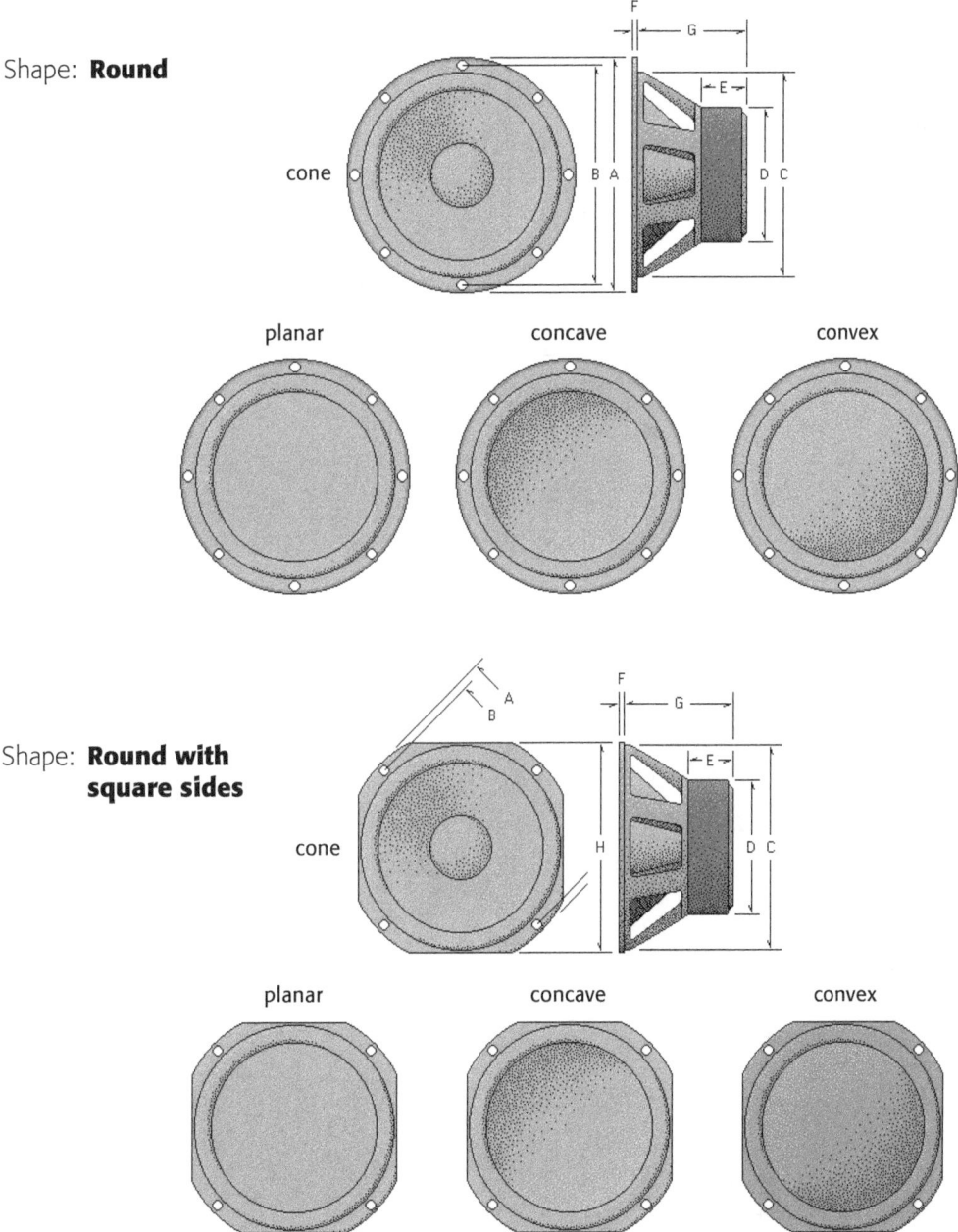

BassBox 6 Pro

Appendix C: The Driver Shapes in BassBox Pro

Shape: **Round with flat top & bottom**

Shape: **Round with mounting tabs**

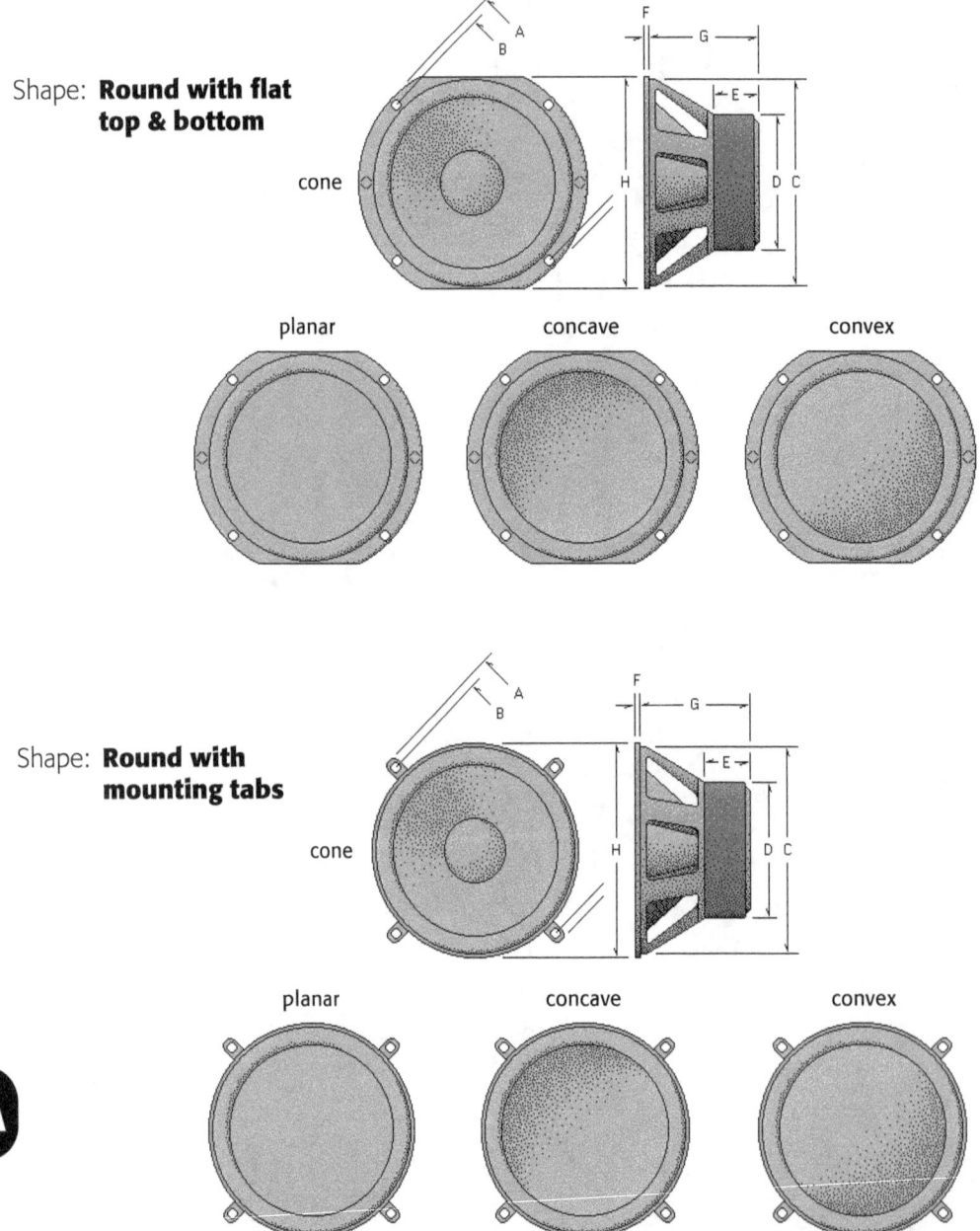

Appendix C: The Driver Shapes in BassBox Pro   **BassBox 6 Pro**

Shape: **Pincushion**

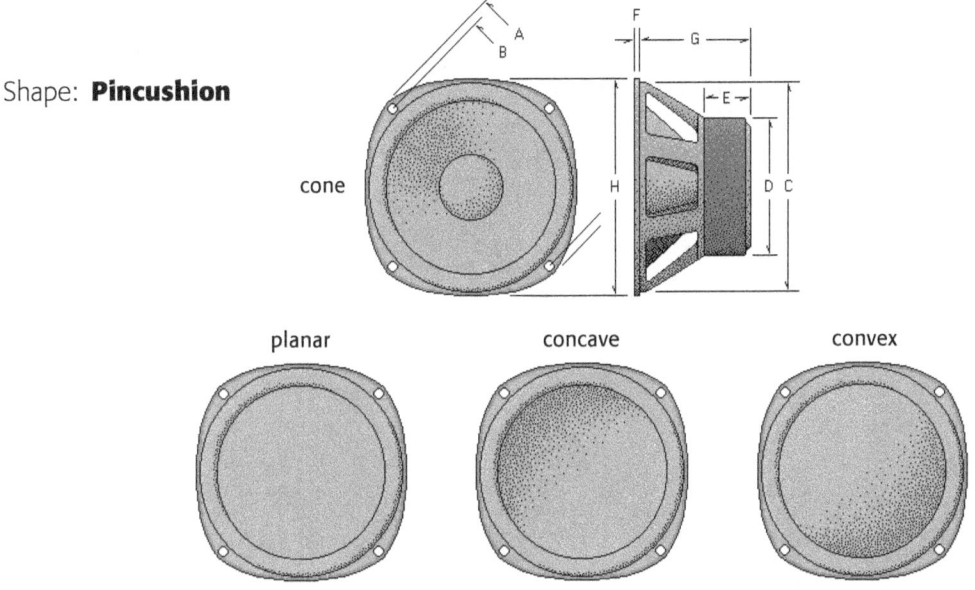

Shape: **Oval**

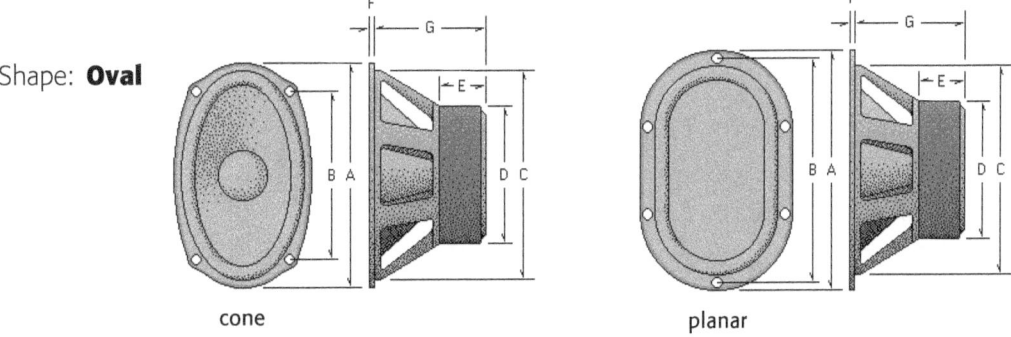

BassBox 6 Pro User Manual © D.E.Harris   **351**

BassBox 6 Pro *Appendix C: The Driver Shapes in BassBox Pro*

Shape: **Square**

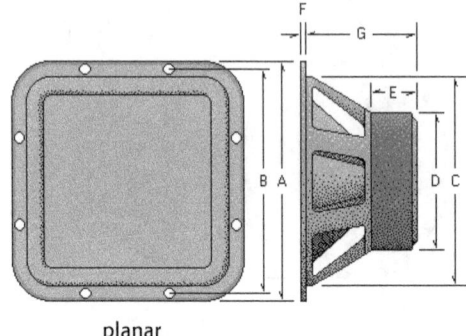

planar

Shape: **Rectangular**

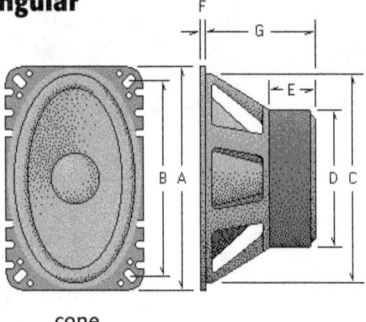

cone

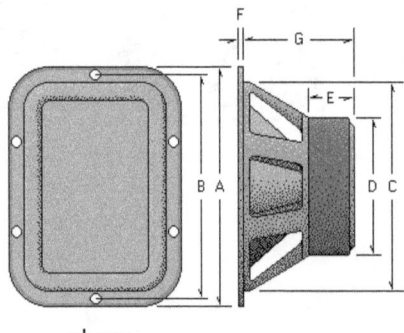

planar

Appendix D: The Box Shapes in BassBox Pro

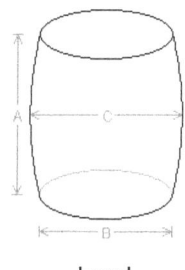
barrel

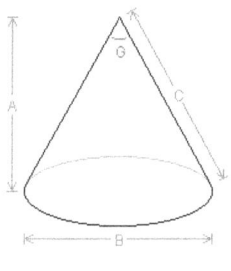
cone

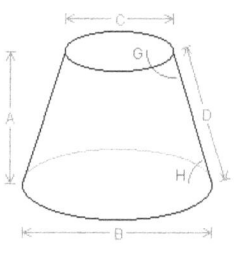
truncated cone

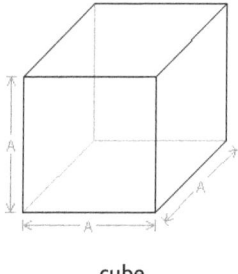
cube

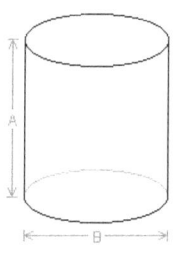
cylinder

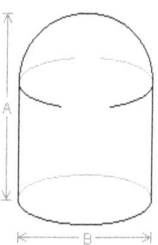
domed cylinder

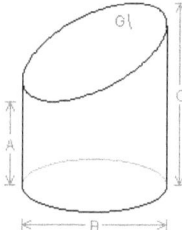
truncated cylinder

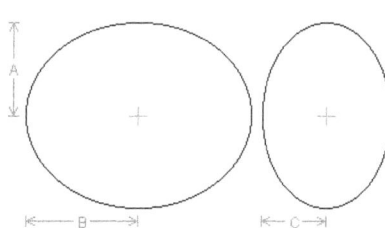
ellipsoid

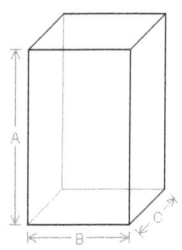
square prism

BassBox 6 Pro *Appendix D: The Box Shapes in BassBox Pro*

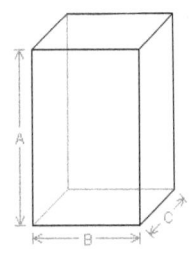
optimum square prism

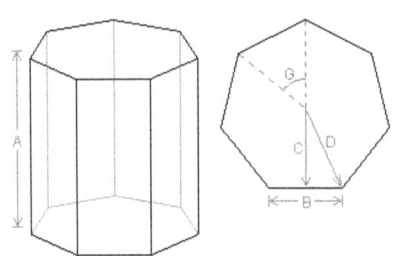
regular polygon prism

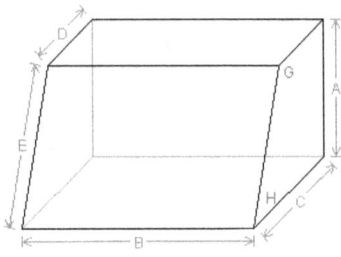
slanted front prism

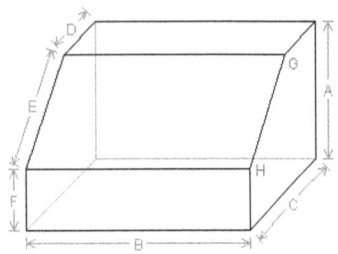
truncated edge prism

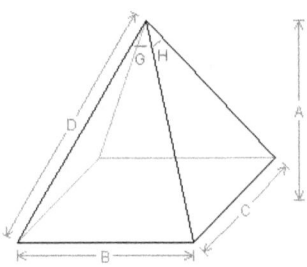
four-sided pyramid

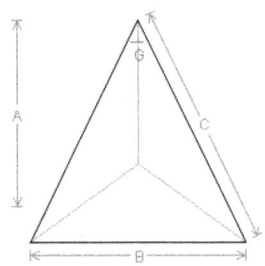
three-sided pyramid

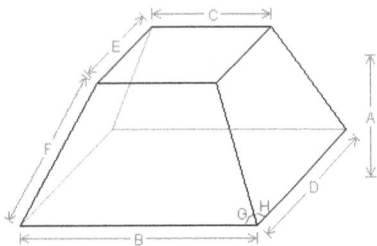
truncated pyramid
(trapezoid)

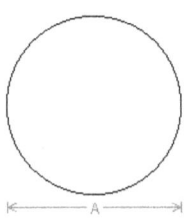
sphere

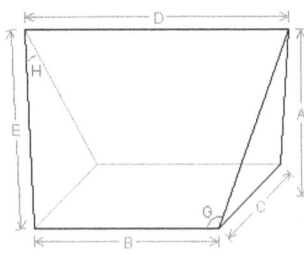
wedge

Appendix D: The Box Shapes in BassBox Pro **BassBox 6 Pro**

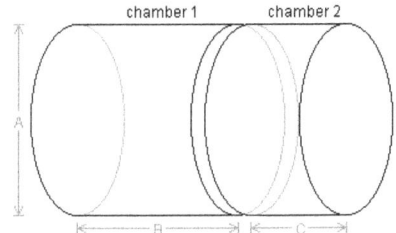
bandpass cylinder – internal, double-chamber

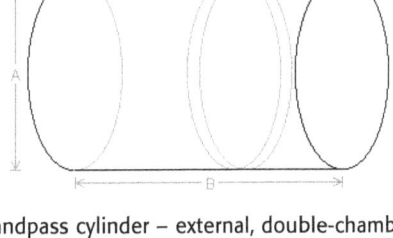
bandpass cylinder – external, double-chamber

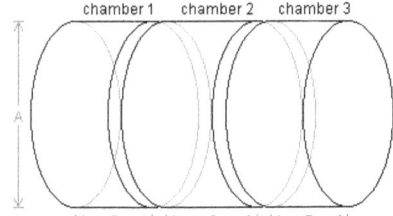
bandpass cylinder – internal, triple-chamber

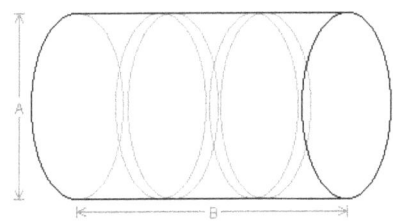
bandpass cylinder – external, triple-chamber

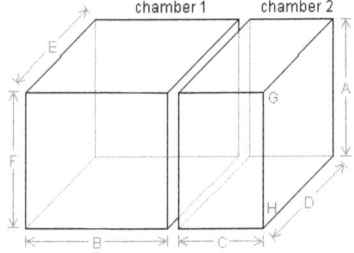
bandpass prism – internal, double-chamber

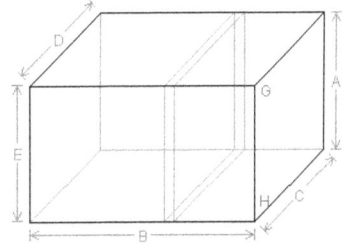
bandpass prism – external, double-chamber

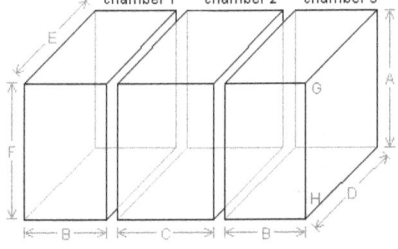
bandpass prism – internal, triple-chamber

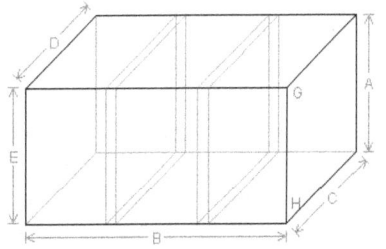
bandpass prism – external, triple-chamber

Appendix E: Suggested Reading

Beginner-Level Reading
Book: *Designing, Building & Testing Your Own Speaker System* by David B. Weems.
Pennsylvania: Tab Books Inc. (McGraw-Hill, Inc.), 1981, 1990 (revised edition)

Book: *How to Build Speaker Enclosures* by A. Badmaieff and D. David
Indiana: Howard W. Sams & Co., Inc., 1966

Intermediate-Level Reading
Book: *The Loudspeaker Design Cookbook*, 5th Edition by Vance Dickason
New Hampshire: Audio Amateur Press, 1995

Book: *Advanced Speaker Designs for the Hobbyist and Technician* by Ray Alden
Indiana: Prompt Publications (Howard W. Sams & Co., Inc.), 1995

Book: *High Performance Loudspeakers*, 5th Edition by Martin Colloms
England: John Wiley & Sons, 1997

Chapter: "Loudspeakers, Enclosures, and Headphones" by Clifford A. Henricksen
Book: *Handbook for Sound Engineers*
Indiana: Howard W. Sams & Co., Inc., 1987

Advanced-Level Reading
Book: *Loudspeakers, An Anthology, Volume 1*, 2nd Edition
(Vol. 1 – Vol. 25 of *Journal of the Audio Engineering Society*)
New York: Audio Engineering Society, 1980

Book: *Loudspeakers, An Anthology, Volume 2*
(Vol. 26 – Vol. 31 of *Journal of the Audio Engineering Society*)
New York: Audio Engineering Society, 1984

Book: *Theory and Design of Loudspeaker Enclosures* by J.E. Benson
Indiana: Synergetic Audio Concepts, 1993

Book: *Testing Loudspeakers*, 1st Edition by Joseph D'Appolito
New Hampshire: Audio Amateur Press, 1998

Additional Reading & References
Article: "Thiele, Small, and Vented Loudspeaker Design: Parts I–IV" by Robert M. Bullock III
Magazine: Issues Four/80 through Three/81 of *Speaker Builder*
New Hampshire: Audio Amateur Publications, 1980-81

Article: "Double Chamber Speaker Enclosure" by George L. Augspurger
Magazine: Dec. Issue/1961 of *Electronics World Magazine*
Ziff-Davis Publishing Co., 1961

Article: "New Guidelines for Vented-Box Construction" by George L. Augspurger
Magazine: Issue Two/91 of *Speaker Builder*
New Hampshire: Audio Amateur Publications, 1991

Article: "The Third Dimension: Symmetrically Loaded, Part 1" by Jean Margerand
Magazine: Issue Six/88 of *Speaker Builder*
New Hampshire: Audio Amateur Publications, 1988

Article: "Passive Crossover Networks: Parts 1–3" by Robert M. Bullock III
Magazine: Issues One/85–Three/85 of *Speaker Builder*
New Hampshire: Audio Amateur Publications, 1985

Paper: "A Bandpass Loudspeaker Enclosure" by L.R. Fincham
Preprint 1512 presented at the 63rd Convention of the Audio Engineering Society
Journal of the Audio Engineering Society (Abstracts), Vol. 27, p. 600 (1979 July/Aug.)

Paper: "An Introduction to Band-Pass Loudspeaker Systems" by Earl R. Geddes
Publication: Vol. 37, No. 5 of *Journal of the Audio Engineering Society*
New York: Audio Engineering Society, 1989

Paper: "Impedance Analysis of Subwoofer Systems" by Arthur P. Berkhoff
Publication: Vol. 42, No. 1/2 of *Journal of the Audio Engineering Society*
New York: Audio Engineering Society, 1994

Paper: "Passive-Radiator Loudspeaker Systems Part 1: Analysis" by Richard H. Small
Publication: Vol. 22, No. 8 of *Journal of the Audio Engineering Society*
New York: Audio Engineering Society, 1974

Paper: "Passive-Radiator Loudspeaker Systems Part 2: Synthesis" by Richard H. Small
Publication: Vol. 22, No. 9 of *Journal of the Audio Engineering Society*
New York: Audio Engineering Society, 1974

Article: "Matching Driver Efficiencies" by John I. Lipp
Magazine: Issue Five/93 of *Speaker Builder*
New Hampshire: Audio Amateur Publications, 1993

Paper: "Complete Response Function and System Parameters for a Loudspeaker with Passive Radiator" by Douglas H. Hurlburt
Publication: Vol. 48, No. 3 of *Journal of the Audio Engineering Society*
New York: Audio Engineering Society, 2000

Appendix F: Driver Parameter Worksheet

The following worksheet lists the driver parameters used by this program. Photocopy it and use it whenever you need to collect the parameters for unknown drivers.

Mechanical Parameters   Units

Fs: _____ Hz

Qms: _____

Vas: _____ _____

Cms: _____ _____

Mms: _____ _____

Rms: _____ _____

Xmax: _____ _____

Xmech: _____ _____

Dia: _____ _____

Sd: _____ _____

Vd: _____ _____

General Information

Manufacturer: _____

Model Name: _____

Serial No: _____

Comment: _____

_____

_____

_____

❏ Underhung Voice Coil   ❏ Magnetic Shielding

❏ Dual Voice Coil

Electrical Parameters   Units

Qes: _____

Re: _____ ohms

Le: _____ millihenries

Z: _____ ohms

BL: _____ _____

Pe: _____ watts

Parallel Dual VC

Qes: _____

Re: _____

Le: _____

Z: _____

BL: _____

Pe: _____

Series Dual VC

Qes: _____

Re: _____

Le: _____

Z: _____

BL: _____

Pe: _____

Combination Parameters Units

Qts: _____

ηo: _____ %

1-W SPL: _____ dB

2.8-V SPL: _____ dB

Qts: _____

ηo: _____

1-W SPL: _____

2.8-V SPL: _____

Qts: _____

ηo: _____

1-W SPL: _____

2.8-V SPL: _____

Appendix G: Acoustic Response Worksheet

The following worksheet lists the data points used by this program for both the measured loudspeaker acoustic response and the interior acoustic response. Photocopy it and use it whenever you need to manually collect the response data.

Hz	Level (dB)	Hz	Level (dB)	Hz	Level (dB)	Hz	Level (dB)	Hz	Level (dB)	Hz	Level (dB)
5		36		135		500		1.9 k		7 k	
6		38		140		530		2 k		7.5 k	
8		40		150		560		2.1 k		8 k	
10		42		160		600		2.25 k		8.5 k	
11		45		170		630		2.35 k		9 k	
12		47		180		670		2.5 k		9.5 k	
13		50		190		700		2.65 k		10 k	
14		53		200		750		2.8 k		10.5 k	
15		56		210		800		3 k		11 k	
16		60		225		850		3.15 k		12 k	
17		63		235		900		3.35 k		12.5 k	
18		67		250		950		3.55 k		13.5 k	
19		70		265		1 k		3.75 k		14 k	
20		75		280		1.05 k		4 k		15 k	
21		80		300		1.1 k		4.2 k		16 k	
22		85		315		1.2 k		4.45 k		17 k	
24		90		335		1.25 k		4.75 k		18 k	
25		95		355		1.35 k		5 k		19 k	
27		100		375		1.4 k		5.3 k		20 k	
28		105		400		1.5 k		5.6 k			
30		110		420		1.6 k		6 k			
32		120		445		1.7 k		6.3 k			
34		125		475		1.8 k		6.7 k			

Index

1-W SPL 187, 200, 321
2.8-V SPL 187, 200

A 245, 246, 249
Absorption 63, 230-233
Acoustic environment 329-330
Acoustic filter 52-53
Acoustic power 27-28, 273-275
Acoustic properties 251-254
Acoustic response
  driver 166, 203-209, 264, 312
  importing data 208-209, 253-254
  open 253
  room/car 19, 166, 251-254, 264
  save 254
Active high-pass 12 dB/octave EQ filter 50-51, 61, 221-222, 225
Add a new driver 172, 309-312
Aim 244
Amplifier 210, 276, 279, 281
Amplitude response 24, 269-272
Architectural acoustic 330
Auto Scaling 339
Automotive acoustic 330
Av 239
Avoiding standing waves 63-64
Availability (driver) 173, 175
Averaged band 207

B 245, 246, 249
B_6 50-51
Back plate 31
Balloon help - see help
Bandpass box 51-55, 61, 65-68, 117-137, 216-218, 221-222, 223, 233, 244
Bandwidth 223
Bessel 188, 189, 191-192, 224
BL 47, 49, 57, 60-61, 180-181, 321
Box 29, 35-45-62, 169, 337-339
  chambers 222
  company 214
  description 214
  construction 63-68

  design 18, 215-229
  dimensions 8, 224, 226-229, 250
  model 214
  priority 219-222
  properties 16-17, 213-250
  selecting 58-60
  shape 227
  type 45-62, 215-218
  volume displaced 243-249
  weight 214
Braces 65, 247-249
Butterworth 224

C 249
Ca 211
Calc 196-197, 241
Calculate all 197-198
Capacitance 211
Car 16-17, 19, 45-46, 251-254, 264
Ce 211
Center plate 31
Chamber 249
Classical box calculations 232, 339
Clear
  acoustic data 209, 254
  all driver parameters 197-198
  design 160, 305
  graph 259
  graph memory 163, 259
  logo 295
Clipboard (Windows) 167
Closed box 45-47, 71-98, 216, 218, 219-221, 233
Closing a design 16, 305
Closing BassBox Pro 14, 15, 167
Cms 60-61, 180-181, 199, 221, 242, 321, 325
Coaxial 183-184
Coherent 189, 192, 272, 275, 277, 278, 280
Coincident 183-184
Color 267, 268, 293, 332, 341
Compact database 161, 316
Company 176-177, 184, 310, 313-315

Compliance 35, 38-40, 56, 199, 221, 234, 242
Compound 187, 244-245
Cone 31, 32
Cone displacement 199, 242, 262, 278
Configuration
  BassBox Pro 329-342
  driver 186-192
Connection terminal 31, 32
Constants 342
Construction 63-68, 214
Copy 16, 167, 267
Cp1, Cp2 211
Critically damped 224
Crossover network 29, 33-35, 44, 210-211
Cs1, Cs2 211
Cursor 260-262, 267, 341
Custom amplitude response 271-272
Cut angles 250
Cutoff rate 42, 45-47, 48-50, 52-55, 56-57
Cutoff region 269

D 249
Damping 39-40, 42, 50, 63, 65, 230-233
Damping factor 210
Damping material 231
Database 171-182, 307-316
  availability options 173, 175
  close 173-174
  compact 316
  driver & co. description 173-174
  drivers found list 173-174, 176-177
  locator size 174-175
  navigation controls 173-174, 177
  repair 316
  search by box 182
  search by company 176-177
  search by model 178
  search by parameters 179-181
  search tabs 173, 175

Index

DCR 211
Default company 308
Delay 283-284, 285
Delete
   an existing company 314
   an object 249
   the selected driver 309
Description
   box 214
   driver 171, 183-185
Design 18
   files 287-290
   panel 15, 18, 287
   properties list 16, 18
   title bar 16
   wizard 18, 169-170, 327
Desired F3 220-221
DF 210, 223
Dia 199
Diaphragm 31, 32, 36, 56, 199, 242
Diffraction 64
Diffraction response shelf 166, 266
Dimensions
   box 224, 226-229, 266
   driver 201-202, 250, 312
   object 249
Displacement limits - see excursion, Xmax and Xmech
Distortion 47, 49, 64, 188, 279
Double chamber 51-52, 222, 244
Double-tuned bandpass 54-55, 217, 218, 223
Driver 29-33, 43-44, 169, 335-336
   acoustic response 166, 203-209, 264
   aim 244
   availability 173, 175, 310
   company 176-177, 184, 310
   configuration 186-192
   database 171, 172, 173-182, 307-316
   description 183-185, 310-311
   dimensions 201-202, 312
   displacement 243-244, 278
   fit testing 339
   locator 173-182, 307-308

   model 178, 183, 310
   mounting 243
   open back 30
   parameters 16-17, 171-172, 179-181, 193-200, 311
   parts 31-33
   piston type 202
   properties 171-211, 290
   sealed back 30
   shape 201
   suitability 60-61, 180-181
   testing 317-321
   type 183-184, 310
   typical parts 31-33
   wiring 190-192, 198
Dual voice coil 31, 198
Duct - see vent
Dust cap 31, 33
Dv 236, 238, 239

E 249
EBP 60-61, 173-174, 180-181, 182, 193
Edit company data 160-161, 313-315
Edit driver data 160, 307-313
Edit menu:
   Clear Selected Design 160, 305
   Clear All Designs 160, 305
   Close All Designs 160, 305
   Edit Driver Data 160, 307-313
   Edit Company Data 160-161, 313-315
   Compact Database 161, 316
   Repair Database 161, 316
   Preferences 161, 329-342
Efficiency 47, 49, 53, 57
Efficiency bandwidth product - see EBP
Electrical configuration 190-192
Empty weight 214
End correction 238
ESR 211
Est 196-197, 202
Evaluating performance 255-285
Excursion 49, 56-57, 199, 261, 273-274, 278

Expert mode 193-197, 336
Exporting a graph 304
Extended bass 50, 56, 220
External box dimensions 228
External network 166, 210-211, 265, 281
External resistance 210

F3 47, 182, 220-221, 225, 269-270
Fb 50, 222, 224-225, 236, 238-239, 281
Fc 321
Features 8-11, 257-268
File menu:
   New Design 18, 159
   Open Design 159, 289
   Save Design 159, 287-288
   Save Design As 159, 288-289
   Print Design 159, 291-304
   Print Setup 160
   Quit 160, 287
Fill 50, 230-233
Filter 211
Flat 37, 46, 270
Flat region 270
Flush mounting 243
Former 31
Frame 31, 33
Frequency 23-25
Frequency response - see amplitude response
Front mounting 243
Front plate 31
Fs 47, 49, 50, 60-61, 173, 180-181, 193, 199, 242, 321, 325
Fx 225

γ (gamma) 232-233, 339
Gasket 31, 33
Graph
   auto-open 340
   clear 165, 259
   close 167, 267
   copy 167, 267
   cursor 165, 260-262, 341

exporting 304
features 257-268
  grid 165, 263, 340-341
  include 166, 264-266
  memory 164-165, 259
  minimize 167, 266
  mode 161, 256-257, 340
  plot color 267, 268, 341
  popup menu 164-167, 258
  preferences 340-342
  printing 267, 291-304
  properties 167, 258-259, 267
  scale 165, 263, 341-342
  size 161, 166, 263-264, 340
Graph menu:
  Display Mode 161, 256-257
  Show Graph 162-163
  Clear All Graph Plots 163
  Clear All Memories 163
Grid lines 263
Grid line darkness 340
Group delay 285

Hearing 24
Help
  balloon help 19, 228, 330
  context sensitive help 20
  getting help 19-20
Help menu:
  On-Screen Manual 20, 164
  BassBox Pro Overview 164
  About BassBox Pro 13, 164
Hertz 23
HF-to-LF Fb ratio 223
High fidelity 37, 220
High output 219-220
Horizontal scale 263
Hv 239

Ignore error messages 294, 333
Impedance 190-192, 281-282
Impedance response - see system impedance
Impedance equalization 211

Import
  acoustic data 208-209, 253-254
  driver parameters 171-172, 197, 290
Include 166, 264-266, 293-294, 296, 333
Individual windows mode 161, 256-257, 263-264
Individual wiring 198
Inductance 200, 211
Inductive reactance 200, 211, 281
Inductive reactance roll-off 270
Infinite baffle 205
Input power 271
Input voltage 271
Interior 243-249
Internal box dimensions 228
Interpolate 207
Isobaric 187, 244-245
Isobaric chamber shape 245

Knee region 270
Known volume 249

L-pad 210-211
La 211
Large-signal
  analysis 8, 194-195
  parameters 194-195
Le 197, 200
Level 203-204, 251-252
LF-to-HF Vb ratio 223
LF Chamber(s) 222-223
Link inputs 341
Linking representatives and distributors 315
Lm 211
Load from database 171-172, 177
Lock 223, 245, 246
Lock Vb 224, 228
Lock vent dimensions 238-239
Logo 295, 334
Loudness 25-28
Loudspeaker - see speaker

Low frequency response 37, 45, 48, 50
Low frequency rise 45-46, 251
Lp1, Lp2 211
Ls1, Ls2 211
Lv 236, 238, 239

M-Vd 202
Magnet 31, 202
Magnetic shielding 185
Magnitude response - see amplitude response
Main window 14, 15-18
Mass 38-39, 47, 199, 242
Materials 64
Maximally flat 37, 42, 50, 220, 221
Maximize 15
Maximum acoustic power 261, 273-275
Maximum electric input power 276-277
Measured acoustic response - see acoustic response
Mechanical configuration 186-189
Memory 164-165, 259
Menu bar 15-16
Menus 159-167
Mid-band thermal power limit 273
Midrange driver 47, 52, 66
Mini preview graph 16-17
Minimize 15, 167, 266
Minimum parameters 194-195, 240, 311
Miscellaneous objects 247-249
Mms 47, 49, 60-61, 180-181, 199, 242, 321, 325
Motor 31
Mounting (driver) 243
Mounting dimensions 201-202
Multiple drivers 186-192, 270, 271-272, 275, 276-277, 278, 279-280, 282

η_0 (eta zero - efficiency) 47, 49, 57, 200, 321

Index

Net values 186, 210
New
 design 18, 159, 169, 288
 logo 295
 object 249
Normal mode 193-195, 336
Normalize 203-208, 251-252
Normalized amplitude response 269-270
Number
 of chambers 222
 of drivers 186
 of objects 249
 of passive radiators 241
 rounding 330

Object 249
Omit acoustic response in graphs 253
On-axis 28, 271, 274
On-axis piston band response 166, 265-266
On-Screen manual 20
Open
 acoustic file 253
 BassBox Pro design file 159, 289-290
 X•over Pro design file 211
Open back 30
Optimum Qtc 337
Optimum square prism 64, 227
Outer shape 201
Overhung voice coil 22, 185
Overlaid plotting 296, 340
Override Vb 228, 229
Overview 15

Parallel double-tuned bandpass 54-55
Parallel wiring 190, 198
Parts list 250
Pascal 25
Passband 52-53, 235, 273, 276
Passive network 211

Passive radiator 41-43, 56-58, 233, 240-242, 278
Passive radiator box 41-43, 56-58, 65-66, 139-156, 217, 218, 221-222
Paths 287-288, 330-331
Pe 200, 271, 274
Performance 255-285, 334
Phase response 42, 45, 48, 54, 55, 56-57, 283-284
Pipe resonance 48, 50, 51-55, 235, 264
Piston band 166, 265-266
Piston diameter 199
Piston type 202
Plot 16-17
 line overlay 340
 line width 340
Polarity 32, 190-192
Port - see vent
Port fit testing 339
Power 200, 271, 276-277
Preferences 329-342
Pressure - see sound pressure level
Printing 267, 291-304, 332-334
Push-Pull 187-188, 190-191

Q 39-40, 50
Q't 223
Qa 232-233, 339
Qec 321
Qes 47, 49, 57, 173, 193, 196-197, 200, 321
QL 205, 225, 337-338
QLv 48, 239
Qmc 321
Qms 196-197, 199, 242, 321, 325
Qtc 39-40, 219, 220, 224, 321, 325
Qts 40, 173, 182, 193, 196-197, 200, 220, 221, 321
Quitting the program 14
Qx 225

Ra 211

Re 47, 49, 57, 200, 321
Rear mounting 243
Recall 259
Repair database 161, 316
Req 211
Reset all messages 331
Reset BassBox screen layout 332
Resistance 200, 210-211
Resonance 35, 38, 42-43, 56, 68, 166, 199, 242, 264, 281
Response - see acoustic response and amplitude response
Restore 207
Rg 210
Rise filter 252
Rm 211
Rms 199, 242, 321, 325
Room 16-17, 19, 251-254, 264
Rp1, Rp2 211
Rx 210

Sample designs 69-156
Sample printouts 298-303
Save
 acoustic file 254
 design 16, 159, 287-288
 design as 159, 288-289
Sealed back 30
Searching the driver database 176-182
Scale 165, 263, 341-342
Screen 31
Sd 199, 242, 321, 325
Sensitivity 186-189, 200
Separate wiring 192, 198, 272, 275, 277, 278, 280, 282
Serial number 13, 185, 214
Series double-tuned bandpass 55
Series wiring 190-191, 198
Series-parallel wiring 191
Shape 227, 249
Shelf filter 252
Single-tuned bandpass 53-54, 216, 218, 223
Single window mode 161, 256-257

Small-signal
  analysis 7, 194-195
  parameters 194-195
Snap main window 332
Sound pressure level 25-28, 271-272
Sound waves 23-25, 36, 42, 49-50, 189, 234
Speaker 29-44
Speaker cables 210
Speed of sound 23, 279, 342
Spider 31, 32-33
SPL - see sound pressure level
Square 202
Standard one-way 183
Standard configuration 186
Standing waves 63-64
Starting BassBox Pro 12, 289
Status bar 15, 18, 20
Status indicators 195-197, 240-241
Store 259
Subsonic 24, 49
Suggest (box) 213, 219-222
Suggest Fb 225
Suggest maximum vent area for Xmax 238
Surround 31, 32-33
Suspension 57, 199
System impedance 281-282
System requirements 11

T-S - see Thiele-Small parameters
Taskbar 12
Technical support 6, 20
Test equipment
Test menu:
  Driver 163, 317-321
  Passive Radiator 163, 317, 322-325
Testing drivers 317-321
Testing passive radiators 317, 322-325
Thermal limit 261, 273-275

Thiele-Small parameters 171-172, 193-200, 204-205
Three-way driver 183-184
Title bar 15
Title block 293, 294
Title window 12-14
Tolerance 195-196, 336
Tools menu:
  Design Wizard 163, 327
  Wavelength Calculator 163, 327
  Start X•over Pro 164, 328
Total 223
Track excursion 261
Triple chamber 51-52, 222, 244
Tweeter 30, 33, 35, 47, 52, 66
Two-way driver 183-184

Ultrasonic 24
Underhung voice coil 22, 185
Units 199-200, 223-225, 245-246, 278, 279, 335-336, 337, 342
Unloaded 43, 46-47, 49, 51, 57
Upper frequency response 51-55
User requirements 11

Vas 60-61, 180-181, 199, 221, 242, 321, 325
Vb 182, 205, 217, 222, 223-224, 228-229, 236
VC wiring option 198
Vd 199
Vent 31
  air velocity 48, 52-53, 235, 262, 279-280
  coloration 50, 51
  cross-section area 234, 238, 239, 279
  cross-section shape 234-235, 237
  dimensions 236
  displacement 246
  end type 238
  minimum vent area for Xmax 238
  pipe resonance 48, 50, 51-55, 68

placement 67-68
recommended size 234
resonance 48, 50, 51-55, 166, 235, 264, 281
Vents 41-43, 48-55, 66, 67-68, 233, 234-239, 279-280
Vented box 41-43, 48-51, 65-66, 99-116, 216, 218, 219-222, 233
Viscosity friction 48, 235
Voice coil 31, 47, 198, 273-274, 281
Vol 243-249
Voltage 271

Wall thickness 229, 246
Wavelength 23, 327
Welcome window 12-13
Windows clipboard 167
Woofer 30, 33, 35-37, 47, 52, 60-61, 66, 180-181
Wv 239

X•over Pro 164, 328
Xmax 47, 49, 60-61, 180-181, 199, 234, 238, 273-274, 278, 336
Xmech 199, 336

Z (impedance) 200

www.ingramcontent.com/pod-product-compliance
Lightning Source LLC
Chambersburg PA
CBHW080233180526
45167CB00006B/2265